Joseph Henry

Joseph Henry about 1829, age 32, thought to be an enlargement of a miniature painted in Albany by Julius Rubens Ames. (Smithsonian Institution Archives, neg. no.78-6067)

The Rise of an American Scientist

Albert E. Moyer

SMITHSONIAN INSTITUTION PRESS
Washington and London

Editor: Jack Kirshbaum
Designer: Janice Wheeler

Library of Congress Catalog-in-Publication Data

Moyer, Albert E., 1945-
Joseph Henry : the rise of an American scientist / Albert E. Moyer
p. cm.
Includes bibliographical references and index.
ISBN 1-56098-776-6 (cloth : alk. paper)
1. Henry, Joseph, 1797-1878. 2. Physicists—United States—Biography. I. Title.
QC 16.H37M69 1997
530'.092—dc21
[B] 97-20686

British Library Cataloging-in-Publication data available

04 03 02 01 00 99 98 97 5 4 3 2 1

The paper used in this publication meets the minimum requirements of the American National Standard for Permanence of Paper for Printed Library Materials Z39.48-1984.

For Lynette, Holly, Emily, Rebecca, and Cheetah

How short the space between the two cardinal points of an earthly career! the point of birth and that of death and yet what a universe of wonders is presented to us in our rapid flight through this space.

—Joseph Henry, 12 April 1878

Contents

Acknowledgments

Marc Rothenberg, head of the Joseph Henry Papers Project at the Smithsonian Institution, suggested in 1991 that I write a biography of Henry. Looking ahead to 1997, he envisioned the book as a bicentennial commemoration of the scientist's birth. Although I was reluctant to take on such a daunting project, Marc convinced me of the wisdom of his farsighted suggestion. Graciously, he then helped with all aspects of the enterprise. His largest contribution came, however, as editor-in-chief for the two most recent of the seven volumes of Henry's papers that the Smithsonian Institution Press has published to date. Indeed, the particular segment of Henry's life on which I concentrate—his crucial first five decades—parallels the chronological range of the first seven volumes of documents in the projected fifteen-volume series.

Marc was not alone in creating the two latest volumes of annotated documents. While I was working on the biographical project—from late 1992 through early 1997—Marc's editorial staff variously consisted of Kathleen Dorman, Deborah Jeffries, Frank Millikan, John Rumm, and Paul Theerman. All these people helped me wholeheartedly, even sharing access to their personal research files. Paul—who moved from the Henry Papers next door to the Smithsonian Archives—was particularly generous. He not only assisted with routine matters such as assembling old photographs but also, near the end of the project, provided a critical reading of the entire manuscript. Kathy was also especially supportive as a guide and reader.

Of course, I owe a similar debt to the editorial staffs that crafted the first five volumes. Nathan Reingold headed the initiative during the early years. Issuing the inaugural volume in 1972, Nate set the high standard of the Henry Papers Project. During his tenure, his aides included: Michele Aldrich, Joel Bodansky, Kathleen Dorman, James Hobbins, John Kerwood, Arthur P. Molella, Stuart Pierson, Marc Rothenberg, Joan Steiner, and Paul Theerman.

The Henry Papers Project is a subdivision of the Office of Smithsonian Institution Archives, which houses many original Henry documents. While searching through the treasures at the archives during my research trips to Washington, I profited from the services and suggestions of director Ethel (Edie) Hedlin and staff members Alan Bain, William Cox, Susan (Libby) Glenn, Pamela Henson, Michael Horsley, Josephine Jamison, Bruce Kirby, Tammy Peters, Mumia Shimaka-Mbasu, James Steed, and Kathleen Robinson Williams. Pam furnished me with an early version of an annotated Smithsonian chronology that she prepared for the institution's sesquicentennial in 1996. Michael helped track photographic prints. Across the Mall, at the National Museum of American History, I also benefited from conversations with Bernard Finn, Arthur P. Molella, Roger Sherman, and Deborah Warner. Roger kindly tutored me on the technicalities of Henry's distinctive approach to electrical science.

Other persons in Washington also advanced the project. Daniel Goodwin, director of the Smithsonian Institution Press, was the model of an encouraging, sensitive editor and publisher. Over lunch at the Castle or beer at L'Enfant Plaza, he provided sound counsel. Press staff, including Cheryl Anderson, Robert Lockhart, Sally Barrows, Janice Wheeler, and Jack Kirshbaum, expertly shepherded the manuscript through editing and production. Also in Washington, at the National Endowment for the Humanities, Daniel Jones offered solid advice on submitting the original proposal for the project as did his counterpart at the National Science Foundation, Ronald Overmann. Thanks to their guidance, the project was generously funded through three-year, simultaneous grants from NEH (Award Number RH-21092-93) and NSF (Grant Number SBER-9310633). (By the way, any opinions, findings, and conclusions or recommendations expressed in this biography are those of the author and do not necessarily reflect the views of either NEH or NSF.) The Smithsonian complemented the grants by providing, through my designation as a research associate, office space and access to the institution's remarkable resources and facilities.

A final person who helped immensely in Washington was Charlotte Webb. Charlotte, a doctoral student in Science and Technology Studies (STS) at my home university, Virginia Tech, spent six weeks for each of two summers sifting documents at the Henry Papers and Archives. The NEH and NSF grants underwrote Charlotte's diggings, and the Smithsonian awarded her a Research Associateship. After Charlotte began grappling with her own dissertation in the fall of 1995, another able STS graduate student, Michael Guarino, stepped in as my campus research assistant.

I also acquired valuable insights and information through conversations with scholars at national meetings of the History of Science Society and through several long-distance exchanges. These scholars included: Jed Buchwald (Massachusetts Institute of Technology); Geoffrey Cantor (University of Leeds); David Gooding (University of Bath); David Hochfelder (Case Western Reserve University); Frank A. J. L. James (The Royal Institution of Great Britain); Richard John (University of Illinois at Chicago); Joseph Mulligan (University of Maryland); Ryan Tweney (Bowling Green State University); and Friedrich Steinle (Georg-August-Universität Göttingen). Through e-mail, Michael Wolfe and James Selkirk, descendants of Joseph Henry's aunt, provided useful genealogical background. On a research trip to Albany, I also gained much from discussions with Stefan Bielinski (Colonial Albany Social History Project, New York State Museum) and Rebecca Watrous (Historic Cherry Hill). At Harvard University, Clark Elliott arranged for me to examine a particularly rare manuscript. And at the Niels Bohr Library of the American Institute of Physics, Joseph Anderson oriented me to the collections of historical books and photographs.

Going beyond written and photographic records, I found it useful to peruse the displays of Henry's original apparatus at the National Museum of American History and Princeton's physics building, Jadwin Hall. At both locations extra bonuses came with backroom tours: Roger Sherman guided me through the dusty cabinet of Henry apparatus upstairs at the National Museum; and Stewart Smith, chair of Princeton's physics department, showed me around a basement cache of old instruments. Although I also enjoyed poking around the nooks and crannies of the original Albany Academy building and the Henrys' Princeton house, I experienced the high spot of my on-site surveys at the Smithsonian Castle during the summer of 1995. The Keeper of the Castle, Richard Stamm, allowed Frank Millikan and me to climb to the roof of the building's main, clock tower—the twelve-story tower that the Henry family had stood atop to track Civil War troop movements. A year later, on 10 August 1996, the tower became the roost of an 821-pound bell commemorating the Smithsonian's 150th birthday.

Virginia Tech contributed substantial support by sharing costs and releasing me from my normal academic duties for two years. Especially helpful in the College of Arts and Sciences were Robert Bates, Gary Downey, Linda Fountaine, Richard Hirsh, Burton Kaufman, Adelene Kirby, Paul Sorrentino, and Daniel Thorp. In the university's Office of Sponsored Programs, Alisa Boothe, Ginger Clayton, and Mary Hunter helped orchestrate the financing of the project.

Finally, but most heartfelt, I thank my wife, Lynette. I also thank our three daughters—Holly, Emily, and Rebecca—who, like planets in resplendent conjunction, are at the moment all teenagers. They, along with members of my extended family, supplied essential moral support. Lynette also provided matchless literary advice.

Prologue: Know Ye Not That There Is a Prince and a Great Man Fallen

The black crepe framed only a placard on the front door, but it seemed to shroud the entire red façade of the Smithsonian Building. Seeing the crepe, passersby on Monday, 13 May 1878, probably did not need to read the placard's blunt message: "Closed on Account of the Death of the Secretary, Prof. Henry." Most residents of Washington knew that this sprawling Smithsonian structure, with its crenellated towers and stained-glass windows, was literally the home of Joseph Henry, the nation's most eminent physical scientist; he, his wife, Harriet, and their children occupied a second-floor apartment of the east wing. Having relinquished a professorship at Princeton to become the first Secretary of the Smithsonian Institution, Henry had devoted the last three decades to organizing and shaping it into a clearinghouse of American science. Whereas many citizens had already heard that the eighty year old scientist was ill, word of his death still threw them, according to the *Washington Post,* "into profound grief." One local resident reported that the announcement "carried a sudden shock to nearly everybody in the city."

The Smithsonian Building—later popularly known as the Castle—had locked its doors Monday afternoon. Late Thursday afternoon, the White House, Capitol, and other government buildings that girded the Smithsonian grounds also closed—for Henry's funeral. Senators and congressmen adjourned their regular sessions to join President Rutherford B. Hayes, Vice President William A. Wheeler, Supreme Court justices, and emissaries from all realms of official Washington at the New York Avenue Presbyterian Church.

A detail of uniformed artillerymen carried Henry's walnut casket into the church, preceded by twelve honorary pallbearers. Chosen from Henry's varied professional associates, the twelve ranged from Washington banker and art philanthropist William W. Corcoran to government astronomer Simon Newcomb and Princeton geographer Arnold Guyot. The pallbearers joined the family and other intimates in the front pews, as did the Smithsonian Board of Regents. Also sitting together were representatives of the

Light-House Board, the National Academy of Sciences, and, in larger numbers, the Philosophical Society of Washington—three groups over which Henry presided at the time of his death. Fewer members represented the National Academy than the local Philosophical Society, but the academy's luminaries stood out among the many mourners: botanist Asa Gray from Harvard, paleontologist Othniel C. Marsh from Yale, and Columbia president Frederick A. P. Barnard. Although the officers of the Light-House Board did not command the attention given the academicians, they did dramatize their grief through a nationwide order closing all the agency's offices and lowering the colors to half-mast on all its vessels. Also present for the funeral were members of the diplomatic corps, including the Japanese and British foreign ministers. Many mourners had to remain outside the church because the remaining seats were filled with representatives from other organizations with which Henry had been affiliated: the Telegraphic Operator's Association, the local Princeton alumni group, the Washington Monument Society, and the trustees of the Corcoran Gallery of Art.

Opening the funeral service, the choir of St. John's Episcopal Church sang a Mendelssohn anthem. One of Henry's long-time friends, Princeton theologian Charles Hodge, offered the prayer, showing much grief. The church's own minister then delivered, according to the *Washington Post,* a "most eloquent and powerful funeral discourse," taking as his text the Biblical passage: "Know ye not that there is a prince and a great man fallen this day in Israel?" After the choir closed the service, the family and near friends entered designated carriages to follow the hearse to Oak Hill Cemetery—two miles away, near Rock Creek in Georgetown. More than a hundred private carriages fell into rank at the rear of the procession, traveling along Pennsylvania Avenue and other streets lined with citizens paying their last respects. William Tecumseh Sherman, commanding general of the army, recalled: "I was present in Washington when this good man died; felt in the very air the evidence of universal grief; saw the President, Congress, and the Supreme Court who had ceased their labors to pay a just respect to his memory and follow him to his grave."

The next day the *Post* judged the funeral, "the most solemn and impressive ever witnessed in Washington, or perhaps in the country." The correspondent to the *Cincinnati Commercial,* however, began his assessment with a qualification: "Since the funeral of President Lincoln, there has not been in America so impressive obsequies as those of Professor Joseph Henry, Secretary of the Smithsonian Institute." The writer became more effusive when he focused on the gathering at the church, labeling it, "the most distinguished assemblage ever congregated on this continent." Also reflecting on "the remarkable number of distinguished men present on the occasion," a reporter for the *Pittsburgh Commercial Gazette* concluded, "one tries in vain to recollect a similar example. Certainly no other man of merely scientific eminence was ever so honored in this country." The correspondent to the *Louisville Courier-Journal* voiced a more definite opinion: "Never was there a greater outpouring of the good and the great."

But the effect of Henry's death extended beyond Washington. Thanks to the telegraph—a means of speedy communication that, the journalists unfailingly pointed out, Henry helped pioneer—essentially every newspaper in the United States logged the scientist's passing. Silver prospectors in Georgetown, Colorado, learned of his death in a note in the *Miner,* and wheat farmers in Omaha, could read a fuller account in the *Daily Bee.* The dispatch in the *Ledger-Standard* in New Albany, Indiana, began: "The death of Professor Henry . . . has caused more universal expressions of regret than that of any public man for years." The headline in the *New York Times* read, "A Loss to All the Nation." Soon the foreign press was running notices. Just across the border, the editor of the *Montreal Herald* suggested that Henry's death "robs the ranks of science in the United States of America of a man who was almost an American Newton."[1]

The day after the funeral, the Smithsonian Board of Regents convened. Presiding was Smithsonian chancellor Morrison R. Waite, chief justice of the U.S. Supreme Court. Among the board members were congressman and next president James A. Garfield, General Sherman, Yale College president Noah Porter, and Harvard's Asa Gray. The regents charged a committee with arranging a public program to honor Henry. By early December the committee's plans became formalized as the House of Representatives and Senate unanimously adopted a concurrent resolution to hold an extraordinary, joint evening session in memory of the scientist.

On 16 January 1879, the main Hall of the House was "thronging" with an "August assemblage" headed by President Hayes. "I see around me, congregated in this, the capitol of a great nation," Virginia senator and regent Robert E. Withers observed, "its highest functionaries in the executive, legislative, and judicial departments of government, distinguished diplomatic representatives of almost every civilized people, the chiefest dignitaries of church and state, men most renowned in peace and in war, those most honored in the world of science, of literature, and of art." Sharing the gavel with the Speaker of the House at this joint session was Vice President Wheeler, himself a Smithsonian regent. After an opening prayer by James McCosh, Presbyterian clergyman and president of Princeton, a series of senators, House members, and regents presented memorial addresses. Asa Gray, for example, sketched Henry's biography, recounting especially his two decades as a professor of natural philosophy (a field becoming better known as physics). Garfield and Sherman added personal reflections on Henry's tenure at the Smithsonian. A high spot of the three-hour session came when regent and Pennsylvania representative Heister Clymer read three memorial telegrams transmitted that evening from Great Britain. Routed through American telegraph magnate Cyrus W. Field, the messages were from executives of two transatlantic telegraph companies and a scientist—Sir William Thomson (later Lord Kelvin), a leading authority in Henry's main, or at least initial, area of expertise, electricity and magnetism. Indeed, Clymer emphasized the poignancy of receiving tributes transmitted by a technology that Henry's researches had helped make possible: "This evening from across the sea there have come to us, by means which his genius and immortal discovery have made possible, messages, telling of the estimation in which the name and fame of Henry are held in the Motherland." For his part, Thomson extolled Henry as a "perennial possession" of the United States, explaining that "Henry's name and works are yours forever, though you now mourn the loss of his life among you."[2]

The congressional ceremony climaxed a succession of memorial gatherings and other formal expressions of grief and homage. Earlier, within days of Henry's death, the Philosophical Society and the Light-House Board had met and issued official resolutions. So too had the Albany Institute—a learned society in the city of Henry's birth, education, and first steady teaching post. The College Chapel at Princeton—where Henry taught in his middle years—also had quickly sponsored two testimonial gatherings. Addresses presented at these and associated meetings constituted a core of memorials that would swell into a large body of remembrances following the congressional ceremony and, even later, after sessions by the American Academy of Arts and Sciences, the National Academy of Sciences, and the American Association for the Advancement of Science, which all delegated members to prepare memoirs in Henry's honor. In addition to printing the speeches presented in the Capitol building, the Senate and House further authorized that these texts be combined with the other principal memorials and published as a special volume. By 1881 the government had printed 15,000 copies of this lavishly bound, 528-page work. Smithsonian officials distributed their allotment widely, to distinguished citizens, professional societies, public libraries, and educational institutions. In reviewing the effusive tributes in the volume—including one by Garfield, now president—the *Philadelphia Public Ledger and Daily Transcript* found striking the unanimity shown by representatives of normally bickering political parties.[3]

Also in the months following the scientist's death, Harriet Henry, her three daughters, and the Smithsonian regents received a cascade of condolences and official resolutions lamenting Henry's loss and testifying to his accomplishments. Under Henry's direction the Smithsonian had established exhaustive, national and international networks of individuals and of learned and technical organizations to share information and receive Smithsonian publications. The nodes in the networks responded promptly to the notice of Secretary Henry's death. First, resolutions came in from regional academies of science in American cities ranging in size from Davenport, Iowa, to New Orleans, and states stretching from California to Delaware. Then, as the months passed, acknowledgments flooded in from around the world. Letters arrived from prominent international organizations as well as regionally active groups like Grey's Hospital Physical Society in London, the Imperial Society of Naturalists of Moscow, the Khedival Library in Cairo, the South African Public Library in Cape Town, and the Ministry of Public Instruction of the Republic of Chile. Even from the far reaches of Adelaide, Australia, came a formal expression of sympathy from a civil servant who identified himself as "Postmaster General, Supt. of Telegraphs, & Govt. Astronomer."[4]

At the joint congressional session, Sherman had ended his memorial meditations by expressing the hope that Henry's example would inspire the nation's youth to pursue the still undervalued profession of science. That Henry, a scientist rather than a luminary from "the more tempting and active scenes of public life," had been honored by "so marked and distinguished an audience in this the Hall of Representatives of the Capitol of our nation" should lead students to emulate him, and by so doing "their stimulated labors in the cause of that science he loved so well will erect to him a monument more lasting than of brass or marble." Apparently, among the nation's legislators, the possibility of a students' incorporeal monument did not preclude the certainty of a more substantial one. In May 1880, two years after Henry's death, Vermont senator Justin S. Morrill, author of the 1862 legislation that led to land-grant colleges and an old friend of the scientist, introduced a bill "for the erection of a bronze statue of Joseph Henry, late Secretary of the Smithsonian Institution." Specifically, the bill allocated $15,000 for the regents to contract with American sculptor William W. Story to create the statue and erect it on the Smithsonian grounds. After a brief debate in which senators raised doubts about awarding the commission directly to Story, the son of a late Supreme Court justice, rather than having an open competition, the bill passed unanimously. Representative Clymer guided the bill through the House, also with unanimous consent. A few days later President Hayes approved the act, thus clearing the way for a major national monument to a scientist—an unprecedented occurrence for a country in which, as the *New Orleans Daily Picayune* lamented at the time of Henry's death, "our monumental brass and marble are devoted almost exclusively to the memory of the heroes of the forum and the field."[5]

Equipped with photographs of Henry, a mask of his face cast at death but broken in shipment, a borrowed academic gown, and a "vivid memory" of once seeing the scientist at the Smithsonian, Story set to work in his studio in Rome, Italy. A year later while visiting the studio, Asa Gray could report that the sculptor "has put his heart into his work and wishes to give Washington a noble statue." As the statue neared completion, Smithsonian officials orchestrated a public unveiling ceremony to coincide with the semiannual meeting in Washington of the National Academy of Sciences. On the balmy spring afternoon of 19 April 1883, an estimated five- to ten-thousand persons blanketed the lawn near the northwest corner of the Smithsonian Castle. As John Philip Sousa led the Marine Band in his own special composition, a march titled "Transit of Venus," the guests of honor filed out of the adjacent, new National Museum Building to the large platform behind the veiled statue. With a few exceptions—Garfield had been assassinated, and his successor, Chester A. Arthur, was traveling in Georgia—the dignitaries marching to the platform composed the same extensive roster of top government officials, scientists, and other distinguished citizens and foreign ministers that had participated in the funeral and congressional memorial service. Also missing was Harriet Henry, who had died a little more than a year before; indeed, her three daughters were still wearing mourning as they took their seats directly in front of the sheathed statue of their father. Government clerks swelled the crowd of onlookers, a presidential directive having closed federal offices earlier that afternoon in deference to the occasion.

Sousa began the main program by leading the band in the "Hallelujah Chorus" from Handel's *Messiah,* and Princeton's A. A. Hodge, the son of the late Charles Hodge, delivered the opening prayer. After making brief comments, Chief Justice Waite, chancellor of the Smithsonian, signaled assistants to lift the white canvas from the statue. "The full glow of the setting sun fell athwart the bronze figure," reported the *Washington Post,* "and the dull metal became transformed and was clothed in a rich, golden glow." The crowd was silent for an instant and then broke into a sustained ovation, with guests on the platform rising. Then as the 200-voice Philharmonic Society joined the Marine Band in performing Haydn's "The Heavens are Telling," the news of the unveiling was flashed to the nation on a special line run to the platform by the Western Union Telegraph Company. A memorial oration from Noah Porter, president of Yale, capped the 90-minute ceremony. "Classic in style, faultless in elocution," is how one guest on the platform described the speech. "Many words of eulogy have fallen above Prof Henry's grave," the guest explained, "but these were not only the purest and fittest, they were also the strongest and most graceful."[6]

As the crowd dispersed to the strains of another march, the statue began its vigil: Rising some nine feet above a seven-foot granite pedestal and cloaked in an academic gown, Henry's figure stood facing his old home, the red sandstone Smithsonian Castle.

"Prof. Henry in Bronze," read the front page headline in the *Washington Post* the next day. For three cents, the cost of the newspaper, a reader could get a full account of the eminent guests at the unveiling, a depiction of the proceedings, unabridged texts of all the presentations, a sketch of the scientist's "Life and Services," and a history of the statue and its maker. The biographical sketch showed that Henry had a distinguished career as a researcher, teacher, and director of the Smithsonian, but curiously it contained little that would lead an uninformed reader to understand why the scientist was the object of such profuse public adulation. The journalists covering Henry's death in 1878 had been correct when they pointed out that never before had an American with a career dedicated solely to science received such a grandiose funeral. From our current vantage, we can add that never before *or since* have Americans memorialized a scientist to the extent that they memorialized Joseph Henry.[7]

How then do we explain "Prof. Henry in Bronze"?

Part of the answer lies, of course, in the public prominence that Henry achieved during his last thirty years as head of the Smithsonian. A fixture in Washington through nine presidential administrations, he exceeded the level of coast-to-coast visibility and Capitol Hill camaraderie experienced by the typical, transient congressman, senator, or cabinet officer. Also, even more than the Supreme Court justices with whom he shared long tenure in Washington, he escaped much of the controversy and notoriety often associated with political or juridical posts. Henry's celebrity as Smithsonian

Secretary, however, only partially explains the extravagant memorials and monuments. Ultimately, his renown as Secretary derived from his earlier scientific contributions—or, at least, people's perceptions of the contributions. That is, as much as the secretaryship embellished his reputation, his previous scientific researches secured and sustained his reputation. His earlier initiatives, particularly in electricity, magnetism, and telegraphy, distinguished him during an era of national scientific emergence and aspiration, and they provided him credibility and authority as he prospered in the Washington limelight.

Although as Secretary he still found time for scientific inquiry, he never again matched the attainments of his academic years. During his three decades at the Smithsonian, he made noteworthy contributions in, for example, optics and acoustics (primarily through his maritime projects, with lighthouses and sound signaling), but he never recaptured the intensity and novelty of his earlier investigations involving especially electromagnetic phenomena—cutting-edge experiments that commanded the attention of contemporary researchers, educators, and commentators in the United States and abroad. A close friend recalled that even Henry realized that, having taken the Smithsonian post, he had abandoned "the paths of scientific glory on which he had entered so auspiciously at Albany and Princeton." Apparently, Henry found consolation in fellow natural philosopher Isaac Newton's similar curtailment of his scientific investigations a century and a half earlier: forsaking his Cambridge professorship, Newton had accepted an administrative appointment in London as head of the mint. In a widely reprinted sketch of Henry's life—one that he himself wrote anonymously in 1871—he affirmed that his years in Princeton, not Washington, were the "most profitable" of his career.[8]

So what did Henry accomplish in science during these "most profitable" years? More to the point, what exactly led to his emergence in Albany and Princeton as the nation's leading physical scientist—a status that helped assure appointment and approval as the Smithsonian's first Secretary? To answer these questions as fully as possible, we need to reconstruct the first five decades of Henry's life, from his birth in 1797 through his settling in as Smithsonian Secretary in 1847. We need to probe his early and middle years, thus tracing how a boy from a family of modest means in a nation of scant scientific resources attained national and international prominence in the esoteric field of natural philosophy. Only then can we more adequately explain the prodigious outpouring of public and private appreciation attendant on his death.

An Unsettled Beginning 1

George Washington was still alive when Joseph Henry was born in Albany, New York, in late 1797. When Washington died two years later at Mount Vernon, Virginia, post riders needed nine days on horseback to relay the news to Albany. By 1878 the dots and dashes signaling Henry's own death would require only a few minutes to travel the wires along essentially the same path.

That Albany ranked as a major American metropolis at the time of Henry's birth accentuates the magnitude of the change that occurred during the eight decades spanning the eras of post riders and telegraph operators. Indeed, the time required for receiving news of Washington's death does not signify that Albany was an isolated backwater along the Hudson River but that it was a main urban center. Residents would be pleased that post riders in questionable weather reached their city in only a little over a week. As the nation's ninth largest municipality, Albany flourished as a center of industry, agriculture, and transportation, and in 1797, it secured its position as a center of government by becoming the permanent capital of New York.

Befitting the capital of a prospering state, the city had its own locally controlled bank and three newspapers. It boasted streets paved with cobblestones and lighted at night by whale-oil lamps. Clean water supplied the town through a long tube of bored tree trunks. The many residents with ancestral roots in either Holland or Scotland could worship at Reformed Dutch or Presbyterian churches—kindred churches that shared a common Calvinistic heritage. Methodists and Roman Catholics had their own churches, and the collective Christian community carried enough clout not only to halt all commerce on Sunday but also to prohibit riding for pleasure on the Sabbath. But religious fervor did not impose too rigid a structure on Albany's citizens, who enjoyed many educational opportunities and recreational outlets. Residents could enroll their children in the private schools of various schoolmasters or borrow books from a 400-volume library for a $4 yearly subscription fee. Citizens could also attend meetings of the Society for Promotion of Agriculture, Arts and Manufactures, visit exhibits at the

Albany Museum, or have their portraits painted in a local studio. For comradeship, friends could tarry in a variety of coffee houses and taverns. Even outside taverns, townspeople enjoyed easy access to liquor, beer, mead, and hard cider. At an ex-mayor's funeral, the guests drank so much from a keg, which the deceased himself had put aside for the occasion, that they had to be ferried home on ox-sleds.

Ox-sleds ranked near the bottom of the list of vehicles available for transportation. After all, Albany was a river town, situated on the Hudson's west bank near the confluence with the Mohawk River. When the Hudson thawed in the spring of 1795, fifty sailing vessels lined the docks, and ninety sloops made regular runs that year. The fastest could complete a round-trip to New York City, 145 miles to the south, in as little as seven days. And in 1797 Robert R. Livingston's small steam-propelled vessel foreshadowed the fast steamboats that Robert Fulton would launch on the Hudson a decade later. On shore the docks connected with an expanding network of post roads and turnpikes that radiated throughout the river valleys and surrounding mountains. Two stagecoaches, each drawn by four horses, ran daily to nearby Schenectady. For about an $8 fare, an Albany politician or merchant could reach New York City in two days.

Being a node in the transportation network, Albany also functioned as a gateway to the "Far West." Many New Englanders were emigrating by wagon or sleigh into the expanses of western New York state. By this time, state officials had expropriated most of these and other territories that the Iroquois Confederacy of Native Americans had traditionally inhabited; chiefs from the Cayuga Nation, for example, met in 1799 with Albany legislators to "sell" all their remaining lands.

As elsewhere in the North, slavery was widely tolerated in the city in which Henry was born. A calamitous fire highlighted the white citizens' scapegoating attitudes toward African Americans. Although the municipality owned hand-pumped "fire-engines" to spray water—and mandated that citizens participate in fire lines using their own buckets to supply the engines with water—major fires destroyed dozens of houses in 1793 and 1797. In the aftermath of the first fire, residents convinced themselves that slaves had been responsible for the blaze. The city's common council ordered that all slaves be placed under a nine o'clock curfew and that all adult, white males take turns serving on a 24-man night watch. In response to the particular fire, moreover, authorities identified arson suspects from the small population of slaves and hastily executed a woman named Dinah; a few months later, they hanged two more female slaves, Bet and Dan, for the same purported offense and then a male named Pomp. In the mid-1790s slavery was so acceptable that the mayor purchased for the city a man named Pompey for 85 English pounds. During 1803 a local paper, the *Gazette*, advertised: "Those who wish to buy one of the most valuable negro wenches, one free from ever having had a husband or child, and one not in the least used to 'black' company, and free from every vice of any moment; will please to inquire of the editors of this paper, from whom they may know the price, and the present owner."

With a population that exceeded 5,000 by the beginning of the nineteenth century, Albany was home to a variety of businesses. These ranged from agricultural pursuits such as the production of maple sugar to industrial endeavors such as the manufacture of glass. But the most visible business by far was politics. In Henry's earliest years, John Jay and then George Clinton held sway as governor in the capitol building completed in 1799. Presiding over a convention in 1801 to revise the state's constitution was Aaron Burr. Two years later, in a duel in nearby New Jersey, Burr would shoot another political leader with close ties to Albany, Alexander Hamilton. Generally, a group of families of Dutch descent dominated the Albany political scene—the Schuylers, Ten Broecks, Ten Eycks, and Van Rensselaers. In 1809 writer Washington Irving, drawing on his experiences in New York City and the Hudson River Valley, would immortalize similar families in his satirical Knickerbocker tales.[1]

In the mid-1770s, two Scottish families had added their names to the growing register of immigrants in the Albany area—the Alexanders and the Henrys. Hugh and Agnes Alexander would eventually share with James and Nancy Henry a grandson, Joseph Henry. But well before Joseph's birth, voyages from Scotland to New York had linked the two families. When the Alexanders and six of their children crossed the Atlantic aboard the *Gale* in May 1774, some of James and Nancy Henry's children were also on board, escorted by an uncle. The next year, at the urging of the uncle, who portrayed America as a paradise, James and Nancy Henry themselves left Scotland presumably with their remaining children and reunited the family in the Albany area.

Joseph remembered his grandfather James Henry as "a man of reading" who was "shrewd" in his understanding of religion and history, particularly theological controversies and Scotland's past. He also recalled that his grandfather had spelled the family name as Hendrie or Hendry before emigrating to America. Apparently, while still a boy, James had witnessed in 1745 the entry into Glasgow of Charles Edward Stuart, the fabled Bonnie Prince Charlie, who was pressing his brief and ultimately unsuccessful military campaign to regain the British throne. Eventually, James Henry married Nancy Herron—whose brother later brought some of the Henry children to America. Landing in New York in 1775, James and Nancy traveled by sloop to Albany, where they found themselves upset and frightened by a gathering of American Indians. After briefly residing in Albany and nearby communities, and still wary of the Indians, they finally settled with their three daughters and two sons in Bethlehem, a rural community just south of the city limits of Albany later absorbed by the town of Selkirk. They were residing on the large farm or estate of established landholder Rensselaer Nicoll, perhaps as tenants. Remaining at this site, James lived into his nineties, long enough to become "the grandfather of a grandfather," dying about 1830.

Joseph's maternal grandparents actually included a step-grandmother. Hugh Alexander's first wife, Eufane Stevenson—mother of Joseph Henry's mother—had died in Scotland in 1765, probably while giving birth to her ninth child. Hugh then married Agnes McClelland and this couple had nine more children, the first five born in Scotland and the rest in New York. The immigrant family apparently first settled in Galway, a village about thirty-six miles north of Albany. Joseph remembered his grandfather Alexander to be a "very ingenious" man who, upon relocating southwest of Albany to Delaware County, constructed a mill with all its complex machinery, even cutting his own grindstones. During the Revolutionary War, American Indians apparently drove him from the mill and he went on to do some type of skilled mechanical work as an "artificer" for the Continental Army. With the war's end in 1783, he manufactured salt at the mineral springs in Salina, in midwestern New York. The family soon moved to Bethlehem, where they participated in the local Reformed Church and crossed paths once again with the Henrys. Finally, in the late 1790s, Hugh, Agnes, and some of their progeny returned to Galway, where at least one of the couple's sons still resided. After Hugh died in 1803, Agnes and various family members remained in Galway, where she would help raise her step-grandson Joseph Henry.[2]

In mid-December 1785 Ann Alexander and William Henry, Joseph's parents, were wed in Albany's Reformed Dutch Church—a centrally located municipal landmark. The bride was almost twenty-five years old and the groom was nearing twenty-two. The newlyweds' first child, a son, was born nine days after their marriage. By some accounts, the precipitant parents set up housekeeping just below the outskirts of Albany—presumably, at or near the farm where William's parents were living in Bethlehem (at a site perhaps immediately south of the Van Rensselaer estate known as Cherry Hill). What is certain, by at least 1788 on, Ann and William were living in a modest house less than a mile further up the Bethlehem-Albany turnpike, in Albany's city limits. According to tax records, their home fronted a stretch of the turnpike eventually known as South Pearl Street (near its intersection with Lydius Street). Ann and William's first son apparently died in his early years, as did a second son born about a year after the first. In 1789 the couple was favored with a healthy daughter, Nancy. Eight years afterward, on December 17, 1797, Joseph was born, and finally, in 1803, his brother, James. One of Joseph's earliest Pearl Street memories was stowing away as a tot in a passing baker's wagon and ending up, for the first time, on Albany's imposing State Street, surrounded by the new capitol and other towering buildings. Apparently, the youngster experienced "an agony of apprehension," worried that he was inextricably caged by the ring of looming structures.[3]

Two of Joseph's closest scientific colleagues later offered glimpses of his mother, Ann. In a memorial essay on Joseph, Asa Gray reported: "She is remembered as a lady of winning refinement of mien and character, of small size, with delicate Grecian features, fair complexion, and when young she is said to have been very beautiful." Simon Newcomb echoed Gray's characterization and added: "Like the mothers of many other great men she was of deeply devotional character. She was a Presbyterian of the old-fashioned Scottish stamp, and exacted from her children the strictest performance of religious duties." She and William had Joseph baptized, one month after his birth, at Albany's First Presbyterian Church—now the family church, located a few blocks from their home. Joseph himself recounted that, when he first learned to talk, his mother taught him a paraphrase of the closing verse of the Twenty-third Psalm; more than sixty-five years later, he still found himself involuntarily repeating the rhyme:

> Goodness and mercy all my life
> Shall surely follow me;
> And in God's house for evermore
> My dwelling place shall be.[4]

William Henry seems to have held various jobs. His occupation through 1790 is listed as "laborer" or "day laborer" on the baptismal records for the Henrys' first two sons and daughter, as registered at Albany's First Presbyterian Church. A few years after his death in 1811, however, a document settling the estate (he died without a will) describes him as "late of the City of Albany cartman deceased"—a driver of a cart, or teamster, being a formally licensed occupation in the municipality. Joseph remembered his father as owning a small sloop on the Hudson River—a memory reinforced in later years when he encountered an Albany resident who had been a sailor under "Billy Henry"; fittingly, the family's home on South Pearl Street was only about one-half mile from Albany's main wharf. A reasonable conjecture would be that William worked at transporting goods along road and river. Whatever his livelihood, he was heading a household of modest means: for the fiscal year of 1799, he paid city taxes of 85 cents, on a house appraised at merely $910 and personal possessions valued at only $62. Most likely, the house was a simple frame structure abutting a larger dwelling. His sole possession to survive, a 1775, translated, Edinburgh edition of the Baron de Montesquieu's *Reflections on the Causes of the Rise and Fall of the Roman Empire,* suggests that he inherited some of his own father's enjoyment of reading. The book carries multiple autographs and statements of ownership by William, along with the purchase price and two dates, 1807 and 1810—the later date inscribed beside the word "Galway," the town where Ann's stepmother and some siblings resided.[5]

William Henry had one other significant attribute, which, lamentably, led to his own fall. He was an alcoholic. Many decades later Joseph revealed this fact to his longtime Smithsonian clerk and confidant, William J. Rhees. In a private conversation Joseph disclosed that his father was "intemperate" and "died in a fit of delirium tremens" and that this "made a lasting impression on his mind."[6] William succumbed to his alcoholism at about age forty-seven, in October 1811, two months short of Joseph's fourteenth birthday. This fatal fit of "the DT's," involving mental confusion, disordered behavior, and uncontrollable trembling, may have continued for several days and probably culminated years of heavy drinking. William's drinking and death no doubt created within the family severe disruptions and stresses that, in turn, affected the children.

The "lasting impression" that it made on Joseph probably had deep-seated consequences of which even he remained unaware. Though psychological interpretations of historical figures are risky, childhood traumas of the type Joseph experienced frequently engender lifelong patterns of personality and behavior.[7] For example, it is common for children of alcoholics to become overachievers. Plagued by feelings of inadequacy and guilt regarding the parent's drinking, the children find that by excelling at some task they not only gain a modicum of personal attention (and, thus, bolster their self-esteem) but also reinstate some sense of family dignity. Faced with the pressures, disruptions, and uncertainties of family alcoholism, these children often implicitly adopt other idiosyncratic but useful behaviors. They learn to keep negative or angry thoughts private and to give voice to only positive expressions. Similarly, they learn to accommodate others' wishes and avoid interpersonal conflicts. The children commonly come to realize that they should simply keep silent about the parent's embarrassing problem.

Ironically, these misbegotten behaviors serve the overachievers well, not only as children but once they become adults acting in wider society. Despite harboring self-doubts and insecurities, the grown-up overachievers gain accolades for being model citizens, effectual and responsible. They frequently gain praise for being compassionate nurturers, although they may be motivated by an external compulsion to please others rather than by genuine, innate altruism.. As children they have learned that by being supportive and accommodating they can gain a semblance of control over an otherwise unpredictable social environment. Not surprisingly, the adult overachievers consolidate control—something that they so lacked as children—by assuming positions of authority. Teaching is a common career choice, with its combined outlet for humanitarian action and leadership.

As we will see, Joseph seems at least in part to have assumed the role of overachiever in reaction to the traumas associated with his father's alcoholism and death. In Joseph's case, the pattern of achievement perhaps is accentuated by his position as oldest surviving son in the Henry household—a position that counterpoised privilege and responsibility. We can certainly speculate that Joseph, being the oldest son *and* the child of an alcoholic, was doubly motivated to be not only achievement-oriented and highly responsible but also exceptionally accommodating and supportive.

Of course, Joseph's personality as an adult would reflect more than having been the oldest son and the child of an alcoholic whose drinking proved fatal. Also involved were his innate temperament and culturally instilled attitudes. In temperament he displayed an introspective, self-reflective bent. As he later told his daughter Mary, even as a youngster he puzzled over "his own frame & being . . . almost to the verge of Insanity." A major cultural influence would be the Scottish Presbyterianism of his childhood, which emphasized piety, humility, personal salvation, trust in divine providence, and selfless service in the world. His mother began a process of Calvinistic religious and moral training that was to be reinforced by his fellow Presbyterians at Princeton and by the wider Christian culture of early-nineteenth-century America. Indeed, it was a culture intertwined with a secular, political culture that similarly emphasized Americans' patriotic obligations to serve their fellow citizens. This Presbyterian and republican milieu no doubt contributed to Henry's value system of duty, rectitude, and compassion.

The "lasting impression" that William's drinking and death made on Joseph also probably manifested itself in his lifelong advocacy of temperance. As a thirty-three-year-old professor at the Albany Academy, he would serve as vice president of the Albany Young Men's Temperance Society. A few years later, advocating a resolute but not overzealous response to the problem of alcoholic beverages, he commented to his brother: "There is no objet of the day in which I feel so much interest as that of the temperance cause and there is none I am convinced which every good citizen is more bound to support. . . . The temperance society appears to me to be founded on philosophical principles. Its objet is not if I understand it right to reform those who have already contracted the habits of a drunkard but to protect the rising generation & those who have not yet acquired the habits. It indeavours to affect this object by rendering so unfashionable the custom of tippling & the common use of spiritous licquors as that no person except the most low and abandoned will be seen drinking in Public." As a professor at Princeton he would work toward this objective, routinely exhorting his students to avoid the pitfalls of drink. Theodore L. Cuyler, a former Princeton student who gained prominence as a Presbyterian clergyman and writer on temperance, recalled that, though Henry avoided public speaking, he would willingly address this compelling issue: "The only speech I ever heard from him was a familiar talk to the students . . . in favor of total abstinence from the use of intoxicants. His main point was that if they drank moderately, they would be tempted to 'take to excessive drinking in case that any misfortune or disaster overtook them,' and he illustrated his position by a most pathetic incident that he had known in Albany."[8] Cuyler's avoidance of specifying the incident, conceivably William's death, perhaps reflected this clergyman's desire not to offend or upset Mary Henry—to whom Cuyler was writing after Joseph's death. Another possibility is that Cuyler simply did not recognize William as the focus of the incident because Henry had concealed his father's name. Desiring not to tarnish his father's name—or, more generally, his own and his family's name—Henry may have cloaked the details when recounting the "pathetic" Albany incident to his Princeton students.

In later years, as might be expected of a child of an alcoholic, Henry often did more than cloak details when reminiscing about his father—he actually dissembled. That is, he hid the reality of his father's alcoholism by creating an alternative version of circumstances. And his motive seemed to be protection of his and his family's reputation. In depicting his childhood to his children, his friends, or the public, he preferred the pretense that he was about six or seven when his father died of some unspecified but presumably natural cause. This version enabled Henry to dissemble even further, providing a socially acceptable explanation of another troubling fact perhaps related to his father's alcoholism—that his parents sent him, at a tender age, to live with his step-grandmother in Galway. In this rendering, Henry implies that, because of his father's untimely death and the resulting financial constraints, his widowed mother was forced to put him in the care of his step-grandmother and an uncle when he was only six or seven years old. In one version, he maintains that the uncle adopted him. Further complicating the story of Henry's early years—and attesting to the dislocations he actually endured as a child—is his lifelong confusion about his age. He believed that he was born in 1799 or even a few years later, rather than 1797; thus, he regularly misjudged himself to have been at least two years younger than he really was.[9]

Whether he went to Galway between six and seven, as he thought—or eight and nine, the more plausible age—his father actually did not die for at least another four or five years. What further entangled family affairs when Joseph was genuinely five, and his sister, Nancy, was thirteen, was the birth of another son, James. Faced with the demands of a younger child and coping with a husband who perhaps already displayed or was developing a drinking problem, Ann seems to have arranged, sooner or later, for her stepmother and one of her brothers to care for Joseph. The brother likely was Ann's twin, John Alexander, who had settled in Galway as an adolescent and now was raising his own family on the outskirts of town. Whatever triggered the thirty-six-mile separation of Joseph from his family, he was in Galway contending with his new situation by July 1808—as indicated by the earliest surviving document written by him, a letter drafted at age ten with his teacher's aid. "Honoured father & Mother," he begins, obviously following a prescribed form, "I embrace this Opportunity to write to you to inform you that I am well at this period Like wise Granmother and all the rest of my Relatives, and I flatter my self that these few Lines will find you enjoy-ing your healths, and my brother & sister it is a general time of health here. I en davour to Make the best improvement in my Learning that I possibly can though at the first it seemed a Little irksome and hard, and I hope to gain the point at Last—pray Dear parents accept of my most humble duty to your selves and kind Love to my Brother & sister. I subscribe myself hoping You will look over the blots from your son. Joseph Henry."[10] Perhaps the stains of ink that he hoped his parents would disregard also signify profounder psychological stains that would carry lifelong consequences. At any rate, it is fair to assume that the tragedy of his father's drinking was compounded by this earlier trauma of being separated from his parents and siblings.

Joseph, who may earlier have had some type of instruction in Albany, now studied under Galway schoolmaster Israel Phelps, a "*learned* pedagogue" and active Presbyterian. Not a motivated student, Joseph briefly fancied that he might become a chimney sweep after watching the impressive acrobatics of an older sweep who scaled the chimney on his grandmother's house. Eventually he began working during mornings in a country store, hence limiting his schooling to afternoons. John Broderick, an Irish immigrant who owned the store and apparently assumed responsibility for Joseph's daily life, proved to be a disciplinarian, administering two "sound floggings" to his helper. Thanks to ordinary boyhood escapades, Joseph was no stranger to the ire of adults; a childhood chum reminded him decades later that, while blindly throwing stones down a hill with other boys one evening in Albany, Joseph almost ended up in "the watch house" when caught by some disgruntled adults who had been grazed by the stones. But Joseph needed no reminder in later decades of Broderick's floggings. The first came on an election day when Joseph joined a gambling game that involved placing money on numbers painted on a cloth and rolling dice to determine the winning wagers. Joseph won. To his chagrin, however, when he boasted about his modest gain to Broderick, he received a whipping. (Later, another bad experience with gambling—a card game for money in which his adversaries accused him of cheating—led him to forswear games of chance forever.) The second flogging occurred, as he later told Rhees, because of "some improper language he used to a colored servant." Given the choice of apologizing to the woman or receiving a whipping, he stubbornly chose the latter, only to find Broderick's beating so severe that he became infuriated and decided to run away. "However, he thought a good deal of it that night & the next morning he determined to remain & try to be better." Joseph's prudent, self-reflective assessment recalls an anecdote about his deliberate nature that he was fond of telling in later years. When his step-grandmother offered him his first pair of boots, he could not decide whether to have the cobbler fashion them with round or square toes. After the boy's many visits to the shop trying to choose, the exasperated shoemaker fashioned a pair of boots with one round and one square toe. (As a clincher to the anecdote, Henry liked to add that a school chum later pilfered the boots, in wintertime—but was easily tracked down by the distinctive tracks he left in the snow.)[11]

Another event that Henry recalled from his Galway years took on such import in his later tellings that it began to sound apocryphal. Like Alice entering Wonderland, young Joseph supposedly followed a wayward pet rabbit into a passageway that led to the boy's own land of marvels—the village library. More specifically, he crawled under a church in pursuit of the rabbit, got distracted by an opening in the floor, wiggled through, and found himself alone in a side room that housed Galway's modest library. Opening the novel *Fool of Quality,* by eighteenth-century British writer Henry Brooke, he was captivated. This lengthy narrative—whose moralistic, Christian message also had once captivated John Wesley—traces the successful efforts of a youth named Henry (but called Harry) to overcome the adversities that befall him after his father dies and bequeaths most of the family's wealth to Harry's older brother. Joseph returned repeatedly through his passageway to devour this and other books—a penchant no doubt heightened by the secret entry. Eventually obtaining permission for front-door access, he exhausted every novel and play in the collection. Having first entered the library as a boy "who did not care to learn anything," he had blossomed into one with "a desire to improve," although he still limited himself to light literature. Unlike other boys who rollicked in sports, Henry relished his reading and soon became "a story-teller to his comrades" who gathered at Broderick's store. All this occurred, Henry recalled, when he was about twelve years old, which translates into the more plausible age of about fourteen.

Close to the same period, he developed a related passion—the theater. On a visit to his mother in Albany, relatives took him to see two plays, his first experiences with drama. Smitten, he returned to Galway and delighted his friends by involving them in reenactments of the performances. He also wrote a play titled "The Fisherman of Bagdad," and then enlisted his friends to stage it. He later had the opportunity to immerse himself more deeply in the life of the theater when, sometime in the next few years, he moved back to his mother in Albany. In the doctored versions of his youth that he later preferred, he alternated between explaining the reunion as precipitated by Broderick giving up the store or his Galway uncle dying. However, while records reveal nothing about the fate of Broderick's enterprise, they do show that his Uncle John lived for decades more; as for other uncles dying about the time of his reunion, the documents are equivocal.[12]

When Joseph returned to Albany in 1814 or 1815 he found an even more cosmopolitan city. The War of 1812 had come and gone, with American and British forces passing through the city, including Commodore Oliver Hazard Perry shortly after his naval victory over the British fleet on Lake Erie. With peace again prevailing by late 1814, more than 200 sloops regularly plied the Hudson River to and from Albany, as did seven steamboats—traffic that would soon increase dramatically with the building of the Erie Canal, an east-west link between Albany on the Hudson and Buffalo on Lake Erie. In the city, some years earlier municipal leaders had changed the name of Cow Lane to Liberty Street, Nail to Lutheran Street, Barrack to Chapel Street, and Swallow to Knox Street. More recently, they had installed iron pipes to transport reservoir water to the residents, who now numbered 10,023 (comprising 4,860 free white males, 5,063 free white females, and 100 slaves). Besides enjoying a new theater on Green Street that raised its curtain in early 1813, citizens had either opened or were building a range of schools: the public Lancastrian School, the Albany Female Academy, and the Albany Academy for boys. In the next few years the state legislature not only mandated a school in Albany for "colored children" but also passed a law to begin on 4 July 1827 phasing out slavery all over New York.

The change that had occurred in Albany's physical makeup was made even more remarkable for Joseph by his perception of the makeup. Oddly, due to the happenstance of departing from his Pearl Street house in one direction but coming back a few years later from another, he perceived the spatial layout of his neighborhood to be reversed during the two phases of his youth. Daughter Mary heard him explain this notion in an 1863 conversation (in which he still mistakenly underestimated the age at which he went to live with his grandmother). "He returned to the city," she noted in her diary, "from a different direction from that in which he left it & the house appeared to be on the wrong side of the street so that now when he looks back all his life before seven years of age seems to him to have been passed upon one side of the street & the rest of his time spent in the same house upon the other."[13]

Job prospects with Albany relatives, coupled with the war's end, probably contributed to Joseph's return to Albany. His income would be important not only for his widowed mother but also for his own livelihood, now that he had reached a responsible age, about fourteen or fifteen in his own estimation. Henry's uncle Hawthorn McCulloch, a prominent citizen and husband of a deceased sister of Ann, hired Joseph to work in his prospering Albany brewery. Soon, however, a cousin on William Henry's side of the family took Joseph as an apprentice in the trade of silversmithing and watchmaking. The cousin, William Selkirk, did his silversmithing along with a younger brother, Joseph's fellow apprentice, in the shop of John Doty. Spending about two years working with jewelry, Joseph gained, as he later told Rhees, "skill in manipulation & some knowledge of metals." Joseph's future brother-in-law, Stephen Alexander, later underscored this point, emphasizing the relevance of Joseph's silversmithing: "This under Providence was the source of much of his facility in combining apparatus and experimenting." But Joseph did not enjoy silversmithing and, as he disclosed to a class of his Princeton students, "he was considered too dull to learn the trade." What he did like, and master, was the theater.[14]

From 1812 through late 1814, as the war with the British sputtered along, the Green Street Theater had flourished, thanks to the enthusiastic patronage of troops garrisoned near Albany. Caught up in the excitement, a group of young Albany residents outfitted another theater and staged their own successful plays. Gifted with handsome features, a good voice, and acting talent, Henry eventually joined and emerged as the leader of this amateur group, known as "The Rostrum." Thurlow Weed—at the time, an aspiring Albany journalist and up-and-coming politician who was Joseph's age—recalled that his friend "became a bright particular star." "His young *Norval*, *Damon*, and even *Hamlet*," Weed further recollected, "were pronounced equal in conception and execution to the personations of experienced and popular actors." So impressed were his friends that they encouraged him to make the stage his profession, a suggestion that became a real possibility when the manager of the Green Street Theater offered him a salaried, permanent engagement. Meanwhile, however, Joseph had discovered another muse—not in art, but science.

Just as his literary muse had appeared by happenstance while Joseph was chasing a rabbit, so too did the scientific Muse materialize by a fluke. Again, a book provided the stimulus. Living in his mother's Albany house, Joseph came across an unlikely volume lying on a table: *Popular Lectures on Experimental Philosophy, Astronomy, and Chemistry.* Written by George Gregory and published in London in 1808, the book sustains a friendly, conversational tone as it presents the fundamentals of physical science for an intended audience of "Students and Young Persons." The owner of the mislaid book was a young Scotsman, one of the boarders that Ann Henry took in regularly to support herself and the family. The book so transfixed Joseph, then about age eighteen or nineteen, that the man gave it to him. Two decades later, on conveying the book to his own son, Joseph inscribed a note on the inside cover that captured his memory of discovering the book:

> This book although by no means a profound work has under Providence exerted a remarkable influence on my life. It accidentally fell into my hands . . . and was the first book I ever read with attention.
>
> It opened to me a new world of thought and enjoyment, invested things before almost unnoticed with the highest interest[,] fixed my mind on the study of nature and caused me to resolve at the time of reading it that I would immediately commence to devote my life to the acquisition of knowledge.

Though this note has the same apocryphal flavor of the rabbit story, Joseph repeatedly maintained in later years that this science book, as he told Rhees, "led to a new epoch in his life."

Henry actually came across only the first of two volumes of Gregory's *Lectures,* the volume covering experimental philosophy and astronomy. In the lead paragraph of the introductory "Lecture," Gregory suggests that the study of physical science offers pupils some payoffs that they presumably desire: not only knowledge to satisfy curiosity but also a means of "fitting yourselves for company and conversation, and of enabling yourselves to proceed gradually in the paths of science, till you arrive at distinction and eminence." Gregory specifically promises his "young friends" answers to basic questions concerning common physical phenomena: the sun's apparent rising and setting, falling bodies, reflection, magnification, thunder and lightning, clouds and precipitation, rainbows, tides, geysers, volcanoes, and earthquakes. For example, after dramatically recounting everyday experiences with thunder and lightning, he leadingly adds: "You have probably never once thought what can be the cause of this thunder and lightning. But will you not be astonished to see it imitated on a smaller scale, the same noise excited, a rapid fire sent forth like that, and producing similar effects?"

Himself a vicar, Gregory professes that ultimately the answers to all questions about natural phenomena trace to "the hand of God." But, he continues, the orderly modes "in which the Divine Wisdom acts and governs the material universe" are open to study as natural laws; though mere mortals "cannot trace effects to their remotest causes," they can discern many of the laws controlling natural processes and events. Unlike their ancient predecessors who wallowed in "fancy and conjecture," Gregory asserts, modern philosophers ground all study of such laws on "the direct and positive test of experience." Early in the seventeenth century, Francis Bacon systematized this modern view, establishing "that it was by experiment alone that any thing in philosophy could with certainty be known." Later in the same century, by putting this precept into practice, Robert Boyle became the "father of modern philosophy." And in the decades around 1700, by masterfully unifying diverse physical phenomena, Isaac Newton further transformed natural philosophy into "a system of truth, illustrated and proved by experiment." Only after completing this introductory paean to the powers of God, experimental philosophy, and the British triumvirate of Bacon, Boyle, and Newton, does Gregory proceed with his chatty "Lectures." In covering experimental philosophy he illuminates everything from gravitational attraction and the laws of motion to hydrostatics and atmospheric phenomena, from light and vision to magnetism and electricity. He ends this unit with a lecture on "The Mechanic Powers," leaving his young readers with the message that experimental philosophy underpins all practical "contrivances"—that is, for all technological mechanisms, it is experimental philosophy "which explains their theory and use, and which extends their application." Henry, his appetite whetted, quickly went on to other books. An older Albany resident recalled the eager scholar frequenting book auctions and reading voraciously.[15]

Supposedly, Henry's discovery of Gregory's *Lectures*—and his resulting reorientation—occurred while he still reveled in dramatic pursuits. Weed picks up the story at this point and explains that, with the invitation from the Green Street Theater pending, the principal of the Albany Academy stepped in and offered Joseph a free course of study at the new school. Uncertain which path to follow, Joseph turned to Weed and two other friends for advice. Weed and one friend counseled that the theatrical career would lead to "assured fame and fortune." The third friend disagreed, arguing that Joseph was still young enough to return to the theater after gaining a solid education—an education that would, in the end, help ensure his success as an actor. The next day, "Joe Henry" supposedly turned down the theatrical offer and began attending the Albany Academy. In a later retelling of this story, Weed added that silversmith Doty influenced Henry's decision when he agreed to release the young apprentice from his remaining indentures only if he chose the academic rather than the theatrical path. But how had the academy's principal, T. Romeyn Beck, known about Henry? According to another story, one passed on by an academy trustee, the school had recently asked a local silversmith to repair a piece of scientific apparatus, an air pump, and the silversmith dispatched apprentice Henry. "The Trustees were so pleased with the skill with which the work was done, and above all with the intelligence and eagerness for knowledge evinced," the story continues, "that they offered him free tuition at the academy, and obtained a release from his engagements with the Silversmith by which he was enabled to accept their offer." Incidentally, by mid-1818, the third friend's advice—to defer the dramatic career—proved to be sound, but for a different reason: the Green Street Theater failed, and the local Baptist Society bought the building.[16]

Neither Weed's nor the trustee's account meshes fully with Henry's own recollection of what happened after reading the Gregory book. According to Henry, though still absorbed in acting when he chanced upon the book, he had by then been forced to give up silversmithing because either his cousin or the shop owner had abandoned the business. Abruptly resigning from the theater group and supporting himself with odd jobs during the day, he pursued his new, academic passion by attending evening classes, supposedly studying under two professors from the Albany Academy. He devoted long hours to scholarship, he later recalled, and found it difficult to juggle his day jobs and night courses. Taking a friend's suggestion that teaching might provide an alternative, Joseph hired a man to tutor him privately in the topics essential for instructing schoolchildren. He also likely studied French in a summer class taught by Albany visitor James Hamilton, a British creator of a popular system of language instruction who had come in 1815 to New York.[17]

Henry further recollected that, with the tutor's help, he solidified his grasp of old subjects and progressed rapidly with new material. The Albany Academy admitted the determined student, but lack of money forced him to postpone his studies. About 1817 James Selkirk, an uncle who lived about seven miles south of Albany in the village of Selkirk, helped find him a teaching position in a country school serving the local district. As a new teacher Joseph earned about $8 per month, an amount that quickly increased to $15 after he proved himself to be competent. As a perquisite of the position, he "boarded round" at the houses of his pupils and particularly the home of a local Methodist minister. Building a reserve of money to support himself and honing his mind in expectation of more schooling, Joseph eventually returned to Albany, matriculating as a regular student in the academy. His program of study would not, however, remain unbroken. When his money ran out he went back to teaching and boarding in his pupil's homes. Theodore Cuyler recalled Henry saying that while staying in the homes of his "rustic patrons" he sometimes "laid down on the floor & pored over Silliman's Journal' by the light of the pine-knot fire"—Benjamin Silliman's recently launched *American Journal of Science* being the only comprehensive scientific periodical published in the United States. Finally, with his savings replenished and his mind further sharpened through reading, Joseph returned to the academy.[18]

To be sure, elements of truth seem to exist in Weed's and the trustee's account—that is, of Principal Beck luring Henry away from the theater with an offer of free tuition after academy personnel inadvertently discovered the young silversmith's academic potential. But Henry's own narrative rings even truer. Beck himself wrote a letter of recommendation for Henry later, in 1826, that adds credence to this version. "Mr. Henry learnt the trade of a silversmith & made good proficiency in it, but having a great desire to advance in knowledge, entered himself in the Albany Academy, studied Mathematics &c then for a support taught school in various parts of this country. With his earnings, he returned & again entered himself in the Academy." Curiously, in this 1826 letter, Beck pegs Henry's current age as about twenty-three. This figure underestimates his true age of twenty-eight by an even greater margin than Henry's more customary two-year reduction. But Henry's and, moreover, Beck's misconceptions clarify why an institution for boys so readily enrolled a young man actually in his early twenties. School records indicate that Joseph studied at Albany Academy from about 1819 through 1822—or from about age twenty-one through twenty-four. However, Henry probably assumed that he enrolled at about age nineteen or, if Beck's misreckoning corresponded to Henry's own misconception, at about age sixteen—well within the usual range for students.[19]

Muddled as his age might be, Henry was at last a regular student in a new-fashioned school. And as it turns out, Beck and his academy staff had a great deal to offer the eager young man who traced his scholastic inspiration to a vagabond rabbit and a forgetful Scotsman.

2 From Student to Surveyor

An early alumnus of the Albany Academy recalled that its capable staff and imposing facilities reflected the community's strong commitment to education. In particular, as the school took shape in the 1810s, most of Albany's leaders "were Dutchmen, and to them the cause of education went hand in hand with the cause of religion—both were equally dear." Housed in a stately, spacious, sandstone building dedicated in 1817 near the city's center and staffed by a respected faculty that included up to six professors and tutors, the academy attracted boys from the municipality's middle- and upper-class families. Boys who were at least eight years old, whose parents could afford the $10 quarterly tuition, could enroll in the academy's English School; this was an elementary division that covered basic reading, penmanship, grammar, and arithmetic along with a smattering of geography, history, and bookkeeping. More advanced "scholars" could pay $12.50 per quarter to enroll in the five levels of the classical and mathematical departments; here they could approach college-level proficiency as they mastered Latin, Greek, belles lettres, geography, geometry, algebra, and natural philosophy. Academy leaders recognized that few comparable schools emphasized subjects involving mathematics and physical science and, moreover, that residents of the region could make practical use of training in accounting, navigation, surveying, astronomy, and mechanics. Therefore, under the guidance of principal T. Romeyn Beck, a physician with wide scientific interests, the school paid special attention to mathematics and physical science. In fact, about the time that Henry enrolled, the trustees approved the addition of chemistry to the curriculum, and Dr. Beck himself initiated a series of affiliated chemistry lectures open to the public at a charge of $5.[1]

Henry likely took advantage of the academy's new offering in chemistry; at the minimum, he owned and apparently used the associated textbook. His personal copy provides a first glimpse of the science he encountered while a pupil there. The author of *Chemical Catechism, with Notes, Illustrations, and Experiments* was Samuel Parkes, a London chemist active in manufacturing. Noticing Parkes's background, Henry probably could have predicted that the author would emphasize "the great importance of chemistry to the arts and manufactures"; in these two practical categories, Parkes lumped everything from smelting iron ore and printing calico cloth to tanning hides and molding candles. Henry probably would have flinched, however, when this British author matter-of-factly declared that he was directing the textbook to children from "the higher and middle Ranks of Life"; Parkes reassured his presumably privileged readers that chemistry would help the "son born to opulence" in improving "the *cultivated* parts of his estate" just as it would help the son who chose "the profession of medicine" to be a better practitioner. Not limiting the benefits of a chemical education to such worldly concerns as manufactured goods and private estates, Parkes also underscored the value of the *Chemical Catechism* for the soul. He said that by revealing the normally hidden "laws of matter," he was giving his young readers "a body of incontrovertible evidence of the wisdom and beneficence of the Deity." Stated differently, he was defending his youthful readers, through chemistry, against "immorality, irreligion, and scepticism." Also, by instructing them in the rigors of experimental investigation, he hoped to dampen the lure of "sophistry," "fanaticism," and "superstition." More generally, he advised that the type of knowledge that comes through the study of natural philosophy and chemistry would lead to "opulence, respectability, and rational enjoyment."

Wending his way through the five hundred pages of *Chemical Catechism,* Henry would repeatedly encounter Parkes's conventional, early nineteenth-century notions of science's material and spiritual utility. He would also find recent, though eclectic, views on chemical theory, experiment, and apparatus. Depending heavily on the interpretive roles of heat and "chemical affinity," Parkes cited the researches of influential chemists from the near past such as Joseph Black and from the present generation such as Sir Humphry Davy. In line with Davy, who at London's Royal Institution was following a varied research program extending from mineralogy to electrochemistry, Parkes also ranged widely as he discussed alkalies, acids, salts, metals, oxides, and other chemical groups. He even had Davy contribute to the book's long, concluding list of "Select Instructive Experiments." Henry's copy of *Chemical Catechism* contains pencil marks alongside some experiments, perhaps revealing the ones he personally watched or tried while enrolled in the academy.[2]

A more personal glimpse of Henry's studies comes from three of his notebooks that contain what look to be classroom drills or supplemental exercises. The first two notebooks cover trigonometric topics; the third ranges over broader mathematical issues and even touches on English usage. About September and November 1821, the dates on the first two notebooks, Henry was using advanced trigonometry to make meticulous ink drawings of spheres and their mappings on plane surfaces, techniques and concepts useful in navigation, surveying, and positional astronomy. Specifically, he devoted the first notebook to "Stereographic projection of the Sphere" and the second to "Astronomical Problems Solved by Oblique Spherics." His mathematical reach extended from the propositions of Euclid's geometry to trigonometric and logarithmic functions such as tangent, cotangent, and log-tangent. Though typically he merely reproduced constructions and analyses from existing textbooks, the textbooks on which he drew were themselves current and demanding. He cited, for example, Thomas Keith's *Introduction to the Theory and Practice of Plane and Spherical Trigonometry and the Stereographic Projection of the Sphere, Including the Theory of Navigation* along with similar books by Samuel Vince, Andrew MacKay, and Nathaniel Bowditch. Henry's copy of Vince's *Elements of Astronomy Designed for the Use of Students in the University* carries, by the way, a short inscription by Henry in which he quotes Sir Humphry Davy: "the Rev. Samuel Vince . . . rose to distinction entirely by the exertions of his own talents."[3]

Mathematics and physical science made up only part of Henry's course of study. Although he devoted most of the third notebook to "Mathematical Problems"—his title for this leather-bound log—he spent eight pages on an English exercise designed to sharpen vocabulary. Likely in response to a professor who was dictating, he transcribed a drill on "Synonymic Elucidation"—that is, a drill on distinguishing similar terms. In the longest entry, he recorded quotations from Hippocrates, Plutarch, Francis Bacon, John Milton, and Samuel Johnson to differentiate twelve related nouns: *saying, saw, by-word, proverb, sentence, sentiment, maxim, adage, axiom, truism, aphorism,* and *apophthegm.* For the briefer set of terms *public house, hotel, tavern, inn, brothel,* and *alehouse,* he dutifully wrote: "all places which receive guests for hire are publick houses[:] an Hotel receives them only to lodge; a tavern receives them only to feed[;] an inn receives them both to lodge and feed; a brothel supplies lude woman; an ale-house is without a wine licence, and sells only beer." The entry on *to honor, to revere, to worship,* and *to adore* ended with a typical, moralistic summary, one that incorporated all four synonyms: "Inteligent benificance is the purest attribute of mind: in our equals it should be honored; in elders it should be revered; in the heroes of our country or of the world it may fitly be worshiped; and in God adored." Earlier in this entry, the discussion of To Honor perhaps struck a personal chord when it turned to parents' responsibilities. Specifically, Henry's own erratic relationship with his parents might have come to mind when, in a comment on the biblical injunction "*Honor* thy father and thy mother," he recorded that "the word *revere* would have been more proper, it being the place of parents to make gifts and to endow their children, and not the reverse."

Henry's main concern in this third notebook was mathematics. The first page he filled with a paean to the life of the mind, especially mathematical inquiry. In this apparent transcription from another source, he began: "Riches and honours are the gifts of fortune, casualy bestowed or hereditarily received, and are frequently abused by their possessors; but the superiority of knowledge and wisdom is a prominence of merit, which originates with the man, and is the noblest of all distinctions." Of all the intellectual pursuits that "call forth the spirit of enterprise and enquiry," however, none is "more eminently useful" than mathematics. After expressing these lofty sentiments, he moved to the mundane but clearly useful topic of figuring simple interest on money. He then progressed to computing the sum of the terms in a geometric progression in an effort to arrive at finite values for series of repeating decimals; he accomplished this by reproducing problems from a recent article on "Circulating Decimals" in the British journal *Annals of Philosophy*. Next, relying on a geometry textbook used in the mathematical department of the academy, he tackled the problem of finding the volume of a section of a solid cone; he signed this derivation, "By Jos. Henry June 27th 1822," thus providing the only date in this third notebook.

Henry devoted the bulk of the third notebook to two, long sets of copied problems and solutions. The latter set was specialized, referring to journals such as the *Edinburgh Philosophical Transactions* and concentrating on "Diophantine" analysis, that is, solving obstinate algebraic equations by assuming certain values for unknown quantities. The first set was more mathematically encompassing, stemming from problems and solutions presented for a wider audience in succeeding issues of the *New York Literary Journal and Belles-Lettres Repository*. Responding to a *Belles-Lettres* exercise, Henry recorded a calculation for "the fluxion of differential" of basic algebraic expressions; in other words, he was becoming familiar with differential calculus. He was also building his skills in natural philosophy or, as it was becoming known, physics—an interest that dated to his original encounter, about five or six years earlier while rooming in his mother's house, with Gregory's *Popular Lectures*. In particular, another *Belles-Lettres* exercise that he transcribed concerned the rate at which a pendulum oscillates under different barometric conditions. A more utilitarian exercise in physics involved analyzing the load supported by a round versus a square beam. Yet another practical problem entailed calculating the height to which a worker must raise a sluice gate in a pond to produce the greatest force of water striking the paddles of a mill wheel.[4]

An anecdote passed down among Albany relatives and friends attests to Henry's diligence as a student. Living probably with his mother or perhaps his aunt Margaret Henry while attending the academy, his habit was to get up during the night and work on unsolved problems that were keeping him awake. For this purpose, he maintained a whitewashed wall in his room on which, by candlelight, he would write with charcoal. When finished, he would rinse the wall with water and then cover it with a fresh coat of lime whitewash, ready for the next night's challenge. No matter the story's truth, by the middle of the school's summer quarter in 1822, Henry had immersed himself deeply in the theories, practices, and applications of mathematics, chemistry, and physics. Through dogged application to courses, textbooks, and journals, the twenty-four-year-old student and intermittent teacher was preparing himself for fuller participation in the regional community of technical educators and practitioners.[5]

For about the next two years, however, he would find that fuller involvement not in one activity but in a variety of ventures. While studying at the academy, Henry had caught the attention of Principal Beck, who apparently employed his bright student to help prepare his lectures. Beck sustained the partnership after Henry finished his course of study, hiring the unsettled young man as a "Chemical Assistant" in his public lecture series during the winter of 1823–24. Orlando Meads, a student during these years, later recalled that the lectures were very popular, attended not only by students but also by "the most intelligent and fashionable in the city"; and though only an assistant, Henry was "already an admirable experimentalist." Henry also assisted the multifaceted Beck in his personal chemical investigations and in preparing his *Elements of Medical Jurisprudence*—a well-regarded guidebook first published in September 1823 with many subsequent American and foreign editions. Henry later told Rhees that Beck maintained a personal commonplace book "of immense use" that he packed with factual information and references to relevant publications; the ambitious assistant would imitate this routine, maintaining for decades to come his own exhaustive reading records and citations of publications. During this period of close association with Beck, Henry even took the first steps toward following him in a medical career. He studied anatomy and physiology under two respected Albany physicians, Alden March and William Tully, graduates of Brown and Yale.[6]

Meanwhile, Beck continued to further Henry's prospects. In 1825 the appreciative young man would write to his mentor: "I cannot suffer this opportunity to pass without expressing by letter what I should never have confidence to communicate by words, the sentiments of gratitude I entertain for the many favours Your Kindness has bestowed on me. To you sir alone I owe what little reputation I may possess and though I may appear to want feeling I shal allways remembered that your countenance gave me support when I had no friends and was almost an isolated being labouring under the disadvantages of at least a doubt."[7] Henry sent this letter of gratitude after Beck had added to his favors by finding employment for Henry as a private tutor, easing his entry into Albany's leading learned society, and helping him land a longer-term job on a major, state road survey.

The tutoring job that Beck arranged was a plum. Not only did it require only three hours a day—allowing Henry time for his other pursuits—but it was with one of the oldest and most prominent families in the Albany area if not New York State. Henry was tutoring the children of Stephen Van Rensselaer and his second wife, Cornelia, daughter of William Paterson, the late governor of New Jersey. Known as the "patroon" because of his large land holdings dating from a seventeenth-century Dutch grant, Van Rensselaer had capped a Harvard education with high positions in New York politics and a general's command in the War of 1812, and currently held a seat in the U.S. Congress. A public spirited patron of education and science, he also served as senior trustee of the Albany Academy in its formative years and as first president of the new Albany Institute. The community's progressive citizenry had formed the institute in 1824 by combining the Society for the Promotion of Useful Arts and the Albany Lyceum of Natural History. Incidentally, Henry remained open to tutoring in other families and instructed Henry James—son of longstanding academy trustee William James and father of psychologist William and writer Henry James. Supposedly, Joseph Henry had involved young James in experiments with miniature hot-air balloons that ended in a freak accident: a severe burn to James's right leg that eventually necessitated amputation. The student and tutor would, however, remain close friends.[8]

The Albany Institute provided a forum for diverse intellectual exchange. An eclectic mix of citizens participated in three departments: physical science and the arts, which built on the long-established Society for the Promotion of Useful Arts; natural history, which absorbed the more recent

Lyceum of Natural History; and history and general literature, a new area. For their meetings and activities, members of the institute used the academy's spacious and centrally located building. Though often minding his congressional duties in Washington, institute president Van Rensselaer would have noticed that his family tutor was an active participant. Indeed, Henry's steadfast service to the institute probably reinforced the favorable impression that Van Rensselaer had formed from Henry's performance as a student under Beck and as a tutor in his own family. When Henry finally quit tutoring following appointment to the state road survey, Van Rensselaer supposedly shook his hand warmly and assured him that, if the need ever arose, he stood ready to aid him with "his patronage or his purse." In future years when angling for jobs, Henry would either rely on Van Rensselaer for a personal recommendation or simply invoke the name of this well-connected general and member of the House of Representatives.[9]

Henry had joined nineteen other backers of the Albany Institute at the main organizational session in May 1824. The minutes for the mid-October meeting describe a typical, early gathering of this local society for the "Promotion of Science & Literature." In particular, they record that Henry was present when a varied array of "donations and deposits were laid upon the Table": clams from Long Island, fossilized ferns, a stalactite and stalagmite, insects, South American nitrate of soda, salt from Salina (further west in New York), a Russian silver coin, an egg shell from an ostrich, West Indian coral, an English chalk fossil, Continental paper money, dried plants, various "petrifactions," and a calculus from the bladder of a horse. Henry himself offered a "Specimen of lithographic printing on Satin," apparently a new technique. Later in the meeting, Beck read a "notice" on the South American nitrate of soda and then presented a committee report recommending that the institute begin publishing its proceedings along with occasional selections from European journals. Henry's own debut for presenting a full "Communication" would come two weeks later. Six weeks shy of his twenty-seventh birthday, he read a lengthy paper titled "On the Chemical and Mechanical Effects of Steam." In ensuing months, he would continue making presentations, including a chemical analysis of a bad smelling mineral, barium sulfate, found at a site in New York. Interested in minerals and other facets of geology, he also accepted election as "curator" in the institute's Department of Natural History, probably at the behest of Beck, who was an officer of the department.[10]

Henry's talk on the nature of steam, which he "illustrated by Experiments," led off a special meeting devoted to the popular topic of steam engines—a meeting that also included a paper on the engine's history and a working model brought from New York City. In the opening section of his address, Henry summarized chemists' views on the "latent heat" associated with steam—that is, the quantity of heat absorbed or released, but not commonly noticed, in the transition between water and steam. Emphasizing concrete results from experiments and observations, he avoided theoretical speculations about the underlying causes of latent heat—except for making passing references to "the matter of heat" and other allusions to the then-common idea of heat as an "imponderable fluid." Rather, he stressed the practical applications that depend on "this great quantity of heat contained in steam," everything from cooking food to heating buildings. Throughout the discussion, he drew particularly on the writings of Joseph Black, the late eighteenth-century, medically trained chemist, active in Glasgow and Edinburgh, who developed the idea of latent heat.

In the remaining sections of this first public address, Henry clarified the dependence of the boiling point of water on pressure, and the "force of elasticity" of confined steam on temperature. This led him to evaluate the two most common types of steam engines—low pressure and high pressure. Citing the recent studies of John Dalton, James Watt, and other leading chemists and steam engineers, he concluded that high pressure steam engines offered the greatest fuel efficiency. Finally, he used a large copper sphere penetrated by a narrow tube with a stopcock to demonstrate a seemingly "paradoxical" phenomena: the ability of a mixture of warm water and air under high pressure to produce snow and ice when suddenly released through a small opening. At an institute meeting four months later, he would present a second paper on latent heat as he expanded this demonstration. For the time being, however, he closed his presentation with a deferential nod to the remaining segments of the evening's meeting: "Gentlemen I will trespass no longer on your time for I have already detourd you too long from the mor interesting subject of the steam engine."[11]

Albany continued to grow as a commercial and political hub through the early to mid-1820s. Among the thousands of residents and visitors passing through its main streets, now embellished with curbstones, were two foreign exponents of American revolution, French-born Marquis de Lafayette, recalling past military glories as he toured the once rebellious states, and Welsh-born Robert Owen, envisioning a social utopia as he made his way to New Harmony, Indiana. With a population already approaching 16,000, the city would expand even more rapidly when the Erie Canal opened in

November 1825. Governor DeWitt Clinton, the leading proponent of the new waterway, rode with other dignitaries as the *Seneca Chief* completed the first trip from Lake Erie to the Hudson River. When the boat reached Albany, cannons spaced along the southern length of the Hudson fired in sequence, relaying their reports of the canal's opening to New York City and back in 58 minutes.[12] Henry missed the Albany celebration. Ironically, he missed it because of the Erie Canal.

In response to residents of southern New York who felt that their tax dollars had helped build a canal that unfairly benefitted central and northern residents, Governor Clinton and other political leaders early in 1825 proposed a Great State Road across the south. Henry would spend the last half of the year, including the day of celebration for the canal's completion, in the sparsely settled counties of southern New York leading one of the road's survey parties. His selection for this responsible, well-paying, state position reflected a growing reputation among influential officials affiliated with the Albany Institute and the Albany Academy, where he had probably studied surveying. Though he had support from Beck and Van Rensselaer, he received the actual invitation through another member of the institute, Alfred Conkling, an Albany lawyer soon to become a federal judge. Henry had become a "favorite" of Conkling, who not only was a member of the institute's Department of History and General Literature but also had attended Beck's series of chemistry lectures in which Henry assisted. It did not hurt Henry's prospects that Conkling also served on the institute's committee on publications with institute stalwart Simeon DeWitt, surveyor general of New York; DeWitt had drawn on Henry for a survey of a farm earlier in the year and, as Henry later recalled, always treated him "with kindness and attention." But what clinched Henry's appointment on the survey was that the governor had appointed Conkling's law partner to be one of three commissioners overseeing the project. Furthermore, partner Jabez D. Hammond earlier had lived in Cherry Valley, New York, home of the commissioners' choice for chief engineer of the survey, William Campbell, himself a corresponding member of the Albany Institute and, later, the successor to DeWitt as surveyor general. Between them, Hammond and Campbell secured Henry's appointment as engineer in charge of a survey crew.[13]

As the crew headed west in late July from Kingston, a town downriver from Albany, Henry was coordinating the efforts of eight other men in two basic tasks. A lead group composed of a surveyor, flagman, two chain men, and an axe man was responsible for charting the compass directions and distances of the cross-country route. Henry, his assistant, and two target men followed behind "to run a line of levels," that is, to measure the gains or losses in elevation for each segment of the proposed road. In the next four months, however, personnel would shift as the men trekked 230 miles west to the town of Bath, near Lake Erie, then reversed direction and added about another 100 miles on an alternative route from near Portland eastward to the village of Angelica. During the last leg of the main, westward trip, for want of a regular survey team, Henry would be forced to borrow a surveyor's compass and combine the surveying and leveling into a single, simplified operation under his personal direction. One crew member remaining through the expedition was Joseph's younger brother, twenty-two-year-old James. Although James started as a surveyor's flagman, he also acted as a targetman, leveled, and worked closely with his brother. Two of the crew, "sons of rich & influential men," tested Joseph's mettle when not only did they rebel against remaining in the field as winter weather set in but also routinely got drunk. Decades later, Rhees recorded Henry telling him that "he would have no drinking men on his party" and, consequently, "was obliged to dismiss them, thereby incurring their opposition & ill-will." The state's flawed system of meeting costs and payrolls also plagued Henry as the engineer leading the party. "I have experienced great inconvenience from want of funds," he wrote to Chief Engineer Campbell in mid-November, "and have been under the disagreeable necessity of borrowing from strangers."[14]

Henry also told Rhees that living and camping in the outdoors improved his health. Standing about five-feet-nine-inches and weighing about 130 pounds,[15] he had overindulged in his Albany studies and work, becoming "very delicate and threatened with consumption." But the outdoors "made a different man of him." Comments that he added to a series of leveling notebooks capture his day-by-day perceptions of this challenging yet restorative adventure. As the party traveled between the Delaware and Susquehanna Rivers, for example, he remarked not only that wolves were plentiful and frequently preyed on sheep but also that rattlesnakes were abundant and that his group had killed one about three and a half feet long. Moreover, determined to meet Campbell's deadline for submitting the survey's data, Henry kept his party working even as severe weather hit—the decision that vexed the two rebellious members of his crew. Near the village of Ellicottville in mid-November, he jotted in his notebook that "this day we leveled about 5 miles in a snowstorm." The bad weather persisted and near Franklinville he complained: "This is a most teidious day for Leveling[;] the Snow drifts and it is so cold that the ink freezes in our pens."

Asides in the notebooks also reveal that Henry broadened his knowledge of geology, topography, and agriculture as he gathered information crucial to deciding the physical and economic viability of the route. In a typical entry, about a new village in Tioga County, he recorded: "Land from the river valued at from 2 to 2 1/2 Dollars per acre a natural soil for wheat." He also used a few of his rare free moments to study sites of scientific interest. In the western part of the state, he first visited a "burning spring" and then went on to other "oil springs," concluding that the inflammable gas bubbling to the surfaces at both sites was "undoubtedly carburetted hydrogen." Henry even took note of remnants of American Indian life. He entered detailed descriptions of a conical hill "remarkable for the trace of an ancient fortification," supposedly "the last residence of the indians in these parts." He also recounted how local tribes could kill "40 deer in one forghtnight" by driving them between two brush-and-log fences arrayed in a "V" that extended two or three miles in each direction.

The expedition also expanded his awareness of daily life among white settlers in the New York outlands. He found it noteworthy that he had traveled 70 miles on one stretch without seeing a church and about 100 miles without meeting a carriage on springs; he added that the rough roads the crew had been following would have proved "fatal" to such a delicate carriage. Traveling through these remote areas, he encountered what he took to be quaint practices. In the region between the Delaware and Susquehanna Rivers, for instance, he recorded that locals still traveled in canoes and made a living by rafting timber down to Philadelphia when the rivers swelled after heavy rains. Despite the remoteness, he did receive a cordial reception. The families in the towns and villages along the route relished the prospect of a state road and, as Henry mentioned in a letter to Commissioner Hammond, treated the party "with the greatest kindness and respect." In fact, the leading citizens invited members of the party into their homes. Years later, Princeton student Cuyler recalled Henry saying that he had enjoyed this hospitality "mightily."[16]

The survey party completed its assignment by the last days of November and headed home, but Henry remained in the field two more weeks retracing the route "in order to make some additional notes." Finally returning to Albany on 15 December, he visited his mother briefly and, as instructed, rushed on the next day to Cherry Valley, about 60 miles to the west, to confer with Chief Engineer Campbell. The sleigh in which he was riding upset twice in the snow but the most disturbing part of the trip was when Campbell greeted him with the announcement that his set of measurements, gathered at great expense to the state, was "worthless." Henry later told Rhees that at that moment, after five months in the field and a harrying sleigh ride, he was so flabbergasted and exhausted that Mrs. Campbell "thought he must be intoxicated." He returned to his room and slept for twelve hours. The next morning, with new resolve, he pressed Campbell on the details of the problem and discovered that an assistant whom he had fired had misreported some figures. Also, he realized that Campbell had made a mathematical mistake in analyzing the data. A forthright man, Campbell apologized on the spot and went on to befriend Henry. He relied on Henry for help in compiling the overall report of the road survey and even hired him to tutor his niece in mathematics. Meanwhile, for his party's component of the state project, Henry had the main responsibility for writing the report and compiling an associated atlas. This work eventually became part of the commissioners' official presentation to the state. Later, Henry would combine his report with fieldnotes from other surveys to prepare an introduction to the *Atlas of the State of New York*—a popular reference book compiled by a fellow surveyor and issued by the legislature in 1829. He also read a version of this introduction at a meeting of the Albany Institute and published it in their *Transactions* under the title "Topographical Sketch of the State of New-York, Designed Chiefly to Show the General Elevations and Depressions of Its Surface."[17]

In one of his leveling books, Henry remarked that "if no state road be constructed this survey will be of great advantage to the country in teaching the inhabitants to explore and awaken their attention to the subject." His comments proved prescient; the legislature never followed through to build the Great State Road. But results such as the *Atlas* did serve to encourage development in the southern counties. For Henry himself, the never-completed project paid handsomely in experience, reputation, and opportunities, not to mention money. Earning $3.50 per day, he ended with the sizable sum of $885.50 after laboring 253 days gathering data in the field and then processing it. However, as the fieldwork drew to a close, in letters to Beck and his cousin Stephen Alexander, Henry divulged that the main reward that he sought was approval of his work: "The last summer has been to me a season of peculiar interest although one of labour and trial. I have devoted myself exclusively to the duties of my office and if my labours should not prove as important as I wish they may, still I hope I have conducted myself so as to merit in some degree the approbation of the commissioners and not to disappoint the moderate expectations of my friends."[18]

This high-minded desire to fulfill the expectations of others, and thereby merit their approval, probably reflected more than either the temperament with which Henry was born or values that he assimilated from the general culture. The desire also probably reflected those attitudes that sprung from the childhood traumas associated with his father's alcoholism. Likely desiring personal affirmation as a child, he had found that he could obtain it by responsibly discharging what he perceived to be his obligations—a behavioral pattern that would resurface throughout his life and often result in over-achievement. Thus, others' "approbation" for properly executing "duties"—not personal or material gain—had emerged as a primary reward for conducting the road survey. And approbation he gained. Beck, Hammond, Van Rensselaer, and others praised the astute, steadfast, selfless surveyor as they joined in a campaign to secure him a permanent position as a civil engineer.

In the United States during the early 1800s, few opportunities existed for a person to study science and mathematics, let alone find employment involving the disciplines. Henry was similar to other Americans with technical interests in that he was partly self-taught—indeed, in a letter in 1832, he asserted more emphatically that he was "principally self educated." But he also had the good fortune to have access to the Albany Academy, where he received the fundamental elements of a European-style scientific and mathematical education; he could have obtained broader classroom training at only a few, select, collegiate institutions such as West Point, Harvard, Yale, and the University of Pennsylvania. Also, when merely a handful of local or regional learned societies offered rudimentary scientific forums—the two best-known being Philadelphia's American Philosophical Society and Boston's American Academy of Arts and Sciences—Henry was lucky to have access to the Albany Institute. But again like most Americans with a scientific bent, he was less fortunate regarding employment prospects. Thus, Beck, Hammond, and Henry's other supporters responded vigorously to the news of a proposed expansion of the Topographical Engineers of the U.S. Army. During this era, Americans did not draw sharp distinctions between basic science, applied science, and engineering, but saw them as blending into one another. And the army was one of the few organizations to offer "scientific" employment through its staff of civil engineers—professionals valued not only for their military endeavors but also for their civilian contributions during this national era of public works and westward expansion.[19]

Beck and Hammond appealed to top political officials in their bid to secure a slot for Henry with the Topographical Engineers. Beck began in March 1826 with a letter to Martin Van Buren, a mover in the Democratic political coterie known as the "Albany Regency," then serving in the U.S.

Senate but soon to become governor and later president. Van Buren knew Henry's uncle, Hawthorn McCulloch, who earlier had employed Henry in his brewery. After sketching Henry's background (and mistakenly listing his age to be about 23 rather than 28), Beck called Van Buren's attention to the aspirant's service as a main surveyor on the state road project; he explained that Commissioner Hammond would contact Secretary of War James Barbour directly to provide details about the service. In the meantime, Beck asked Van Buren to intervene for Henry. "I have no doubt of his perfect competence, of his natural & acquired abilities," Beck attested, "& his conduct & deportment are so modest & correct, that none who know him can avoid being prepossessed in his favour." Van Buren passed the letter to Barbour along with a one-sentence note vouching that Beck was "a gentleman of high respectability." Within a few days, Barbour heard from Hammond, himself a former member of the U.S. Congress and four-term state senator. Recommending that Henry be added to the army's corps of civil engineers, Hammond affirmed that he "is a young man of excellent character and correct habits and highly distinguished for his advances in Science and particularly for his mathematical acquirements." Hammond closed by mentioning that Van Rensselaer, currently a delegate to the House of Representatives, would be calling on the secretary to petition personally on Henry's behalf. By the time Van Rensselaer made the visit, Barbour apparently had consulted about Henry's case with the lieutenant colonel in charge of the army's Topographical Bureau. The secretary had to disappoint Van Rensselaer, informing him that any new positions would be filled from within the military ranks—from present officers or new graduates of West Point, the nation's leading school of engineering and, perhaps, science.[20]

Henry was determined, however, to follow a career in civil engineering, particularly surveying. He had, as he recalled in later decades, become "enamored with the profession of an engineer" and found it "very pleasant & congenial employment." Though the army position dangled beyond his reach, his reputation from the road survey continued to spread and he began to receive tenders of other commissions. Offers were pending to become either an engineer on a canal project in Ohio or a director of a mine in Mexico. But in April a job opened closer to home. The professor of mathematics and natural philosophy at the Albany Academy resigned because of a personal scandal involving "a female of *low character.*" Beck urged Henry to apply for the vacant position. Henry demurred, as he later disclosed to Rhees, not only because it was "against his wishes for he preferred engineering" but also because a close friend actively sought the position.

Through Beck's influence, however, the academy trustees elected the reluctant Henry to the position, with duties commencing the following September at an annual salary of $1,000.[21]

Silversmith, actor, man of science, teacher, physician, and engineer had all been options, but an unexpected academy resignation settled the twenty-eight year old's uncertainty about a career. Hammond had written in support of his ex-employee's candidacy: "Altho' I do not claim to be a competent judge of Mr. Henry's scientific acquirements I hope you will excuse me for embracing the present occasion to state that he executed the business assigned him last summer by the Commissioners of the State Road with great accuracy, skill and ability and in a manner highly satisfactory to the Commissioners and I believe to the public. In relation to his moral qualities I must be permitted to add that I never knew a person more free from faults or more amiable in his disposition." Half a century later, in his private conversation with Henry, Rhees recorded that when the academy trustees elected Henry, "he took the place almost with sorrow. He attributes much of his success in life to the fact that he never left one position for another without much misgiving—but a determination to do the best he could in his new sphere." In an era of U.S. history when few persons made science their occupation, when even fewer went into natural philosophy, and when still fewer broke into science from modest socioeconomic backgrounds, Henry was beginning his life's labors as a rarity among rarities.[22]

Teaching at Albany Academy 3

On 2 May 1826, four days after the academy trustees approved his appointment, the incoming professor of mathematics and natural philosophy boarded a packet boat and headed west on the Erie Canal. The boat stood out from others not only because of the cooking stove and kitchen paraphernalia that the crew had rigged on the vessel's fore and aft decks but also because of some of the unusual belongings that the twenty passengers had stowed in the lone, large, central cabin: textbooks, chemical apparatus, and hammers and chisels. Half a year after Governor DeWitt Clinton had formally opened the canal, Henry was beginning a group tour of the belt of notable geological sites along the waterway. He and his comrades would spend up to six weeks living, eating, and sleeping on board as the boat made a leisurely round-trip between Troy on the Hudson River and the canal's western terminus near Buffalo on Lake Erie. With ample provisions, including a donated barrel of beef and another of beer, they would be gliding through a conveniently exposed cross-section of the geology of New York State.[1]

Heading the tour was geologist Amos Eaton. A fifty-year-old protégé of Clinton and Stephen Van Rensselaer, Eaton had established his scientific reputation primarily through surveys of the canal's geology. More recently, in Troy he had helped organize the Rensselaer School, a scientific and technical teacher-training school that later evolved into Rensselaer Polytechnic Institute. For the canal tour, he had assembled a party composed primarily of students from his school, including eight of the ten young men who, a week earlier, had made up the first graduating class. Among the nonstudents were two other young men who shared with Henry an Albany background: Dr. James Eights, son of a respected physician, and George Clinton, son of the governor. Clinton later recalled that, as one of the "party of lads" on the tour, he found Henry to be exceptionally "charming" in all his interactions with the group. Eaton wrote to his wife that the "new students," including Henry and Clinton, "are all good fellows."[2]

Henry likely joined the tour out of personal interest. The previous year he had become a curator in the Albany Institute's Department of Natural

History, the division that covered geological subjects. His inclusion on the tour also probably reflected Van Rensselaer's or Beck's desire to broaden the new professor's understanding of geology before he assumed his teaching post in September. Both men had early notice of the canal tour; Eaton had consulted with his patron, Van Rensselaer, in planning the excursion, and Beck's younger brother was a Rensselaer science professor who originally had intended to help lead the tour. In fact, the brother, Lewis C. Beck, was only a year younger than Henry and had become a close associate while in Albany; he served, for example, with Henry as a curator in the Albany Institute's natural history branch. He knew Henry well enough to write to him early in the tour requesting a favor: that Henry do various chemical tests on natural springs along the way. Lewis Beck also enjoyed close relationships with Clinton and Eights. "Remember me to," he added in a playful note at the bottom of the letter to Henry, "Mess. Clinton, 88s, & Prof. Eaton."[3]

Eaton conducted the tour as a school, announcing detailed rules of conduct for his canal-faring pupils. He also required every participant to "collect and label a complete set of the geological specimens of the canal line" and to maintain "a complete journal of every important occurrence, with a description of every interesting natural or artificial object."

True to character, Henry was a conscientious pupil. In his journal he adopted Eaton's often idiosyncratic geological labels to create a precise, daily record of significant rock formations. On the fifth day of the tour, for example, he wrote that, after leaving the boat at an inn, the party walked a few miles through the forest to view a waterfall on Starch Factory Creek. Along the creek, they observed "slaty graywacke which resembles the second greywacke or calciferous slate," a formation overlaid with "millstone grit, saliferous rock, Grayband, & Feriferous slate"; this arrangement of the strata led the party to agree that the lower formation "must be the first gray wacke of Eaton." Eaton's tendency to assign his own labels to formations probably reflected not only vanity but also a desire to establish American geology's distinctiveness in an era when European systems of nomenclature dominated. Later on the tour, at a site where Eaton labeled a rock "Feriferous slate," Henry commented: "Mr. E's views of the strata at this place is undoubtedly correct but the names which he has given to the new strata may meet with some opposition." Henry could see, nevertheless, that much work "remains to be done in examining the rocks of our country." Over the next two years he would join with Clinton and Eights in helping Eaton publish in Silliman's *Journal* a report on North American "Geological Nomenclature."[4]

Henry also listened attentively to Eaton's formal lectures and, in the process, encountered a geological theory that, while controversial, was popular in the United States and Europe. Called diluvialism or Neptunism and associated with German geology teacher Abraham Werner, the theory traced the earth's strata to the action of water and oceanic floods, particularly the great biblical deluge of Noah. Thus, Henry recorded that Eaton repeatedly stressed the chronological classification of formations into three types: antediluvial, diluvial, and postdiluvial. On the return trip, Henry had an opportunity to hear the ideas of another well-known naturalist. Constantine Rafinesque, a professor from Transylvania University in Kentucky, happened to be in the region and joined the tour at Rochester. Henry offered a mixed appraisal of the newcomer's work: "His fort appears to be zoology & botany and in these sciences he is very industrious but his usefulness is said to be much impaired by his proness to make new genera & species." Eaton was not alone in his enthusiasm for naming.

Instructed to include in his journal a record of all interesting events and objects, Henry inserted personal reactions to side excursions the group took while spending a week in the Buffalo area near the canal's western terminus. After traveling eleven miles in a hired wagon to Niagara Falls, he recalled his first visit to the famous site half a year earlier while leading the road survey—a disappointing visit in which the falls "did not answer my expectations." "Perhaps I did not view them with propper poetical feelings," he remarked, adding "but in this visit I have been mor fortunate." A few days later, he and his companions also visited the nearby ruins of Fort Erie, site of deadly combat twelve years earlier between the American forces that captured the fort and the British troops that then set siege. Shown a depression in the soil that indicated a mass burial of British soldiers, Henry responded: "The sinking of the ground showes that the mortal remains of these gallant soldiers are fast mingling with their mother earth & the demolishing hand of time is nearly as rapidly leveling the monuments of their labour and the reliques of their valour." Apparently, he still retained the poetic feelings experienced at the falls.

The return passage on the canal went more swiftly than the outbound journey. Professor Rafinesque's reflections enlivened the eastward trip as did Eaton's anecdotes about his old friend, writer Washington Irving. Ten days out of the home port of Troy, however, Henry left the boat and boarded a stagecoach to Albany. Still in a geological frame of mind while on this overland excursion, he took time in Cherry Valley to collect specimens and sketch the rock strata—a copy of which he sent to George Clinton who was still with the tour. But the primary reason for his early departure and stage ride home was to expedite a pending visit to the United States Military Academy at West Point.

Henry was probably rushing to West Point, about 90 miles below Albany on the Hudson River, to arrive in time for a prearranged review of the military academy's science facilities and programs. With his own teaching assignment looming, he could profit from exposure to the military academy's technical offerings, which were among the best in the nation. The timing of the visit, which lasted for at least two weeks in June 1826, allowed him to be present during the annual, extended gathering of the school's board of visitors. He could thereby take advantage of the activities associated with the board's regular inspection of operations, including some social interactions; in a log of the trip, Henry recounted, for example, touring a nearby Revolutionary War fort and attending the formal dinner that the board held for the academic staff.[5]

Although he opened his log by reporting on an iron foundry near West Point, he concentrated on the teaching of chemistry. He recorded that the "chemical laboratory" was a room in which students sat behind a railing while the instructor used the other half of the room for displaying apparatus and performing demonstrations. While inspecting the laboratory, he found striking the professor's reliance on a particularly innovative instructional device: "One article very necasary in teaching chemestrys is found in this room viz a black board on which the student is taught the attomic theory and all algebraical formula in chemestry. Indeed it appears to be one of the principles of teaching in this institution that every thing as far as practical should be demonstrated on the black board. The student is even required to draw all articles of chemical aparatus & explain them in this way." For Henry, who in his younger days purportedly had made do with writing on a whitewashed wall with charcoal, the West Point blackboards seemed decidedly progressive.

Henry also noted particular teaching initiatives of the two instructors in charge of the overlapping fields of chemistry, mineralogy, and geology. Professor John Torrey and Assistant Professor Jonathan Prescott each broached a topic to which Henry himself would devote considerable attention during the next few years. Torrey alluded to making an improved version of a scale of chemical equivalents—a mechanical device, with sliding logarithmic scales, for finding weight relationships between basic elements and compounds. Within a year, Henry would be collaborating with his friend Lewis Beck on constructing and selling just such an improved version.

The team would depart from the traditional design by incorporating a more recent view that hydrogen has an atomic weight of one and that all other chemical bodies, whether simple or compound, have weights that are integral multiples of hydrogen's. Though they sought a copyright for their design and sold many dozen copies of the scale, the manufacture and distribution proved difficult. Eventually, the entrepreneurs abandoned the project.[6]

Professor Prescott also dealt with a topic that would later preoccupy Henry. While giving an examination to his chemistry students, he covered the joint field of galvanism and electromagnetism. In the decades bracketing 1800, Luigi Galvani, Alessandro Volta, and others ushered in the science of galvanism with their development of electrochemical cells that could produce sustainable electrical currents. Earlier in the eighteenth century, Benjamin Franklin and other "electricians" had found that whereas electrostatic friction machines and Leyden jars could produce and store large charges, they could generate only small, transient currents. "Galvanic" cells or batteries, however, generated much larger currents that flowed in a continuous but gradually diminishing manner. Electrochemical batteries thus extended the boundaries of electrical science by providing early-nineteenth-century researchers and teachers with reservoirs of electrical current. Henry summarized in his West Point log a discussion of a decisive 1820 finding that joined the previously separate sciences of electricity and magnetism into the combined field of "electromagnetism." In particular, he recorded a theoretical explanation of Hans Christian Oersted's celebrated discovery that an electric current from a battery can produce a magnetic effect. Oersted had found that a current-carrying wire, when positioned above and then below a magnetized compass needle, causes the needle to aim in opposite directions perpendicular to the wire, thus indicating a circular magnetic pattern surrounding the wire. As long as the current persisted, the needle held its particular orientation. The explanation that Henry recorded treated the current as an "electric fluid" that flowed from the battery's negative pole and spiraled around the wire; supposedly, this fluid attracted the north pole of the needle, aligning it with the direction of the fluid's circular motion.[7]

Over the next few years, Henry would place electromagnetism at the center of his varied research program. At West Point in June 1826, however, he still had found the topic novel enough to describe in his log a version of Oersted's seminal discovery and a variant of the Danish scientist's initial theory of spiraling electricity. In a related vein, he also recorded advice on giving classroom demonstrations of "all the electromagnetic phenomina." This additional guidance came from James F. Dana—a Harvard-trained chemistry professor from Dartmouth College about to assume a post at the New York College of Physicians and Surgeons. Dana, a member of the board of visitors, detailed for Henry the type of battery to use and the size of magnetic needle necessary "so that the class might see every movement" in demonstrations involving electromagnetism.

West Point did not provide Henry's first exposure to electricity and magnetism. Perhaps his earliest formal contact occurred in Albany when initially enchanted by Gregory's *Popular Lectures on Experimental Philosophy, Astronomy, and Chemistry.* He certainly had become acquainted with the basics of the subjects by 1823 while assisting T. Romeyn Beck in his chemical lectures and investigations. But at the military academy in mid-1826, faced with the prospect of braving his own roomfuls of students needing to "see every movement," he was perhaps more attentive to details. This attentiveness continued after he left West Point and traveled fifty miles further south to New York City, an extension of his trip most likely intended for additional scholastic enrichment. On visiting the shop of an instrument maker, he transcribed detailed instructions for creating strong magnets. This artisan, who made and sold a variety of technical instruments, suggested a convenient way to produce a group of bar magnets by placing them in a circle and sweeping them all with a horseshoe magnet; Henry revealed that he had already been working with magnets when he noted that he had previously been using this technique. The instrument maker then offered advice on fashioning an exceptionally strong magnet, one capable of lifting at least 60 if not 150 pounds. "In making a powerfull magnet," Henry jotted in his log, along with the artisan's other instructions, "the shape should be a horse-shoe and each bar should be as thin as a saw blade, magnatised separately and united all the north poles together with screws or rivetts of lead." The concept of a "compound" horseshoe magnet—one that gained its strength from separate magnetic components—would be one that Henry would also extend in the next few years.

While in New York City, Henry spent an evening visiting "Peals museum in Broadway." This was a newly opened branch of painter Charles Willson Peales's main museum in Philadelphia, a popular and respected enterprise offering a distinctive mix of art and science. In later years, Henry recalled that "almost from childhood," he had wanted to visit the notable museum. The Broadway branch, run by Peales's son Rubens, juxtaposed three disparate displays or activities in three different rooms. In successive rooms, Henry viewed an exhibit featuring natural history, a collection of fine art paintings, and a chemistry demonstration. For the latter, on the evening Henry was in the audience, "Mr. Peal exhibited a number of experiments with oxygen & hydrogen gases." Producing bright flames and loud explosions, the experiments evidently were engrossing. The Peale plan for mingling an arresting array of natural history, art, and scientific experiment would also resurface in Henry's coming years—but not for another two decades.[8]

When the Albany Academy's fall term opened in September 1826, the trustees held a formal ceremony to induct the new professor of mathematics and natural philosophy. In addition to the main event, which was Henry's own inaugural speech, one trustee offered welcoming remarks and another gave the concluding prayer. In his opening comments, reported afterward in the *Albany Argus and City Gazette,* Gideon Hawley encapsulated the trustees' philosophy toward Henry's area of mathematics and natural philosophy, a subdivision of the academy's "English department" rather than its more traditional "classical department." Hawley explained that, though the trustees valued classical training for traditional professional careers, "they cherish with peculiar favor" the English department, which is "best fitted to prepare the greatest number of pupils for the useful pursuits of active life." Hawley noted that this preference for practicality sprung from the trustees' belief that the academy should be responsive, as a local institution, to the needs of Albany's citizens.[9]

Hawley boasted that Henry's preparedness for his new position proved the Academy's effectiveness; after all, Henry had "received the greater part of [his] education under its instructors." Always an attentive student, Henry apparently had assimilated the academy's guiding pragmatic philosophy: the central theme of his inaugural speech was that mathematics and natural philosophy offer significant practical benefits in everyday life. Of course, this utilitarian message permeated early nineteenth-century views of science; Henry had also encountered it in books such as Gregory's *Popular Lectures* and Parkes's *Chemical Catechism.* Ultimately, the message reflected the persistence of seventeenth- and eighteenth-century ideas about science—particularly the Enlightenment trust in the power of humans to achieve progress by using reason to discern natural laws.

Henry structured his address around mathematics, first distinguishing between "pure" and "mixed" mathematics. The former is an abstract study without worldly connections, he remarked, and the latter entails applications to natural or social realms and gives rise to pursuits ranging from astronomy to surveying and political economy. Through a sketch of the history of mathematics—which he portrayed as a story of "progress towards its present state of perfection"—he arrived at the topic of modern calculus and its applications to physical science. In exploring the ties among calculus, physics, and astronomy, he elaborated three popular tenets regarding the nature of science: the efficacy of Baconian scientific method, the accessibleness of natural law to human reason, and the discernability of God's handiwork in the physical universe. Natural philosophers have been successful in applying calculus to physics and astronomy, Henry asserted, because they have been "building upon principles established by the inductive methods taught by Bacon." Moreover, this controlled use of calculus led Newton to establish a "sublime system" of terrestrial and celestial physics based on one law of universal gravitational attraction, thus demonstrating that an "entirely complete" theory was within the pale of human reason. Similarly, French mathematical physicist Pierre Simon de Laplace later refined Newton's insights, creating a work that honors "not only the authour but the whole human race." Finally, Newtonian physics has led investigators to discern cycles in the heavens that are so subtle that they require thousands of years to complete; these ordered nuances, Henry suggested, reveal a deity's "infinite power."

The application of the admittedly rarefied principles of calculus to the mechanics of the earth and heavens did not constitute Henry's main example of mathematics's utility. He soon disentangled himself from this complex topic, assuring his audience that, "since one great object of science is to ameliorate our present condition by adding to those advantages we naturally possess[,] it is fortunate that all problems valuable for their practical application can be deduced from principles of a much more elementary kind." In fact, the academy's core of mathematics courses suffices to prepare students to pursue all science-based practical fields—everything from navigation and surveying to chemistry and mineralogy. The area that best exemplifies the usefulness of mathematics and natural philosophy, however, is that of "the mechanic arts." And epitomizing all the inventions and machinery associated with the mechanic arts is the steam engine, "that legitimate & gigantic ofspring of mathematical & chemical science." Henry reminded his Albany audience, living along the Hudson River, that the steamboat held particular fascination not only because of its "unparalleled usefulness" but also because of its being "an American invention completed & first tested in our own state."

Henry continued the motif of pride in the United States and New York as he enumerated further ways in which mathematics and science lead to useful results. The country's flourishing program of internal improvement—

canals, roads, and pending steam railroads—illustrated "how mathematical & mechanical knowledge may be applied not only to the necessities of ordinary life but to increase the power & promote the wealth of a nation." The achievements of New Yorkers, in particular, have caused citizens of other states to look to them for examples of profitable works and a supply of experienced engineers. Moreover, these and other American engineers have rightfully differentiated themselves from their transatlantic cousins: "The tedious methods & expensive plans of European engineers have been found but ill to agree with the state of finance or the active character of the American people. Since with us great results must be produced with small means." And this distinction between American and European engineering affected what Henry intended to teach. "To teach therefore the practicle and scientific principles of engineering as adapted to this country," Henry judged, "becomes an important duty of the mathematical professor of this academy." These nationalistic feelings about teaching engineering perhaps reflected Henry's experience on the state road survey and his exposure to West Point's distinctive educational program. For whatever reason, he was confident that engineering was an enterprise dependent on science and, in the United States, an enterprise constrained to operate pragmatically, achieving maximum results with minimum resources.

Having devoted most of his speech to fields illustrating "mixed" mathematics, Henry closed with a few comments on teaching "pure" mathematics. He grounded his comments on the seasoned but popular psychological theory that a student's mind contained separate "faculties" controlling distinct mental powers, each of which must be trained through "mental discipline." Henry had recently heard a discussion of the theory when, on the canal tour, Eaton had defined a mental faculty to be "that power which produces any results different from any other result," such as love or anger. Henry now argued that, by studying pure mathematics, a student received "discipline" for the "reasoning" or "inventive" faculties. Significantly, the student retains these powers as part of his "liberal education" and can draw on them in other life experiences, even if he forgets the associated mathematical content. Henry invoked the name of the respected eighteenth-century Scottish philosopher Thomas Reid to buttress his main assertion that, of all academic subjects, pure mathematics was the best equipped to strengthen the reasoning and inventive faculties. He even professed that the study of Euclid's geometry encourages in the student a spirit of intellectual independence. Specifically, understanding of the logical, unprejudiced nature of geometrical demonstrations leads the student to adopt the same standard for making judgments in other realms of thought: "He finds himself emancipated from bondage & exults so much in his new state of independence that he spurns all authority & would have demonstration for every thing." Henry adds, however, that experience soon teaches the student that, in the everyday world of making judgments, airtight demonstrations are not always possible and that "he must rest contented with mere probability or at most with analogical reasoning."

In sum, as an incoming professor Henry believed that instruction in mathematics served not only to prepare academy students for practical pursuits such as engineering but also to ground their liberal education by strengthening their reasoning faculty. Though insistent that mathematics was important in the curriculum, he closed his inaugural speech by acknowledging that mathematics was not above other subjects, "for they all contribute by various means to sweaten, to adorn & to embelish life."

For the new professor these pronouncements about pedagogic philosophy would not prove empty rhetoric. Through a combination of the trustees' intentions, Principal Beck's direction, and Henry's implementation, the curriculum in mathematics and natural philosophy actually emphasized practical applications. It also emphasized the psychological theory of mental faculties and the associated mental-discipline approach to teaching. At least by the fall of 1829, when the trustees reorganized the academy and updated its governing statutes, Henry felt that the curriculum embodied his pedagogic philosophy. By then he had been teaching for three years and had contributed to the restructuring.

The call for restructuring grew from the trustees' realization that Beck was being forced to spend too much time teaching elementary courses for the youngest students. The same problem plagued Professor Henry, who had "to attend daily to 160 recitations"; with this heavy teaching load, he was unable to "acquit himself in such a manner as even to satisfy his own laudable ambitions." In his second year, for example, Henry was responsible for seven basic classes in arithmetic along with two in Euclid's geometry and one each in trigonometry, algebra, bookkeeping, and using globes. The trustees responded by splitting off the elementary courses in the curriculum and placing them in a newly created Fourth Department. This move left the academy's three principal professors responsible primarily for the three remaining, advanced departments. Henry anchored a five-year sequence of courses in the "Mathematical Sciences," which included natural philosophy. Professor Peter Bullions oversaw "Classical Studies," with its eight years of traditional offerings in Latin and Greek. And Beck held sway in "English and Natural

Sciences." This five-year progression of courses meshed geography, history, rhetoric, and belles lettres with chemistry, mineralogy, and natural history, capping them all with moral philosophy and natural theology. To teach natural theology as a culminating course was to attest that science augments Christian religion and that the study of "design" in nature can reveal God's existence. Championed by Englishman William Paley and other influential thinkers in the eighteenth and early nineteenth century, the central claims of natural theology had a wide following in Beck and Henry's day as evidenced by the religious passages in books such as Gregory's *Popular Lectures on Experimental Philosophy, Astronomy, and Chemistry* and Parkes's *Chemical Catechism*.[10]

In Henry's view the courses in his department followed a progression dictated by the "principles of Education"—that is, by the principles stemming from the psychological theory of mental faculties. He briefly stated the essence of these established principles, declaring "it is evident that the different faculties and powers of the mind are not simultaneously but successively developed and consequently that some periods of life are better suited to particular acquirements than others." Boys between the ages of about six and twelve, in particular, have minds that are "susceptible of acquiring habits of action of industry and of mental expertness." Young students, then, could take only mathematics courses in the elementary department where, through drill and memorization, they could develop habits of correctly adding, subtracting, multiplying, and dividing simple numbers. With age the power of forming mental habits begins to fade and the reasoning faculty starts to develop. The boy of twelve or thirteen, therefore, could take courses in the department of mathematical sciences, first studying algebra. Eventually, the student reaches the critical age at which he can appreciate the logic of Euclid's geometrical demonstrations. "At the age of fourteen, and scarcely before," Henry insisted, "the student may begin Euclid and from this period to the end of his course his pursuits would be almost exclusively an exercise of the reasoning faculties." These later, more advanced pursuits included plane trigonometry, mensuration as applied to surveying and navigation, analytical plane and spherical trigonometry, the application of algebra to geometry and conic sections, trigonometry as applied to nautical astronomy and topography, natural philosophy, astronomy, and differential and integral calculus.

In teaching these courses—and at various times he taught them all, and more—Henry relied on the traditional mainstays of the mental-discipline approach to teaching: repetition, memorization, and recitation. For example, to present arithmetic to a boy who was ten or eleven years old, Henry concluded that a book was of little use: "I may say from experience that by *oral* communication & by questioning on the subject of the lesson a teacher may give a puple of that age a thurrow knowledge of the principles of this science but I believe it is only by oral instruction & that often repeted that he can be made perfectly to understand them." He added that, in conducting these recitations, he always used the blackboard—the innovative contrivance that he had chanced on at West Point—to illustrate the principles; he then assigned the students examples to work for the next day. Once, to illustrate the advantages of drilling the academy's beginning students in basic arithmetic—rather than discoursing to them on abstract computational rules—he organized a demonstration. Taking half-a-dozen boys that he had trained for two years, he gave them an advanced lecture on geometric shapes and then examined them orally on the content of the lecture. The boys' grasp of the demanding material impressed the senior professors assembled for the demonstration; Henry interpreted the outcome as confirmation of his belief that early training of the "doing faculties" prepares the mind for proper, later use of the "thinking faculties." In his more advanced courses, although recitation and repetition continued, textbooks figured more prominently. He taught the crucial "first book" of Euclid's *Geometry* with particular care: "About four propositions are given each day as a lesson and the text of those passed over are repeatedly reviewed. In this manner the first book is read through four or five times."

Teaching in this intensive manner, Henry estimated that merely his "ordinary duties" consumed about seven hours each day. Moreover, he was teaching up to fifteen courses each year and, during his six years at the academy, its enrollment rose from about 140 to 240 pupils. To handle such enrollment, the three senior professors could seek help from only four other tutors and instructors and one occasional lecturer, Lewis Beck. On top of Henry's normal teaching load, he also offered occasional public lectures. Furthermore, as his reputation grew at the academy and institute, he felt obliged to respond to citizens expressing various scientific and technical interests. Within a three-month stretch in 1831, he received a Native American battle axe accompanied by an account of its discovery, a comment on a design that he had suggested for the steam engine of America's first public railroad, and a request for his opinion regarding the best location for a steamboat's paddle wheels. The following year, he appeared as an expert witness before a New York legislative committee trying to arbitrate a dispute involving a survey of state lands. Being a professor entailed considerably more than seven hours of daily interactions with students.[11]

The students appreciated Henry's efforts. George Carpenter, a pupil from 1826 who later tutored at the academy, recalled: "In disposition he was most kind, in manners courteous and agreeable; his face was remarkably handsome, his finely chiseled features being lighted up with mild, blue eyes. He was a favorite with the pupils, always ready and willing to aid them in their studies, winning their esteem and affection." H. N. Smyth reminisced that Henry's willingness to help extended even to "a dull boy" such as himself, who was having trouble with Dr. Beck's course in English literature during the academic year 1830–31. "My composition was heart rending," he recollected, "and as they were submitted to and supervised by Doct. Beck little consideration for my short comings was extended. I had however the sympathy of Prof. Henry, who sat by me and aided me with ideas and phrasiology." As a young student, Smyth felt that Henry "was the realization of all that was manly, generous, sympathetic and kind as an instructor." Yet another student, Alexander Bradford, fondly remembered that Henry "rose with the sun to instruct his pupil, eager after knowledge." Local citizens also seemed to appreciate his public presentations. After Henry moved to Princeton, he told his new students an anecdote about his first, open Albany talk on galvanism. Though the talk had been "attended by a large Assembly of the Elite from Albany & its surroundings," a merchant friend in the audience nonplussed Henry. The merchant told him afterward that whereas the presentation was interesting and good, it contained "nothing especially new." A few weeks later, however, a lecture on the same topic by a traveling speaker from New York City completely enchanted the merchant. Henry recalled that his own talk conveyed the latest information clearly but with "little ornamental flourish"; in contrast, the visiting speaker gave an amateurish discourse but packaged it in eloquent language. "So young gentlemen," Henry reportedly advised his Princeton students, "you see if you would be considered *profound,* you must not be *too plain.*"[12]

Henry believed that the courses in his department contributed to the students' general mental development as well as fostered particular practical skills. That is, the courses not only embodied the principles of the psychological theory of faculties but also exemplified the precept of science's utility. Besides the offerings involving surveying and navigation, there was an entire course on "Technology, or the application of Mathematical and Mechanical Science to the Useful Arts." Furthermore, there was explicit instruction on architecture and civil engineering. At one time or another, Henry taught each of these subjects.[13]

About the winter of 1832, Henry advanced a detailed defense of the utility of "Mechanical Philosophy"—a label that he used loosely to refer to areas of physical science involving especially the principles of mechanics (how material bodies or systems react under the action of forces). He presented the defense in the introductory lecture of a public course that probably he and the Beck brothers were offering. Beyond carrying his regular classroom load, Henry frequently helped teach special courses on topics popular among Albany's citizens. This one concerned chemistry, but Henry briefly shifted to the subject of mechanical philosophy to exhibit the academy's recent acquisition of instruments for illustrating principles of physical science. He had overseen the purchase of these devices, financed by Stephen Van Rensselaer and other Albany notables, after reviewing collections at leading institutions such as Yale College. The thirty-two new items ranged from a hydraulic ram to a hygrometer. Before displaying the instruments, however, he presented "a brief exposition of the importance of the study of Mechanical Philosophy[,] a branch of Physical science intimatly connected with all the arts and with the developing of the human mind and progress of civilization."[14]

Henry explained to his audience that mechanical philosophy had not received its deserved attention either in Albany or the nation. Part of the problem was that this branch of physical science was not as flashy as its cousins such as chemistry. But the main problem was that the general citizenry did not appreciate mechanical philosophy's tangible, practical benefits. The disregard was evident particularly among those workers in the mechanic arts who actually built machinery. Indeed, in the United States, the "operative mechanic" viewed the "theoretic inquirer" with "contempt" and "prejudice." To underscore his belief that nearly all technological advance hinged on the principles of mechanical philosophy, Henry marshaled a long list of examples, starting with the plough, windmill, waterwheel, and steam engine. He felt strongly that the United States could learn from England, the "queen of the mechanic arts," the advantages of an "intimate combination of scientific knowledge with practical detail. . . . Without a knowledge of science, the artizan is little more than a labouring machine."

Having made a case for the dependence of inventions on physical science, Henry further claimed that, ultimately, a nation's political economy rested on mechanical philosophy. Similarly, but with less clarity, he suggested a link with ethics: If "a hoard of savages" completely lacked useful inventions and a knowledge of their underlying scientific principles, then they would be unable to embrace even "the most sublime system of ethics." With more clarity—and forcefulness—he also surmised that mechanical philosophy "furnishes the most powerful extrinsic argument which can be adduced in support of the truth of revealed religion." Any person versed in inductive reasoning would inevitably agree with William Paley that an appreciation of the "mechanism of nature" engenders a "sense of the existence and attributes of the Deity." By this date, Henry owned a copy of Paley's influential *Natural Theology* and probably the first volume of his *Principles of Moral and Political Philosophy.*

Henry closed the lecture by briefly mentioning two other practical aspects of mechanical philosophy. A fuller appreciation of the principles of physical science would keep individuals and the nation from wasting time on misguided inventions that inherently contradicted the principles. Probably with contrivers of perpetual motion machines in mind, Henry hoped to repress "the premature zeal of ignorant although perhaps ingenius inventors." And frustrated by persons applying for unjustified patents, he hoped to convince "such inconsiderate inthusiasts of their real ignorance." The benefits of anchoring inventions to science would redound to the nation's wealth and power, he maintained, raising the United States to a level as high if not higher than the leading European countries.

Although throughout his lecture he had stressed the utility of mechanical philosophy, he ended by insisting that persons should also be able to pursue the topic "for its own sake." That is, he reserved a place for mechanical philosophy as a purely speculative endeavor aimed only at extending the boundaries of science. Indeed, investigators of this bent sometimes displayed heroism comparable to "that exhibited in the field of battle or on the stormy deep" as they engaged in "daring enterprises" in their pursuits of knowledge. (While dramatizing the battle scars of earlier, heroic alchemists—"A lost eye, a dismembered limb, or a scorched and crisped visage"—Henry perhaps had in mind a laboratory accident of his own. About a year before, an exploding gas retort cut his face so badly that, after spending over a week to recuperate, he was left with a cauterized scar on his face.[15]) Henry acknowledged that skeptics would be unable to appreciate the significance of this often heroic, speculative search. For these doubters, he offered the assurance that even the most seemingly impractical investigation will have "in some manner an indirect bearing on the wants or luxuries of society." To reinforce this point, he cited the apparently useless investigations of electricity by the ancient Greeks; these investigations led, centuries later, to "the most astonishing results ever recorded in the history of discoveries or inventions"—Benjamin Franklin's successful efforts to protect houses with lightning rods. Science's utility, though sometimes apparent immediately, often requires time to manifest itself.

In teaching natural philosophy to the academy's more advanced students, Henry covered not only Franklin's subject of electricity but also the related subject of magnetism. By making this natural extension to magnetism, Henry was embarking on an odyssey of teaching and research that decades later would culminate in some admiring commentators even listing his name alongside Franklin's in the pantheon of American scientific achievers. The commentators would suggest that what the esteemed Franklin had done for electricity, Henry had done for magnetism, specifically electromagnetism.

As he did with all scientific topics, Henry depended heavily on classroom demonstrations in presenting electrical and magnetic principles. When the academy expanded its collection of philosophical instruments in 1830, he had included standard items for displaying the effects of static electricity as pioneered by Franklin and other eighteenth-century electricians. He could demonstrate for his students a large "electrical machine" that used friction on a rotating glass cylinder to generate a charge, a brass "Electrometer" for measuring the quantity of charge, and a "battery" of nine joined Leyden jars for storing electrostatic charge. The next year he added a "Galvanic Deflagrator," a large electrochemical battery composed of groupings of many zinc and copper elements immersible in a few troughs containing acid. Robert Hare, a noted chemist at the University of Pennsylvania, had introduced this popular battery a decade earlier, a battery so powerful that its discharge could "deflagrate" or inflame objects. In turn, Hare's design built on the initial development of electrochemical cells by Galvani, Volta, and others.[16]

Whereas the hand-cranked electrostatic generators were straightforward to use, the electrochemical batteries depended on sensitive mixes of acids and metals that were constantly depleting each other, causing the batteries to have short lives. They also involved critical connections between the zinc and copper galvanic elements and associated wires. With his solid background in chemistry, Henry had the skill to tease reliable currents out of various combinations of acids and metals. Complementing this skill was his familiarity with soldering and manipulating wires and metals that he gained as a silversmith's apprentice. His lecture notes reveal that he was using this amalgam of practical knowledge to display for his Albany students not only more manageable magnetic and static electrical effects but also the less tractable galvanic phenomena. Furthermore, he was illustrating Oersted's

1820 discovery that an electric current can influence a magnet—an effect that Henry had seen being taught at West Point a few months before assuming his Albany post. To be sure, Henry supplemented his lecture-hall demonstrations with expositions of theory. He covered, for example, French physicist André-Marie Ampère's popular theory that all magnetic effects were ultimately due to electrical currents in the magnet's underlying particles—minuscule currents that were circulating in planes perpendicular to the magnet's north-south axis. Nevertheless, he set experimental demonstrations as the anchors of his lectures. The most repeated verb in his lecture notes on electromagnetism is "show."[17]

In September 1827, as he began his second year of teaching, Henry talked about his electromagnetic interests in a letter to Lewis Beck. He greeted his friend by quipping that although, as customary in dealing with his students, he had been "busily engaged in applying *birch* to one end and *arithmetic* to the other," he had found time to complete two pieces of routine business: to research a chemical issue that concerned Beck and to mail various batches of their improved "Scale of Chemical Equivalents" to potential customers. Having settled these matters, he devoted the rest of the letter to electromagnetism. More exactly, he devoted it to possible electromagnetic interpretations of the aurora borealis, which in recent weeks had been "displaying its fantastic flirtations amoung the clouds and stars of the north." Henry recounted that some uninformed Albany citizens dismissed the aurora as a "lunar rainbow" or a "reflection from a cloud." Other residents, noticing that the recent displays began soon after a heavily attended public hanging of a murderer, believed that it was "a sign in the heavens" that full justice had not yet been served. Though amused by these conjectures, Henry opted for an electromagnetic explanation—an informed electromagnetic explanation.[18]

In speculating on the electromagnetic links of the aurora, Henry cited articles in two British periodicals, the *Edinburgh Journal of Science* and *Annals of Philosophy* (a London publication, which in 1827 consolidated with the *Philosophical Magazine*). These well-regarded scientific publications were part of the up-to-date collection of basic European and American journals maintained in the library of the Albany Institute or in members' private collections. Henry habitually studied these journals, meticulously transcribing to notebooks a running record of important passages and ideas in all his areas of interest, not only electricity and magnetism. Adept by now at reading French, he included translations along with commentary. His appointment in 1829 as institute librarian (and resignation as a curator in the natural history division) facilitated the self-imposed routine that he had seen his mentor Beck follow to such advantage. The institute's library was located at the academy, where Henry already tended the school's library.[19] In systematically copying extracts from the institute's and academy's periodicals—and books—Henry was keeping abreast of current thinking in a variety of areas. For the aurora's seeming alignment with the earth's magnetic pattern, he mentioned to Lewis Beck an article by Christopher Hansteen, prominent Norwegian expert on terrestrial magnetism. And for a laboratory experiment that likely simulated the aurora's magnetic alignment, he alluded to an anonymous 1821–22 "Historical Sketch" of the new-sprung field of electromagnetism. Henry apparently did not know that the author of this comprehensive review was Michael Faraday, chemist at the Royal Institution in London, who would emerge as a leading authority in electromagnetism—and would anticipate Henry in a major discovery.[20]

The experiment from Faraday's anonymous article was one that Ampère had devised to show that, in line with his theory, a circular electrical current would mimic a magnetic compass needle when exposed to the earth's magnetism. The experiment called for bending a wire nearly in a circle and arranging the two ends, one above the other, as vertical pivot points. When a galvanic current passed through the wire, Henry told Beck in his paraphrase of Faraday's article, the circle would orient itself in a plane perpendicular to the earth's north-south magnetic meridian. Thus, in Henry's own interpretation of the experiment, the top portion of the circle would simulate the "auroral arch." Henry considered this "a striking analogy," particularly in light of Ampère's theory of the earth's magnetism, "which he supposes to be merely caused by currents of electricity circulating in planes at right angles to the magnetic north & south." Henry was referring to a further summary in Faraday's article of the French physicist's well-known theory. "But be this as it may," Henry closed, "I am sure that there 'are more things in heaven & earth' connected with electro-magnetism than are yet dream't of in philosophy.'"

As a young Albany actor portraying Hamlet, Henry would never have dreamed that a dozen years later he would be delivering to his own friend a scientific rendering of Hamlet's dialogue with Horatio. Indeed, this onetime Shakespearean had become such a disciple of science that, a month before writing to Beck, he had intended to do electrical experiments on the corpse of the recently hanged murderer; though Henry had borrowed special galvanic troughs to produce suitable currents, for some unknown reason, he was "disappointed in performing the experiments."[21] A month after his letter to Beck, however, he succeeded with a group of less macabre experiments. Presented at a meeting of the Albany Institute in October, these experiments and the accompanying paper became the basis of Henry's first publication on electromagnetism. The institute issued the paper a half year later, in June 1828, in the inaugural number of its *Transactions,* sending copies to learned societies and editors of leading scientific journals in North America and Europe. In the paper Henry showed how to modify standard electromagnetic instruments to enhance, and make more readily visible, subtle effects involving galvanic currents and the earth's magnetism.[22]

Henry's motive for modifying the instruments was pedagogic. He pointed out that, though electromagnetism was one of the most exciting subjects and one of the ripest for discovery, it was also one of the least understood sciences in the United States. Neither popular lecturers nor most science instructors had incorporated the subject in their presentations. "A principle cause of this inattention to a subject offering so much to instruct and amuse," he deduced, "is the difficulty and expense which formerly attended the experiments—a large galvanic battery, with instruments of very delicate workmanship, being thought indispensable." Recently, however, Englishman William Sturgeon had partially remedied this problem. Sturgeon built on the finding, from Dominique Arago and Ampère, that a current-carrying wire when wrapped around an insulating glass tube containing steel needles could create a magnet; this led him to introduce a full-fledged electromagnet that required only a modest galvanic battery to produce a strong force. He made this breakthrough in the mid-1820s by wrapping a thick iron wire with a loose helix of bare copper wire; he also bent the iron into a horseshoe shape, thus increasing the magnet's effectiveness for lifting by placing its north and south poles in the same plane.

The young Albany professor may have first seen a simple Sturgeon electromagnet in January 1827, nine months before presenting his paper to the institute. The opportunity arose in a popular lecture, part of a series on electromagnetism, given in New York City by Dana, the professor who had offered advice on demonstrating electromagnetic effects earlier at West Point. In later years, though Henry himself never acknowledged attending the lecture—only his daughter and others made the claim, well after his death—he did, in a public statement, judge that Dana's version of the magnet, like Sturgeon's original, was a comparatively "imperfect" instrument of "feeble" power. Be this as it may, Dana's lecture notes show that he gave his audience a detailed rationale for spiraling a wire around a horseshoe of soft iron; the multiple turns of the wire and the wire's orientation at near right angles to the iron presumably combined to enhance the horseshoe's magnetism.

Whereas Dana paralleled Sturgeon's design and used only a single strand of wire to form the helix along the length of the horseshoe, he speculated about the ability of multilayered, concentric wires—which he approximated by a solitary, flat spiral or "volute" of wire—to concentrate the magnetic effect in their mutual center. Dana also mentioned Ampère's justification for using a helix of wire to magnetize enclosed needles, summarizing Ampère's belief that all magnetism was due to tiny, circular electrical currents. As fate would have it, Dana's death a few months after his lecture deprived Henry of any possible further information. However, Henry did enjoy access to Sturgeon's original article as abstracted in the November 1826 issue of *Annals of Philosophy,* one of his favorite sources. Indeed, in articulating his pedagogic reasons for building an improved set of instruments, Henry not only borrowed the rationale from Sturgeon but also paraphrased him. Sturgeon had called for making the interesting but little understood science of electromagnetism more accessible by increasing magnetic power while decreasing battery size. Specifically, in words that Henry would restate, Sturgeon had sought to overcome the difficulty, expense, and small scale of prior electromagnetic demonstrations.[23]

In his October 1827 presentation, Henry pointed out that Sturgeon's resulting set of instruments, "with large magnets and small galvanic elements," was suitable for use in either "the private study or the public lecture room." However, he felt he could go further. Whereas Sturgeon's instruments improved prospects for displaying electromagnetism, they still illustrated only a limited group of effects. In particular, Sturgeon's magnets were inappropriate for showing either the earth's magnetic action on a conducting wire or the interplay of two conducting wires. "To form therefore a set of instruments, on a large scale, that will illustrate all the facts belonging to this science, with the least expense of galvanism," Henry decided, "evidently requires some additional modification of the apparatus, and particularly in those cases in which powerful magnets cannot be applied."

In taking this pedagogic initiative, Henry was not only echoing Sturgeon but also paralleling the path he had set for American engineers in his inaugural speech: to produce great results with small means. Moreover, he was embarking on a project leading to breakthroughs in electromagnetism that would result in widespread professional and popular recognition.

Honoured father & Mother
I embrace the Oppertunity to write to you
to inform you that I am well at the [illegible]
Like wise Granmother and all the rest of my
Relitives, and I flatter my-self that these
few Lines will find you enjoy-ing your
healths, and my brother & sister. it is a
general time of health here. I endeavour to
Make the best improvement in my Learning
that I possibly can. though at the first it
seemed a Little irksome and hard. and
I hope to gain the point at Last
pray Dear parents accept of my most humble
duty to your selves and kind Love to my Brother
& sister &c I subscribe myself [illegible]
You will look over the blots from your son,
Galway July 26 Joseph Henry
1808

Letter that Henry, in 1808 at age ten and a half, sent from Galway to his "Honoured father & Mother" in Albany. (Smithsonian Institution Archives, neg. no. 93-9592)

Albany Academy Building, where Henry studied and taught. Located near the capital city's center and dedicated in 1817, the reddish-brown sandstone building now houses state education offices. After Henry became professor of mathematics and natural philosophy in 1826, he supposedly set up his laboratory in a basement room at the building's northeast corner *(lower right)*. An inscribed pedestal with a large bronze statue of this "Pathfinder in Science" now graces the approach to the front steps. (Photo courtesy of Albany Academy Archives. Smithsonian Institution Archives, neg. no. 96-4474)

Left: Henry's house while a professor at the Albany Academy. Located near the academy at 105 Columbia Street, the redbrick row house was torn down in 1899. (Smithsonian Institution Archives, neg. no. 46639-B). Right: Record-breaking electromagnet, capable of supporting more than 2,000 pounds, that Henry built for Benjamin Silliman Sr. and his Yale students in 1831. Donated to the Smithsonian in 1896 at Mary Henry's urging, the historic magnet was included among the items to be evacuated from Washington if the city were threatened with attack during World War II. (Smithsonian Institution Archives, neg. no. 78-6063)

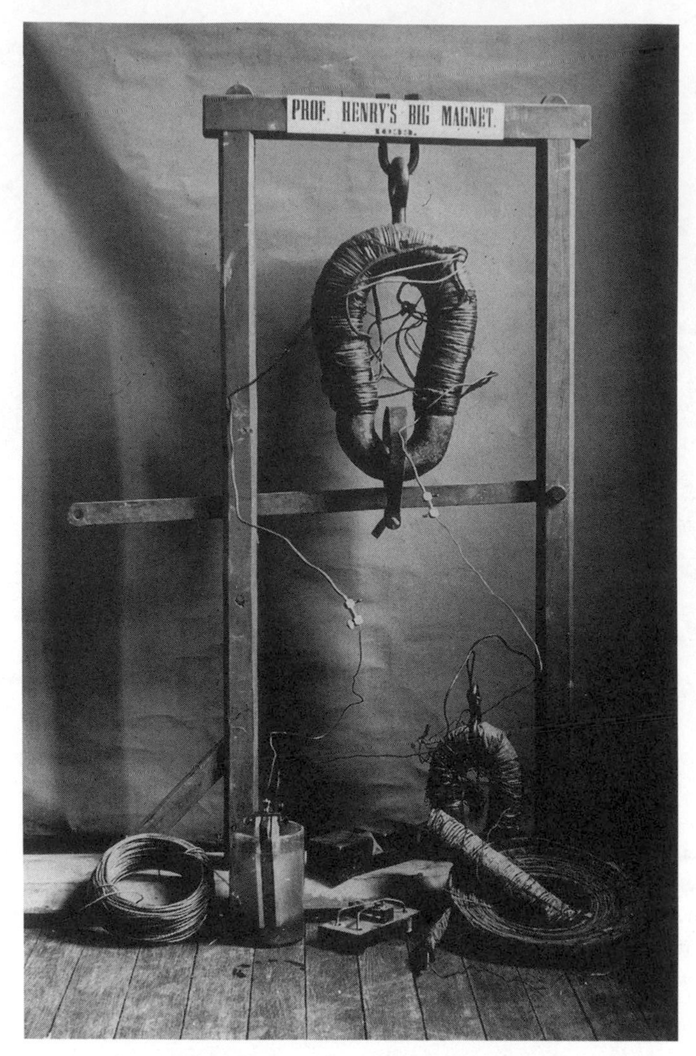

Key scientific instruments. (a) The smaller horseshoe electromagnet *(bottom right, background)* is the test device with nine switchable coils that Henry described in his first article in Silliman's *Journal,* 1831, and used in Albany to clarify his distinction between "intensity" and "quantity" magnets. Later, in 1832, when he wrapped the magnet's iron armature *(pictured to the left of the instrument)* with insulated wire and rigidly attached it to the ends of the horseshoe, he joined Michael Faraday in successfully being able to create an electrical current through mutual induction. (b) The larger horseshoe, in the five-foot-tall frame, is "Big Ben," which Henry built in 1833 for his Princeton classes and used to achieve a new lift-record of 3,600 pounds. (c) The "flat spiral" of copper ribbon *(lower right, foreground)* figured into Henry's investigation of self-induction, a discovery that he began to pursue in earnest at Princeton in 1835. (d) The coil of insulated wire and the cylindrical battery *(lower left)* were included in this 1880 photograph, an engraving of which appeared in *Scientific American,* as examples of "modern" equipment. Most of the items pictured in this photograph are still on display in Princeton's Physics Department. (Smithsonian Institution Archives, neg. no. 78-6069)

Above: Site of many of Henry's experiments, Princeton's "Philosophical Hall." The top floor, renovated by Henry in 1835, contained his lecture hall, laboratory, workshop, and office. (Smithsonian Institution Archives, neg. no. 46639-A). Below: The Henrys' home in Princeton, completed in 1838 near the site of their original house and next to the Library. Both buildings in the photograph still stand, although the home is now located across the quad after being moved twice. The photograph dates to before 1870, when workers first moved the house. (Smithsonian Institution Archives, neg. no. 82-3192)

The Henrys playing croquet on the Smithsonian grounds in the early 1860s. The most likely identities of the family members are *(from left to right):* Harriet and Joseph with children Helen, Caroline, and Mary. Photograph by Titian Ramsay Peale. (Smithsonian Institution Archives, neg. no. 75079)

A Mathew B. Brady photograph of Henry, with the natural philosopher's autograph, about 1863. (Smithsonian Institution Archives, neg. no. 76-5847)

Above: The Smithsonian Building from the southwest, looking toward the Capitol. (Smithsonian Institution Archives, neg. no. 91-10382). Below: Unveiling of sculptor William W. Story's statue of Henry, April 1883, as photographed from the Smithsonian Building. In 1934 workers moved the bronze figure from its original location in front of the building's west wing to a central site near the main door. Until a crane rotated the statue in 1965 to face the Mall, Henry's gaze had been fixed on the Castle. (Smithsonian Institution Archives, neg. no. 10723)

Master of Electromagnets 4

A coil of wire underpinned all of Henry's modifications of electromagnetic devices—in his 1828 paper and in subsequent, more consequential Albany papers. He realized that if a single circle of current-carrying wire could react to the earth's magnetism—as he had mentioned in his letter to Lewis Beck—then a coil of many strands would multiply the effect. Thus, a multilayered coil would enable lecturers to display geomagnetic and other electromagnetic effects in a highly visible manner while using only a small, inexpensive galvanic battery.

His inspiration for the "multiplying coil" came from an instrument devised by Prussian physical scientist and journal editor Johann Schweigger. Shortly after Oersted's announcement of a current's influence on a compass needle, Schweigger realized that he could amplify the effect by surrounding a similar magnetized needle with a vertical coil of wire. Others improved the device and soon many scientists had adopted it, using it to test for the presence and strength of a current. Though Schweigger labeled it a multiplier, it became known as a galvanometer—Ampère's name for his own, more rudimentary instrument for gauging a current.[1]

Henry might have learned about what he called "Prof. Schweigger's Galvanic Multiplier" through a mention by Faraday in his "Historical Sketch." Another source might have been a pamphlet that Ampère prepared in 1822 on recent discoveries in electricity and magnetism; Henry heavily annotated his copy. Most likely, however, he obtained the information from a new book on electromagnetism by American colleague Jacob Green. In this 1827 publication Green freely used copious "extracts" from the writings of various experts on electromagnetism to present the "essential facts" from this "new and unsettled" branch of "chemical philosophy." These authoritative writings included works by Faraday—not only an 1827 manual titled *Chemical Manipulation* but also probably his earlier "Historical Sketch," the first half of which Faraday devoted to the "facts" of electromagnetism. Intending to reach the "practical chemist," Green avoided mathematical and theoretical issues, though he did bow to Christian natural theology by positing that God animates even the material universe. Green was a chemistry professor at Jefferson Medical College in Philadelphia, who previously taught at the College of New Jersey (which would be renamed Princeton College) during his father's presidency of the school. He had earlier lived in Albany, remained a close associate of T. Romeyn Beck, and was an active contributor to the Albany Institute; he would become an intimate colleague of Henry. In his guidebook Green emphasized that the multiplier was a "very important acquisition to the apparatus of chemical research" because it allowed practitioners "to detect the smallest stream of electricity." He also specified how to build a particular version of Schweigger's multiplier—an enhanced version by Oersted—including the technique that first made the multiplier's coil possible: insulating the wire. Whereas many contemporary electrical instruments still relied on bare wires, the coil used wire "wrapped or wound around throughout its whole length, in silk thread." The silk provided electrical insulation, allowing successive turns of thin copper wire to be "wound one over or above the other" without making electrical contact and shorting out.[2]

Henry fixed his attention on this innovation, giving in his 1828 paper a similar account of his own multiplier: "the coil is formed of about twenty turns of fine copper or brass wire, wound with silk, to prevent contact, and the whole bound together so as to have the appearance of a single wire." He went on in his paper to apply similar "multiplying coils" to various electromagnetic instruments previously dependent on single strands of bare wire, such as the standard instruments for illustrating the effect of terrestrial magnetism on a current-carrying wire and the effect of two current-carrying wires on each other. Except for references to directions of flow for galvanic currents, Henry's paper mimicked Green's book in its avoidance of electromagnetic theory. He limited himself to the practical, pedagogic goal of devising instruments to illustrate important principles "on a large scale . . . with the least expense of galvanism."

Over the next year and a half he devised a more consequential application for "the coil." Although he introduced it without fanfare, this application would eventually establish his reputation as an innovator in electromagnetism and would influence his scientific career for decades to come. Not content merely to supplement Sturgeon's set of instruments, whose facility derived from the English scientist's rudimentary electromagnet, Henry applied his special coil to the electromagnet itself. Instead of following the usual procedure of loosely wrapping a few feet of bare wire around a horseshoe-shaped, shellacked piece of iron—the shellac providing electrical insulation—Henry "tightly wound" the horseshoe "with 35 feet of wire, covered with silk, so as to form about 400 turns." Supplying the insulated coil with current from only a small galvanic source, he found that his magnet exhibited strength much greater than that from a comparable Sturgeon magnet using a larger battery.

Without fanfare, Henry demonstrated his enhanced electromagnet in the spring of 1829 at a regular meeting of the Albany Institute.[3] His incessant reading and note taking had enabled him to keep abreast of the latest American and European developments in electromagnetism, but his deep immersion in Albany's regional scientific and academic communities apparently inhibited him from seeking a broader forum for his results. A new professor with an exhausting teaching load and divergent research interests, he seemingly neither presumed nor had time to announce more widely his improvement of the electromagnet. In view of his erratic education and disjointed record of employment, he must have been pleased just to be addressing as a peer the regionally prominent men who constituted the Albany scientific and academic circles. Indeed, among the politicians, physicians, lawyers, merchants, and educators who made up the eclectic membership of the Albany Institute, Henry must have enjoyed his status as one of the most imaginative and productive members. And he must have been encouraged to learn that word of his talent was spreading to other American intellectual centers. When he first met Benjamin Silliman a year later, in May 1830 while visiting Yale College to get ideas on apparatus, the eminent chemistry professor, public lecturer, and editor of the *American Journal of Science and Arts* expressed great pleasure, saying that he already knew of Henry through reputation. Similarly, while visiting Columbia College a few weeks after meeting Silliman, Henry recorded that physical science professor James Renwick: "Spoke very highly of my experiments on magnetism &c &c."[4] Though Henry recognized that his enhanced electromagnet held promise for the lecture hall and the research bench, he still neither conceptualized nor reported his findings in ways that fully reflected the practices and norms of the growing network of European electromagnetic researchers. What awakened him to the full significance of his enhanced electromagnet for the broader professional community was the announcement of a similar magnet by Dutch physicist Gerrit Moll in the October 1830 issue of David Brewster's *Edinburgh Journal of Science.*

"I have lately had the mortification of being anticipated in part by a paper from Prof. Moll in the last No of Brewster's Journal," Henry wrote to

Silliman in early December 1830. Noting that he had delayed publishing descriptions of his own electromagnets for almost two years, Henry implored Silliman to issue as quickly as possible a paper that he would immediately write. Within two weeks, he assured Silliman, he would send this paper on "the development of great magnetic power with a small galvanic element." Silliman, who had not communicated with Henry since their meeting in New Haven earlier that year, responded that he had already arranged, coincidentally, to reprint Moll's article in the pending, January issue of the *Journal;* this timing made it essential that Henry's paper also appear in the same issue. Though he had already finalized most of the issue, Silliman told Henry that he was willing to "reserve a form for you at the very end where you can come in by way of appendix." After finishing the paper "in great haste," Henry rose early two days before New Year's Eve to speed it in the morning mail; in a postscript to the cover letter, he jotted: "Please excuse this scrawl as I am writing without a candle just after daybreak."[5]

As Silliman had warned, the reprint of Moll's article appeared in the January issue. The article showed that Moll also had increased the number of turns of wire around Sturgeon's type of horseshoe magnet. He left the wire uninsulated and loosely coiled, however, as he sought mainly to increase the strength of Sturgeon's design by enlarging the magnet's iron core and the galvanic battery; on two of his test magnets, nevertheless, he did take the precaution of "coating the horse-shoe with silk." The article further showed that Moll expressed surprise, in light of his "astonishing" results, that an exhaustive search of British and Continental journals turned up no other attempts to improve Sturgeon's design. Given Moll's declaration about his fruitless search in European journals, it is ironic that Henry's paper appeared later in the same issue of the *American Journal of Science and Arts.* As Silliman had promised, he secured a place for the paper as an appendix—the sole entry in a separate section after the "Miscellanies," at the very end of the two-hundred-page issue.[6]

Silliman soon wrote back to Henry: "It is a highly important & interesting paper & will undoubtedly be printed in the foreign Journals & I shall be much gratified in quoting the results in my chemical book." After the paper appeared, a colleague at Union College in nearby Schenectady called it "a glowing prize for Mr. Silliman" and claimed that it was "worth all the attempts at original experimental investigation that have ever been recorded in his journal"; he also joined Silliman in predicting that it would "undoubtedly attract that attention abroad which it so richly merits." Not only would Silliman include the findings in his popular textbook, *Elements of Chemistry,* published later that year, but also European scientists would have access to Henry's findings through Silliman's fairly broadly distributed *Journal.*[7] Though his debut lacked grace, Henry was now entering an international stage.

In his paper Henry took advantage of the reasonably wide circulation of Silliman's *Journal* first to summarize his earlier findings from the Albany Institute's *Transactions.* Again, he explained that he aimed to eliminate from electromagnetic experiments the necessity of "a large and expensive galvanic apparatus." Although Sturgeon's electromagnet provided an important first step toward achieving this goal, Schweigger's spiral coil of silk-wound wire held additional promise. After recapping his initial applications of the coil, Henry moved to his latest extension: the electromagnet that he had shown to the institute in the spring of 1829. This modified magnet, he pointed out, successfully met the criterion of increasing the size of the instrument while reducing the galvanic source. He emphasized, however, that since devising this prototype two additional ideas had enabled him to advance even further. First, he realized that he could achieve a greater magnetic power by substituting a larger horseshoe constructed from a longer and thicker iron bar; in particular, he improved his results by using a bar ten inches long with a one-half-inch diameter. Second and more important, he discovered that he could double the power of even this larger magnet by overlaying the first coil with a second coil having an equal length of insulated wire, and independently connecting both coils to the original galvanic source. That is, he soldered corresponding wires from each coil to the zinc plate of a small battery and the remaining wires from each coil to the copper plate; later generations of scientists would call this a "parallel" connection, where the available current from the battery divides between the two branches and later recombines. (In contrast, a "series" connection would have the outgoing wire of the first coil joined to the incoming wire of the second coil, with the remaining two end wires completing the circuit to the battery; the entire current would flow sequentially through each component of the combined coil.)

He mentioned that his idea of using multiple coils came from "reading a more detailed account of Prof. Schweigger's galvanometer." The technique strengthens the magnet, he speculated, because the multiple coils conduct "a greater quantity of galvanism." As he later elaborated to his natural philosophy class at Princeton, an array of smaller coils seemed preferable to one, larger coil because a galvanic current loses "free passage" when confronting the increased resistance of a lengthy wire; he pointed out that, in fact, when he began his work, "it had not been possible to send a current through a long wire." Another advantage of multiple coils arose from, he further reasoned in his initial paper, the property of a galvanic current to produce a magnetic effect perpendicular to the current; by placing separate coils along each segment of a large bar, each segment would receive the maximum magnetic effect because the circular windings would be almost at right angles to the bar. This last explanation was reminiscent of Dana's rationale, in his New York City lecture four years earlier, for wrapping his simple horseshoe electromagnet with a single wire spiral. More to the point, it was reminiscent of Ampère's even earlier theory that all magnetism was due to circular electrical currents. Indeed, Henry would later tell his students at Princeton that not only had Ampère's theory influenced him but also that he had been the first in the United States to study the theory seriously. One student recorded that Henry arrived at his design for the electromagnet by "carrying out an original inference from Ampère's Theory." Another young man—not a Princeton student but a later Henry protégé—recalled Henry even identifying the moment in Albany when he first had the basic realization that Ampère's theory suggested tightly wrapped, insulated wire. Apparently, while sitting with a friend in his study and becoming momentarily lost in thought, Henry suddenly jumped to his feet and exclaimed that he had the important insight. Henry's explanation for specifically using multiple coils also calls to mind the advice that he had received from the New York instrument maker on constructing strong conventional magnets: build up a "compound" horseshoe magnet by layering a sequence of thin bars that were separately magnetized.[8]

With preliminaries out of the way, Henry devoted the bulk of his paper to a main series of twenty-three experiments designed to clarify the properties of the two types of electromagnets: single coil and multiple coil. A slightly younger Albany colleague, Dr. Phillip Ten Eyck, helped with these experiments—as did, to a lesser degree, Lewis Beck. Though Henry referred to various trial magnets, he primarily relied on a large, painstakingly constructed magnet that allowed him to vary the coils and measure the corresponding magnetic strengths. The magnet—a twenty-one-pound iron horseshoe with its poles joined by a seven-pound "keeper" or "armature" (thus forming an "armed" magnet)—hung from a frame with its lateral limbs pointing down. A pan to hold weights hung from the armature. Nine tightly wrapped coils of insulated copper wire, each having a length of sixty feet, covered two-inch segments of the horseshoe. Henry carefully tagged the two ends of each coil so that, by soldering various ends either together or directly to the galvanic plates, he could produce a variety of longer or shorter coils and of single or multiple connections to the galvanic plates. The experiments confirmed his preliminary finding that an increase in the number of separate coils soldered directly to the galvanic plates increased the lifting power. Indeed, as he demonstrated at the Albany Institute a week before mailing his paper to Silliman, on separately soldering all nine coils to a slightly larger battery, the magnet supported 750 pounds. In contrast, Moll's largest magnet, approximately a twenty-six-pound iron horseshoe with a four-pound armature, supported only 154 pounds. Calculating that his magnet held over thirty-five times its own weight, Henry suggested that "it is probably, therefore, the most powerful magnet ever constructed."[9]

In his experiments, Henry varied not only the combinations of coils but also the types of galvanic batteries, and the lengths of wire connecting the coils and batteries. This systematic approach, which conformed to the standard empirical practice of varying experimental parameters, enabled him to go beyond his preliminary findings. Specifically, it allowed him to announce the "remarkable result" that, under certain circumstances, the use of longer wires between the battery and the coil seemed to enhance the magnet's strength. The circumstances were, first, that a "trough" battery with many pairs of linked galvanic plates be substituted for the simple battery of two galvanic plates and, second, that a large continuous coil be substituted for the magnet's multiple coils. Trough batteries, according to the conventional view, produced a current with a large "projectile force" or "intensity" similar to that generated by an electrostatic machine. Applying one of Robert Hare's theoretical notions, Henry speculated that this force allowed the "galvanic fluid" to travel long distances; while traveling these distances, the fluid lost velocity, thus moving at a slower rate perhaps best suited for producing the greater magnetic effect.[10] "From these experiments," Henry summarized, "it is evident that in forming the coil we may either use one very long wire or several shorter ones as the circumstances may require: in the first case, our galvanic combinations must consist of a number of plates so as to give 'projectile force;' in the second, it must be formed of a single pair."

Beginning later in the decade, Henry's curious results found ready interpretation in an analysis of electrical circuits worked out by German physicist Georg Ohm. The analysis involved the relationship between what scientists came to call the "potential difference" of a battery, the "current" flowing in a circuit, and the total "resistance" of the circuit—including the resistance of external loads such as a coil around a magnet, the resistance internal to the battery itself, and the resistance of the intervening wire. Expressed algebraically (in units of electrical measure adopted only near the end of the century), the relationship holds that potential difference (measured in "volts") equals the current (measured in "amperes") multiplied by the total resistance (measured in "ohms"). Ohm had proposed the basic analysis in 1826, but Henry and most other natural philosophers and chemists were still unaware of his work. Through trial and error leavened with a bit of rudimentary theory, Henry had ferreted out a specific implication of Ohm's analysis: for maximum transfer of energy between a battery and external load (in Henry's case, for the maximum magnetic strength), the internal resistance of the battery must equal the remaining resistance of the circuit (of the connecting wire and coils). Without realizing it, Henry had been roughly matching internal resistances and external loads.[11]

On the one hand, a trough battery has a large internal resistance because it compounds, through series connections, the individual resistances of its many cells, each made up of a pair of dissimilar metal plates and a highly resistive solution of dilute acid; this compound resistance roughly matches the large external resistance of long connecting wires and a sizable continuous coil on the magnet. On the other hand, a simple combination of galvanic plates has a small internal resistance; this approximates the low external resistance that results from using multiple coils connected separately, or to use the later technical phrase, in parallel. In subsequent writings, Henry would label the continuous-coil magnet an "intensity" magnet; he was extending the label commonly applied to trough type of batteries (with many cells connected in series) that produced an electrical force of great "intensity" (what later became known as high potential difference, measured in volts). Similarly, he would name the multi-coil magnet a "quantity" magnet; he was adapting the label commonly applied to a simple pair of galvanic plates (or combinations of plates connected in parallel) that produced a significant "quantity" of electricity (what later became known as current, measured in amperes). Regarding these labels, Green, for example, in his 1827 guidebook excerpted a recent description by Faraday on arranging batteries "to produce intensity or quantity at pleasure." Copied from Faraday's *Chemical Manipulation,* the London scientist's 1827 book-length manual instructing students how to do experiments, the excerpt also emphasized that each type of battery had advantages depending on the intended electromagnetic result. Earlier, in his "Historical Sketch," Faraday had reported experiments by Moll distinguishing between the ability of a trough battery to produce chemical effects and the ability of a pair of galvanic plates to produce magnetic effects.[12]

When Henry referred to a "very long wire" he meant 1,060 feet (323 meters) of copper wire strung around the upper perimeter of the academy's assembly hall—a large, high-ceilinged room that spanned the front central section of the building's second floor. That he could use a galvanic trough to trigger an intensity magnet over this distance led him to a portentous realization—one that would in later decades not only embroil him in a public controversy but also furnish what Americans popularly perceived to be his main "scientific" contribution. In a brief phrase in his paper, he mentioned that the magnet's ability to respond over a great distance is "directly applicable to Mr. Barlow's project of forming an electro-magnetic telegraph."

As Green had recounted in his book on electromagnetism, many researchers realized that Oersted's breakthrough held promise for devising a functional instantaneous telegraph, a longstanding goal of electricians. That is, they realized they could probably build a telegraph incorporating the magnetic effect of galvanic fluid on a compass needle. British researcher Peter Barlow, however, cast doubt on the widespread hope and related trials. Barlow showed that the magnetic capability of galvanic fluid diminished precipitously after the fluid had advanced only 200 feet through a wire, supposedly ruling out any chance of an electromagnetic telegraph. Though Henry did not elaborate in his paper on his striking reversal of Barlow's verdict, he did arrange a telegraphic device in the academy's large assembly hall, later known as the chapel, and lectured on the topic in the course he taught for members of the senior class. As his students from 1831 and 1832 later attested, he positioned a regularly magnetized bar horizontally on a pivot with its north end between the poles of a single-coil intensity electromagnet; when he triggered the electromagnet through the long, intervening circuit of wire, the north end of the conventional bar magnet swung toward the electromagnet's south pole, causing the other end of the bar to strike a bell. By reversing the battery's connections, he could switch the polarity of the electromagnet, thus restoring the bar to its original position away from the bell. Eventually, he increased the wire's length, which he had strung in successive loops around the assembly hall. John Newland, one of Henry's pupils who helped with these "mechanical arrangements" at the Albany Academy, recalled that in 1831 he saw Henry transmit a galvanic current through a wire one and one-half miles in length. George Carpenter, an assistant who wrapped the long wire with yarn to insulate it, remembered that the wire extended to Henry's basement laboratory to dramatize further the distance between the sender and receiver. Though many investigators had experimented with telegraphic devices—especially devices that used a galvanometer's needle to detect electrical signals—Henry had devised the first telegraph that used an electromagnet to receive signals transmitted over large distances; this design eventually would figure prominently in the development of the American telegraph system, especially the instruments of Samuel F. B. Morse.[13]

Henry seemingly understood the practical implication of his telegraph. New York State geologist James Hall remembered that when he visited the Academy in August 1832, Henry displayed the setup with the bell and mentioned its potential for signaling over long distances.[14] During the period around 1830, however, Henry emphasized only its lecture-room possibilities. He also stressed the pedagogic rather than the practical implications of his strong, multiple-coil quantity magnets. In fact, following the publication of his article in January 1831, he devoted much of his limited free time to building electromagnets for colleagues at other schools to use in their lecture rooms. When he first sent the article to Silliman, he offered to construct for Yale a major magnet that would support more than 1,000 pounds. When Henry shipped Silliman's electromagnet, at the end of March after the Hudson River had thawed, the instrument's strength exceeded expectations. The accompanying description, which Silliman published in the April issue of the *Journal* under Henry's and Ten Eyck's names, reported that in tests the 82.5-pound magnet supported more than 2,000 pounds. Silliman added an editorial note proclaiming that Henry held the distinction of having built the world's strongest magnet, "eight times more powerful than any magnet hitherto known in Europe." But both Henry and Silliman concentrated on the lecture-room applications of this powerful instrument. In describing the device to Silliman, Henry detailed the most theatrical ways to exhibit it. For example, disconnecting the battery and allowing the magnet to release a 1,500-pound load "never fails to produce a great sensation among the audience as before the fall they can scarcely believe that the magnet supports the weight." Later, Silliman wrote back that the "beautiful magnet" was a success at Yale: "*Here* it excites unqualified admiration & an audience of 300 looked at its performance last week with astonishment."[15]

Reading about the extraordinary Yale magnet, other Americans wrote to Henry for information on building their own. Parker Cleaveland, a professor at Bowdoin College in Maine, followed Silliman's lead and arranged for Henry to construct one for his lecture room. Fulfilling a promise made the previous summer, Henry also soon sent a smaller electromagnet and related apparatus to Jacob Green, the Philadelphia professor who had compiled the survey of electromagnetism. When a canal engineer wrote from Kentucky requesting information on Henry's strong electromagnet, Henry took time to provide background on recent developments: "The whole subject belongs to a new department of science lately discovered called electromagnetism which has excited much interest in Europe but is very little known in this country." In a mildly boastful aside, he further remarked that, although the Europeans were already enthusiastic about the new field, the "striking results" of his own electromagnets "have had a tendency to render the subject more popular among men of science in America and doubtless more important improvements will be made."[16]

Henry faced many obstacles and expended valuable time in building the large magnets for Silliman and Cleaveland. Special materials were scarce in Albany, a problem exacerbated when the Hudson was frozen and river transportation halted. For large pieces of iron suitable for the magnets' cores, he either had to make do with the miscellany available locally or had to purchase directly from the operators of an iron furnace and forge in the state's northern region. For sheets of zinc to make batteries and for copper wire, he had to send to New York City. Fabricating the magnet's various components proved time-consuming. He told Cleaveland: "I can get nothing made in Albany in the philosophical line except I stand continually over the workman during the operation or unless, which is most often the case, I do the work intirely myself." On the Yale magnet, to cover the copper wire with insulating cotton thread, two persons using a spindle from a spinning wheel could dress about 30 feet of wire in ten minutes. On the Bowdoin magnet, Henry used another technique for insulating the wire. After wrapping the iron horseshoe with "strips or ribbons" of old silk cloth, he applied several coats of varnish to the wire and then wound it on the magnet with more silk interspersed between alternate turns. "This process was a very tedious one," Henry wrote to Cleaveland, "and occupied myself and two other persons every evening for two weeks." (Later family tradition held that, to insulate the coils in his earliest electromagnets, Henry improvised with ribbons of white silk cut from one of his wife's petticoats. This might have been the case with the Bowdoin magnet.) He passed on the cost of materials and workers' labor to Silliman and Cleaveland, totals of $35 and $45, but he refused reimbursement for his own toil. He told Silliman that he would have liked to absorb all the expense himself, but could not afford it. "My experiments," he confided, "have already cost me considerable." He added that, of course, his personal expense on the Yale magnet "is sufficiently paid by the honour of constructing it." Similarly, to Cleaveland he wrote: "I consider myself sufficiently paid in the additional knowledge and experience I have acquired in the construction [and] use of the instrument." In an expansive gesture, Silliman included an extra five dollars in his payment, labeling it "a trifling aid towards your future experiments." Also, he suggested that, if it would enable Henry to continue his investigations, he would help underwrite, and line up others to help support, Henry's work.[17]

Although it is unknown whether Henry accepted Silliman's offer, he did continue with his electromagnetic investigations. First, however, he had to get past his demanding teaching schedule, not only the seven hours he daily spent meeting regular classes but also a special course of lectures that he and Beck were offering on chemistry and physics. Probably taking advantage of a brief summer break from teaching, he published in the July 1831 issue of Silliman's *Journal* a letter announcing an "Electro-magnetic Engine." "I have lately succeeded," he began, "in producing motion in a little machine by a power, which, I believe, has never before been applied in mechanics—by magnetic attraction and repulsion." He was correct about his originality. In the early 1820s Faraday had achieved steady mechanical motion by building a device in which a current-carrying wire revolved around a regular magnet; Barlow and others elaborated on Faraday's idea. But Henry explored new territory by devising a machine that depended on the continual reversal of polarity of an electromagnet—a design laden with implications for later generations of direct-current motors.[18]

Devising his machine and telegraphic bell ringer about the same time, Henry designed both around the idea of a pivotable bar moved by changing magnetic poles. Specifically, the key to Henry's machine was an electromagnetic bar, balanced on a fulcrum like a seesaw or a steam engine's horizontal working-beam, with each pole positioned over the north pole of a vertical magnet of the regular type. He recounted that he had wrapped the bar with three separate coils having their corresponding wires united into two leads. The two leads of this responsive quantity electromagnet divided into left and right pairs of long horizontal wires whose ends, depending on which side of the bar was down, extended to the terminals of a battery on either side of the apparatus. Having arranged the two batteries to produce currents flowing in opposite directions in the coils around the bar, the bar alternated in its magnetic polarity depending on which battery was in contact. He positioned the two sets of extending wires so that the instant a contact occurred with one of the batteries, the flow of current magnetized the bar in such a manner that its poles would react with the conventional vertical magnets and flip the bar to the other side; this motion would break the contact with the first battery but create the contact with the second battery, thus reversing the bar magnet's polarity and causing the magnet to snap back the other direction. By intermittently replenishing the acid of the battery, Henry could maintain "a uniform motion, at the rate of seventy-five vibrations in a minute . . . for more than an hour."

"Not much importance, however, is attached to the invention," Henry contended, "since the article, in its present state, can only be considered a philosophical toy; although, in the progress of discovery and invention, it is not impossible that the same principle, or some modification of it on a more extended scale, may hereafter be applied to some useful purpose." The machine's status as a scientific novelty did not inhibit Henry from, as he later recalled, "applying a flywheel to it and afterwards giving it a rotary motion"; but he did not deem these alterations "of sufficient importance to merit a publication."[19] In public, as he had done with the telegraph, Henry contented himself with describing the unadorned apparatus and announcing "a new effect produced by one of the most mysterious agents of nature." Also, in line with his view of basic science as the progenitor of practical applications—the crucial first step that determined subsequent steps—he contented himself with merely pointing out the possibility of the device's future utility.

As he explained a few years later in a letter to Silliman, he had a sound technical reason for refusing to pursue the practical implications. In particular, as a source of mechanical power, the battery-dependent machine was undeniably more expensive to use and less convenient to operate than a coal-fired steam engine. He told Silliman that, following his first article on electromagnets, he had been "pestered with letters from almost every state in the Union," but he always responded the same way: "I have uniformly referred them to the description of my little machine in the Journal and stated that I freely renounced all right to the invention as I consider the machine in the present state of the science only a philosophical toy." By the date of this letter, September 1835, he could add that he considered the proliferation of similar engines in countries such as Germany, France, and Italy as nothing more than reinventions of his device and, therefore, still philosophical toys. The same was true, he further informed Silliman, of the new motor involving rotary motion built by Thomas Davenport—a Vermont blacksmith whose interest had been aroused by the power of Henry's electromagnets, whose learning had been aided by Silliman's textbook account of Henry's experiments, and whose effort to promote the motor had been met with a cool, personal reception from Henry, who pronounced the device a mere "curiosity." It was also true, Henry told Jacob Green in early 1834, of the rotational machine that Scottish physicist William Ritchie recently exhibited in London. For Henry his reciprocating engine was primarily a device to illustrate the scientific principles of electromagnetism; it enlivened his lectures both at the academy and the institute. His close colleagues apparently viewed it in a similar light. In 1834 Green borrowed the engine for use in a talk "to a *popular* audience"; he and Henry jokingly called the eye-catching device a "sheeps tail," probably after the rapidly wagging bar. More pragmatic inventors such as Davenport, however, felt that electromagnetic motors were capable of performing practical tasks and went on to pioneer and patent successful designs.[20]

In following decades as the telegraph transformed communication and as electric motors recast industrialization, Henry would maintain that when first building such devices he had consciously chosen not to exploit them for their technological or commercial value. In an 1876 letter to a friend tracing his early research, the aged scientist recalled: "At the time of making my original experiments on electro-magnetism in Albany, I was urged by a friend to take out a patent, both for its application to machinery and to the telegraph, but this I declined, on the ground that I did not then consider it compatible with the dignity of science to confine the benefits which might be derived from it to the exclusive use of any individual." "In this perhaps I was too fastidious," he added, knowing in hindsight that Morse had used some of the Albany findings to create a commercially successful electromagnetic telegraph. Earlier, while in the thick of a mid-century dispute with Morse over the invention of the telegraph, he had similarly written: "I have sought . . . no patent for inventions, and solicited no remuneration for my labors, but have freely given their results to the world, expecting only in return, to enjoy the consciousness of having added, by my investigations, to the sum of human knowledge, and to receive the credit to which they might justly entitle me."[21]

These high-sounding, retrospective words capture an aspect of Henry's attitude during his Albany years toward using technical knowledge for personal gain, but they do not fully reflect his outlook. During the very months that he was first demonstrating to his students the telegraph and reciprocating engine, he was working closely with regional entrepreneurs on the magnetic extraction of iron from crushed ore. And he was pressing for payment and a possible patent. More than an idealistic rejection of monetary gain, therefore, lay behind his hesitation in similarly promoting the technological or commercial implications of the telegraph and engine. To be sure, he did share the common belief that natural philosophy was above payment and patents. However, he also believed that both devices would remain into the foreseeable future mere philosophical toys. In other words, he was unable to envision their more immediate technological significance. Indeed, though he eventually grasped the full potential of the telegraph, he continued into later years to argue that electromagnetic engines could never successfully compete with steam engines because of the higher comparative costs of batteries over coal.[22]

Early as the summer of 1831, he was employing his strong electromagnets to help extract iron from crushed ore. Specifically, he was using his electromagnets to magnetize more effectively the magnetic steel "points" on separation machines that iron-ore workers were already using in the Adirondack Mountains of northeastern New York; the points, which workmen had to regularly "retouch," projected from a revolving cylinder and attracted iron from the ore. Henry's contact with the ore operators perhaps grew out of his efforts to obtain substantial bars of iron for building large electromagnets. He eventually furnished the operators with an electromagnet that greatly facilitated the process of retouching the points—a powerful magnet that soon caught the attention of blacksmith Davenport and triggered his career as an inventor. He also contemplated his own design of an extraction machine that depended directly on electromagnets. As in his other involvements with electricity and magnetism, this ore initiative consumed much time. Repeatedly, he had to prod artisans into completing promised work and to spend periods ranging from hours to days actually monitoring them. Eventually, he realized that he needed to guard his own interests in this industrial venture. When one ore operator, wary of competing manufacturers, reminded him to keep private his insights on improving an existing extraction machine, Henry used the reminder to assert his claim for payment. He replied:

> You need be under no apprehension on this score for although I have heretofore been perfectly free in giving to the public any Knowledge I might possess of use and although I have considered it almost below the dignity of science to ask pay for my Knowledge yet I now conceive that I have been rather too free for my own interest and that the subject of magnetism is worth something more to me than mere fame. As however I do not wish to devote much time to the subject of your machine without a fair prospect of remuneration, I have thought best before commencing to make particular inquiries in relation to the validity of your patent or of any that I might obtain for the application of Galvanism to the same purpose.[23]

Perhaps jaded by earlier dealings with the ore men, Henry let lapse his conviction that a true scientist should remain above exploiting knowledge for financial gain. As he mentioned to the ore operator, he did initiate "particular inquiries" on patenting his magnetic designs. The "best authority in the state" however, tendered misleading advice: that once a person publishes a scientific discovery, it becomes public knowledge and, therefore, no one can claim a patent on practical applications of that knowledge. Henry construed this to mean that, because he had already published his findings on electromagnets, he could not acquire a basic patent "for the application of Galvanism to the separation of iron." The authority, however, provided an out by mentioning that a person could obtain patents for "alterations and improvements" in the original machine. With his eye on such an auxiliary patent, Henry offered to proceed with modifications of the existing magnetic separator in addition to constructing his own electromagnetic machine. Several months later, in a letter to Silliman, he remarked that he was still "considerably engaged" in the project, but yet lacked a "definite arrangement" with the ore operator. "I think the apparatus could be made a very valuable affair," he commented, "were it properly formed on scientific principles and the management of it given to a person who understood the subject of magnetism & mechanics."[24]

Henry apparently did not envision comparable prospects for his lecture-room telegraph and reciprocating engine. Indeed, his misperception of patent law likely reinforced his moral and practical aversion to developing the telegraph and engine—devices that he probably assumed were already in the public domain because of his publications earlier in the year. And although he imagined great prospects for the extraction apparatus, he apparently had no further opportunity to apply "scientific principles" to the apparatus. His initial contribution, nevertheless, made an impact on the ore industry. A few years later, after Henry had moved to Princeton, a newspaper ran a story on how he had successfully adapted his electromagnets to the separation machine. In his public utterances, Henry would have voiced pleasure with the moral that the author drew from the story of these "first fruits of the electromagnetic theory." Indeed, Henry himself perhaps suggested the moral to the journalist: "This discovery has become invaluable to the furnacemen of the North, where the principle is now recognised; and it shews how much, after all, the money making, and money getting portion of society are deceived, when they think science of no consequence to any but men of leisure. Science puts money in *their pockets,* and creates the dividends upon some of their most valued stocks." In private, however,

Henry retained a less magnanimous opinion of the furnacemen capitalizing on his electromagnetic techniques while putting no money in his pockets. He mentioned to one of his Princeton students that he, not the iron manufacturers, deserved the patent right to the successful techniques. By these Princeton years, Henry was still intrigued by the relation between basic science, the mechanical arts, and money; but private talk of patents and personal gain had given way to more customary, public pronouncements on the greater gains of society.[25]

5 The Hunt for Electromagnetic Induction

Henry had worked in makeshift quarters as he explored the extension of his electromagnets to the telegraph, reciprocating engine, and ore extraction. He had set up his laboratory in the academy's basement, in an ample but dimly lit room with only two small windows that faced the students' playground, part of the parklike grounds around the building. While he pressed on with his studies in this secluded Albany setting, other scientifically minded persons were beginning to read and react to his three articles in Silliman's *Journal.* Specifically, they were responding to his reports on electromagnet design, the large Yale magnet, and the engine. As early as July 1831, he could note with evident pride that the descriptions of his electromagnets in the first two articles "have since been republished both in France and in England so that the invention is now fully before the scientific world." In a similar but more lighthearted vein, by August 1832 Amos Eaton could jest with Henry about the wide effect of his enhancements of electromagnetic instruments; quoting English writer Alexander Pope, Eaton joked that the improvements have "'damned you to everlasting fame,' in the remotest parts of the Earth, wherever the word, *Science,* has been told."[1]

Both Henry and Eaton overestimated Henry's foreign influence, but they were not far off the mark. London's widely read *Philosophical Magazine* carried a report on the Yale electromagnet in the October 1831 issue. About the same time in London, an installment on electromagnetism in the popular *Library of Useful Knowledge* included a summary of Henry's Albany and Yale multicoil devices; the summary derived from earlier reports in the more specialized *Journal of the Royal Institution.* In the March 1832 issue of *Philosophical Magazine,* electromagnet pioneer William Sturgeon followed with a detailed description of Henry's multicoil design, proclaiming that it accounts for "a magnetic force which completely eclipses every other in the whole annals of magnetism." The influence of Henry's work was even more widespread at home. Besides hearing from Cleaveland at Bowdoin College, he received requests for further information from a varied group that included a Presbyterian minister, a high school principal, a professor at Amherst

College, and a Kentucky resident "in pursuit of the perpetual motion." When the announcement of the Yale electromagnet appeared in Silliman's *Journal,* Yale lecturer Charles U. Shepard appended in a footnote a summary of an experiment "to learn the chemical effects of this instrument." Successfully constructing and testing their own electromagnets based on Henry's design, and publishing their results in Silliman's *Journal,* were two established chemistry professors: Harvard's John W. Webster and University of Pennsylvania's Robert Hare, well-known inventor of powerful batteries. Henry's counterpart at nearby Union College, professor of mathematics and natural philosophy Benjamin Franklin Joslin, soon followed Webster and Hare in publishing a description of his own modifications of Henry's multiple coil electromagnet; instead of wrapping the iron core with individual loops of wire, he substituted an insulated sheet of copper. By late 1834 a New York instrument maker was building and selling, with Henry's encouragement, a version of his multicoil electromagnet, a "fine affair" mounted in a mahogany frame.[2]

Meanwhile in Europe, another established professor was less successful in recreating Henry's potent electromagnets. He was none other than Gerrit Moll, the physicist at the University of Utrecht in the Netherlands who earlier had scooped Henry by being the first to publish refinements of Sturgeon's electromagnet. He complained of his inability to duplicate Henry's magnets in a letter to Faraday, who by this date, June 1831, had emerged at the Royal Institution in London as an internationally active chemist. "I have been toiling very much these days," Moll wrote, "in endeavouring to repeat the American electro-magnetic experiments, but without success. I could not convince myself that by increasing the number of coils the power of the temporary magnet was increased in the least degree." He then retraced his failed efforts, expressing frustration regarding Henry's recommendations. "I shall be very happy to learn whether you have been more successful in repeating the American experiments," he ended. Five months later Moll had still not heard from Faraday regarding the American experiments, and he asked again. By then, however, Faraday had put to historic use not only Moll and Henry's general style of strong electromagnets but also Henry's particular design with multiple, tightly wrapped, insulated coils.[3]

Faraday had shared with other active investigators the hope of finding the converse of Oersted's famous effect. That is, whereas Oersted had shown that electricity can produce magnetic effects, Faraday had soon joined Augustin Fresnel, Ampère, Auguste De La Rive, and others in seeking to show that magnetism can produce electrical effects. Because Oersted had found that an uninterrupted current in a stationary wire sustained an adjacent magnetized needle in a deflected position, the searchers assumed that a stationary magnet would produce a steady current in an adjacent stationary wire. Fresnel, for example, led the search when, using a helix of wire wrapped around a fixed bar magnet, he attempted in vain to produce a current that he could detect through the subtle test of electrically decomposing water into hydrogen gas and oxygen gas. This test involved a familiar electrochemical process that Faraday labeled "electrolysis"; if a current were present—presumably, a steady current—it would pass through the water via a submerged "cathode" and "anode," producing telltale tiny bubbles of hydrogen at the cathode and oxygen at the anode. At the end of August 1831, more than a decade after Oersted's initial findings and Fresnel's fruitless experiment, Faraday succeeded. He identified what became known as electromagnetic "induction," a term that he extended from the well-known interactions between static electric charges. As the century unfolded, induction turned out to be not only a foundational concept in the physics of electricity and magnetism but also the much acclaimed principle behind the technology of electrical transformers and generators—two mainstays of modern industrialization.[4]

As Faraday revealed in his research diary—and later in a paper he read before the Royal Society of London—his crucial breakthrough came while experimenting with an electromagnet made from an iron bar bent into a ring and then welded together. He wrapped wire transversely around segments of the ring in a distinctive multicoil manner—that is, with separate, tightly wrapped, insulated coils. "Wound many coils of copper wire round one half, the coils being separated by twine and calico—there were 3 lengths of wire each about 24 feet long and they could be connected as one length or used as separate lengths. . . . On the other side but separated by an interval was wound wire in two pieces together amounting to about 60 feet in length, the direction being as with the former coils." Faraday's finished ring was comparable to a horseshoe electromagnet with a curved armature joining its poles; the half of the ring with three coils of wire corresponded to the main horseshoe while the half with two coils corresponded to a rigidly fixed, curved armature.[5]

The ring configuration was uniquely his own, but the general style of the electromagnet, strong and controllable, reflected the influence of Moll's and Henry's innovations.[6] And though Faraday had used tightly wound helical coils of insulated wire in experiments with regular magnets since the early to mid-1820s (the same period when he settled on a ring as a good shape for a regular magnet), he left no record of using multiple coils until his late August experiments on the ring electromagnet.[7] This was either after Moll's letter alerted him to Henry's technique of "increasing the number of coils" or, more likely, after he came across Henry's findings himself. Indeed, Moll probably first learned of the "American electro-magnetic experiments" through the installments that he was receiving of Faraday's own *Journal of the Royal Institution*. This London publication contained, in the inaugural, May 1831 volume (which gathered the earlier installments together), an extended summary of Henry's main article on large, multiple-coil electromagnets; the second or August volume carried an overview of the article on the Yale magnet. The former summary highlighted Henry's innovation of multiplying the number of coils and provided details on how Henry and Ten Eyck apparently produced "the strongest magnet ever constructed." The latter overview similarly specified how Henry and Ten Eyck constructed the even more powerful Yale magnet by boosting the number of coils to twenty-six. Both synopses displayed technical clarity and likely benefited from the editorial hand of Faraday himself, a mainstay of the short-lived *Journal of the Royal Institution* (a colleague sympathized with Faraday about "all the plague & interruption which the journal must be to you"). Whatever his role in preparing the two summaries, Faraday explicitly referred to the Yale electromagnet in his Royal Society paper when he announced his findings on electromagnetic induction. In addition, when he met Henry a few years later, he happened to mention that during the earlier period he also had been familiar with Henry's third main publication, on the reciprocating engine. He commented to Henry that, as soon as the July 1831 issue of Silliman's *Journal* reached England, he had demonstrated a model of Henry's machine at the Royal Institution (where, supposedly, William Ritchie got the idea for his own device). In his specifications for the machine, Henry had described using an electromagnet with three coils.[8]

Although Faraday had more urgent concerns, the articles might even have contributed in a small way to his decision to interrupt a longstanding, time-consuming, Royal Society project on improving optical glass. "I would wish, under present circumstances," he wrote to the Secretary of the Royal Society in early July, "to lay the glass aside for a while, that I may enjoy the pleasure of working out my own thoughts on other subjects." In any case, Henry's writings at least likely lead Faraday to adopt multiple coils and also triggered, or reinforced, his basic strategy to seek the inductive effect using a strong electromagnet. The most influential of Henry's publications might have been the one on the Yale electromagnet or, to be more precise, the extended footnote that Silliman appended to the original article. After Silliman declared Henry's newest electromagnet the strongest in the world, Yale natural history lecturer Charles Shepard briefly described his effort to tease a "chemical effect" from the extraordinary magnet. Shepard was a twenty-seven-year-old specialist in mineralogy who had assisted in Silliman's laboratory since 1827 and had helped set up and test the Yale electromagnet. Though ultimately unsuccessful in observing a "chemical effect," Shepard used the large electromagnet in a perceptive variation of what informed readers would recognize to be Fresnel's abortive attempt to produce a current using a stationary regular magnet—a presumably steady current that Fresnel had tried to detect by decomposing water through electrolysis. "As its magnetic flow was so powerful," Shepard reported about Henry's Yale electromagnet, "I had strong hopes of being able to accomplish the decomposition of water by its means." This note—which called attention to Shepard's search for chemical decomposition and, hence, by implication, a current induced by magnetism—was portentous. It appeared with Henry's article in April 1831, two months before Moll's letter to Faraday and four months before Faraday's historic use of an electromagnet to induce a current. Moreover, the overview of the article that ran in the August volume of Faraday's *Journal* devoted the last of its dozen sentences to Shepard's experiment with the Yale electromagnet: "It was attempted to decompose water by this magnet, but without success."[9]

Faraday could have configured his adaptable, multicoil ring electromagnet as either an intensity or quantity magnet. He opted for the former method. After conducting a preliminary trial in which he connected only one of the three main coils to the battery—he now joined all of the coils into a continuous strand (i.e., combined them in series) and ran the ends to the battery. The battery was a trough device with ten pairs of four-square-inch plates. Later scientists would come to refer to this portion of an electromagnet, involving the basic windings and battery, as the "primary" circuit. Faraday's choice of an intensity magnet for his primary circuit perhaps reflected either the ready availability of the multiplate battery at the Royal Institution or a desire to tap the battery's high intensity (i.e., its high potential difference as measured in volts). On the other half or armature segment of his ring—the "secondary" circuit—he similarly connected the two remaining coils in series and continued the end wires into an external circuit that passed over a magnetized needle three feet away, far enough so as not to be directly swayed by the electromagnet. In the preliminary trial involving only the single coil in the primary circuit, when he either made or broke the connection with the battery, he witnessed a slight, transitory oscillation in the needle positioned under the external wire in the secondary circuit. Straightaway rewiring the primary circuit to convert the three coils into a continuous strand, he reported in his research diary: "Effect on needle much stronger than before." Later, he confirmed this evidence of a current by drawing a small spark in a gap in the external wire during the instant he connected the battery. Reminiscent of Shepard at Yale, however, Faraday failed in a test to find "evidence of chemical action" by using the current to decompose a solution of copper and lead. (For both experimenters, electrolysis probably miscarried as a test because the induced current was simultaneously too evanescent and too weak to produce discernible gas bubbles.)

Worried that he was observing merely a spurious effect, Faraday devised by mid-October what he considered a decisive test. This experiment involved a freestanding, regular bar magnet thrust into a paper cylinder encircled with eight separate helices of insulated copper wire having Henry's preferred type of parallel connection and running to a galvanometer. "The 8 ends of the helices at one end of the cylinder," Faraday recorded in his diary, "were cleaned and fastened together as a bundle. So were the 8 other ends. These compound ends were then connected with the Galvanometer by long copper wires." When he inserted the bar magnet into the array of helices—or withdrew it—the galvanometer's needle moved in a fleeting response. In an aside evocative of Henry's comments, he noted that when he tried connecting the eight helices in series rather than parallel, the galvanometer responded only feebly; that is, he realized that to produce the greatest deflection from the array of coils, "it is *best* in pieces and combined at the end." The multiple coils provided Faraday, just as they had for Henry, a means not only of amplifying feeble effects but also of conveniently alternating configurations of coils to achieve the clearest results.

The experiment with the bar magnet and cylinder confirmed, Faraday concluded in his Royal Society paper, that he had indeed achieved "the production of electricity from ordinary magnetism." He added that, whereas the currents detected in the various experiments were all moderately weak, the use of more powerful electromagnets could strengthen them—strengthen them to the point that they could produce not only "a brighter spark" but also the elusive "chemical action." Specifically, because all the effects could be created using iron-core electromagnets, "there is no doubt that arrangements like the magnets of Professors Moll, Henry, Ten Eyck, and others, in which as many as two thousand pounds have been lifted, may be used for these experiments." Faraday was referring to the record-setting magnet that Henry, with the assistance of Philip Ten Eyck, had built for Silliman at Yale. Later in the paper he referred more simply to the electromagnets "of Moll, Henry, &c." as "very powerful." Although he never directly attributed his own electromagnet to either Moll or Henry—nor its tight wrapping of insulated wire to Henry—he at least cited their achievements. He omitted even mentioning, however, Henry's contribution to the multicoil design—a design that Faraday seemingly incorporated in his pace-setting investigation.[10] Faraday could possibly have isolated electromagnetic induction without using multiple coils; after all, in his preliminary trial with his electromagnet, he achieved a weak effect using only a single coil in the primary circuit. At the least, however, the multicoil technique aided his investigation by accentuating effects: first, in his main experiment when he produced a strong intensity electromagnet by rewiring the three coils into a continuous strand and, second, in his decisive follow-up experiment when he surrounded the regular bar magnet with a sensitive array of eight coils connected in parallel. At the most, Shepard's perceptive but disappointing chemical application of Henry's promising multicoil electromagnet rekindled Faraday's search for a magnetically induced current.

Indeed, before turning to electromagnets and multiple coils, Faraday had repeatedly failed to coax electricity from magnetism. As early as the mid-1820s, in a notebook that he labeled "Chemical Notes," he recorded an experiment with a regular magnet wrapped with a single copper coil connected to a galvanometer. Apparently, he was seeking a current resulting simply from the proximity of the magnet and coil, both immobile relative to each other. He was in good company. As mentioned earlier, Fresnel, Ampère, and others had investigated similar, static arrangements of bar magnets and coils following Oersted's suggestive findings; they were speculating that a conventional magnet would induce a steady current in analogy to the constant magnetic effect that Oersted had elicited from a current flowing evenly through an immobile wire. A synopsis of a later talk by Faraday at the Royal Institution, after he had made his breakthrough, characterizes this unsuccessful experiment with a stationary magnet and coil as a classic one—an experiment that had been attempted "by various persons hundreds of times, in the hope of evolving electricity from magnetism." Evidently, no more than a few months before identifying electromagnetic induction, he was reexamining the "Chemical Notes" containing his version of the experiment from the mid-1820s.[11] Perhaps, with the earlier attempt in mind, he realized the significance of Henry's new multicoil electromagnets and Shepard's recent initiative with one of the powerful instruments. The upshot was that Faraday soon fashioned two ways for updating the classic experiment. The first involved brute force—the greater strength of a three-layered intensity electromagnet—and led to his germinal experiment with the ring electromagnet. The second involved finesse—the greater sensitivity of multiple coils connected in parallel to a galvanometer—and led to his corroborating experiment with the bar magnet.

Similar to Shepard capitalizing on the Yale electromagnet to try to decompose water, Faraday was drawn to an electromagnet of Moll's and particularly Henry's style probably because of its strength.[12] The instrument allowed him to repeat the classic bar-magnet experiment, but now he could substitute the stronger armature segment of his ring electromagnet for the weaker, regular bar magnet of earlier times. As he had done years before with the bar magnet, he wrapped a coil around the electromagnet's armature so that the coil and armature were stationary relative to each other. It just so happened, however, that Moll's and Henry's new type of magnet (along with its armature) had the extraordinary property of surging, almost instantaneously, into activity following connection of its primary circuit to a battery. And this switching on (and off) of the electromagnet—necessitated by the instrument's dependence on a chemical battery—turned out to be the previously missing key to the creation of electricity in the detecting coil. Faraday's scientific descendants would come to speak of a "magnetic field," saying that only a *changing* field induces a current. For the induced current to flow in the electromagnet's secondary circuit and for Faraday to notice it, he had to have preconnected the armature's coil to a detecting device and be in a position to watch it. He had, in fact, completed the connection before activating the battery, most likely in imitation of the mid-1820s arrangement of a bar magnet and other fixed components but also perhaps to avoid premature depletion of the multiplate battery (chemical batteries being notoriously short-lived). Moreover, in line with his earlier, abortive attempt, he positioned the components in close enough proximity that he could observe the magnetized needle while activating the battery.

After first detecting a current using a comparatively strong intensity electromagnet—and after a few intermediate experiments—Faraday returned to the arrangement of a coil around a weaker regular magnet. Now, however, he realized that he needed to look for spurts of current during the insertion and withdrawal of the bar magnet—the physical action corresponding to switching the electromagnet on and off. Moreover, apparently following Henry's lead, he knew to use an array of eight detecting coils connected in parallel to a galvanometer; this distinctive arrangement enabled him to discern more easily the weaker transient current in an experiment involving only a feebler regular magnet. And though it used merely a traditional bar magnet, this was not a minor, follow-up experiment; rather, it was a foundational experiment that had frustrated "various persons hundreds of times" and that Faraday now considered a decisive confirmation of the inductive effect. (It is interesting to speculate that if the investigators over the previous decade had possessed equally responsive apparatus, they might have detected the feeble, transient current accidentally—by, for example, happening to jar the normally stationary bar magnet. This conjecture assumes, of course, that the unsuspecting investigators would not have unthinkingly dismissed the fleeting current as spurious.)

In sum, Faraday's isolation of induction depended heavily on Moll's and Henry's enhanced electromagnets—especially Henry's, with their tightly wrapped, insulated coils. His success also depended, in particular, on Henry's versatile and sensitive multiple-coil design.

That Faraday did not explicitly acknowledge any debt to Henry for the multiple coil design typified the responses of many Europeans to Henry's findings on electromagnets reported in his three articles in Silliman's *Journal* and excerpted in various European publications. In Paris, Claude Pouillet built on Faraday's announcements and, in the process, incorporated Henry's (and Moll's) electromagnet design without explicitly acknowledging the source. Demonstrating a large electromagnet and its inductive capabilities for the first time at the comparatively late period of June 1832, Pouillet nevertheless continued through later years in various editions of his textbook to lay claim to the originality of his electromagnets and date them, probably inadvertently, to 1831. The Europeans' pattern was simply to absorb Henry's enhancements and adaptations into everyday practices. London's William Ritchie, for example, gave more formal recognition to Henry's electromagnet than many Europeans, when in an 1833 article he passingly referred to "an electromagnet according to the American method." In 1841 Henry accurately characterized this tendency in a complaint to Swiss colleague Auguste De La Rive, who unwittingly credited the invention of the first electromagnetic machine to Ritchie rather than Henry. "The affair is not of much consequence," he assured De La Rive, "but we find in this country that unless we are somewhat tenacious in relation to our rights we are liable to have our discoveries and inventions claimed by others owing to our distance and the want (before the late triumph of steam) of rapid communication." Ironically, because Americans depended primarily on European publications for information, Henry's own compatriots sometimes repeated De La Rive's type of misattribution. Thus, in 1838, when electrical inventor Charles Page publicly credited the first electromagnetic machine to Sturgeon, Henry quickly dispatched a letter to him saying, "I have all ways supposed myself to be the first in this invention," and asked Page if he had information that would show that someone else had been first.[13]

Particularly after Henry's 1837 trip abroad, however, British scientists such as Sturgeon, Brewster, and James Prescott Joule were citing Henry's contributions to electromagnet design and application. Even a respected, instrument maker whom Henry met in London, writing in the influential *Philosophical Magazine,* assigned full priority to Henry's machine. Also, after the trip, Henry came to realize that the previous failures to cite his work were not personal slights but manifestations of a general disinterest in what most Europeans perceived to be inconsequential American efforts. In Ritchie's case, Henry later noted: "The neglect was probably due more to the feeling which then existed in England in regard to the character of American scientists than from want of courtesy to myself since I was indebted to him for much kind attention during my residence in London in 1837."[14]

Faraday did not knowingly slight Henry. He honestly felt that the crucial elements of the discovery of induction were his alone; this became evident in a misunderstanding involving two Italian researchers about priority in the discovery. In mid-December 1831, before his findings appeared in print, Faraday had sent a letter to a colleague in Paris summarizing his investigation. Working from this sketch, Leopoldo Nobili and Vincenzo Antinori extended Faraday's work and published their own results in a paper dated January 1832—but in a journal that happened to be backdated to a time before Faraday's announcement. Though the Italians acknowledged their debt to Faraday, an inattentive editor of a popular British journal and others inadvertently ascribed priority to them. This led Faraday to publish a sharply worded statement in March 1832 clarifying the misunderstanding and confirming his own priority. He attested: "I never took more pains to be quite independent of other persons than in the present investigation; and I have never been more annoyed about any paper than the present by the variety of circumstances which have arisen seeming to imply that I had been anticipated." On another front, but one that was less combative, he also felt compelled to defend his priority over Ampère. On learning of Faraday's results, Ampère used hindsight to suggest that he and De La Rive had anticipated electromagnetic induction. Edinburgh professor James D. Forbes commiserated with the beleaguered Faraday, saying of Ampère's retroactive claim that "it is wonderfully easy to connect crude & unintelligible fragments when another has furnished the key." "Do not doubt my dear Sir," he added, "that you will get all the credit due to you for your fine experiments on this subject[.] All the nibbling of the Français will not do you much harm."[15] Convinced that he deserved full credit for his experiments on induction—and probably assuming that Henry's multicoil electromagnet design was part of common knowledge—Faraday seemingly never imagined that he needed to acknowledge explicitly Henry's contribution to the experiments.

Faraday continued with his initial experiments through the fall, eventually recording in his diary, "R. S. Paper Read 24 Nov. 1831." Eight days before the Royal Society of London heard the paper, Henry wrote to Cleaveland at Bowdoin College: "I have lately had forged a large horseshoe weighing 101 lbs which I intend fitting up for some contemplated experiments on the identity of electricity and magnetism." When he published a related set of results the following July in Silliman's *Journal,* it became clear what he meant by "identity." The "contemplated experiments" were to show the production of electricity from magnetism, the long-sought complement of Oersted's discovery that Henry had no way of knowing Faraday was about to announce; by finally producing this complementary effect, Henry felt that he would be establishing that electricity and magnetism indeed display a "reciprocal action" if not an "absolute identity." In the July article's introduction, he recounted that he had been working on the project since at least the late summer of the preceding year. This was three months before mentioning it to Cleaveland but four months after the publication, in his own article on the Yale magnet, of Shepard's footnote calling attention to the "powerful" magnet's potential for evoking "chemical effects." Unlike Faraday, Henry left no record of an earlier interest in personally probing the possible chemical or electrical manifestations of magnetism. He also left no mention of Shepard's influence. In his July article, however, Henry described the genesis of his experiments:

It early occurred to me, that if galvanic magnets on my plan were substituted for ordinary magnets, in researches of this kind, more success might be expected. Besides their great power, these magnets possess other properties, which render them important instruments in the hands of the experimenter; their polarity can be instantaneously reversed, and their magnetism suddenly destroyed or called into full action, according as the occasion may require. With this in view, I commenced, last August, the construction of a much larger galvanic magnet than, to my knowledge, had before been attempted, and also made preparations for a series of experiments with it on a large scale, in reference to the production of electricity from magnetism.

A decade later, while lecturing to his students at Princeton, he recalled that simple reasoning led him to turn to a large electromagnet. In particular, because he knew that a "small portion of electrical effect produced great magnetic effect," then it followed that a "great magnetic effect or power would produce small electrical effect."[16] And though he left no record of thinking it, perhaps Shepard's failure to exhibit the electrical effect with the Yale magnet, already the world's largest electromagnet, convinced Henry that he needed to try an even larger magnet, the projected, prodigious, 101-pound device.

With the return of students for the Albany Academy's fall session in 1831, however, Henry suspended construction of the huge magnet. He still had not completed it by the time he wrote to Cleaveland. And when Silliman asked in March 1832 if he had any results to publish, Henry replied that the project was on hold:

In answer to your inquiries relative to my farther experiments in magnetism I have little definite to communicate at the present time. I have not abandoned entirely the subject but have been prevented by circumstances from prosecuting a series of experiments which were commenced on a very extensive plan. I had partially finished a magnet much larger than any before made and had constructed a kind of a reel on which more than a mile of copper wire was wound. I was obliged to abandon the experiments on account of the room in which my apparatus was erected being wanted for the use of the Academy and it has not been convenient to resume them during the winter.

All he could report to Silliman was completion of a magnet for Cleaveland and work on the extraction of iron from crushed ore. He had never consummated his initial round of "contemplated experiments on the identity of electricity and magnetism." In fact, he would not be able to return to the experiments until the closing weeks of June 1832, and then he would have to resort to using an earlier, smaller electromagnet. As he expressed it in his July article in Silliman's *Journal,* he had been "accidentally interrupted in the prosecution of these experiments" and was unable to resume the investigation "until within the last few weeks, and then on a much smaller scale than was at first intended."[17]

When he did come back to the project, however, he worked with dispatch. At the end of June, he explained to a colleague that he had been delinquent in answering a letter because of the push to complete the project; he had been "engaged in a series of experiments which I wished to announce in the next no. of the American Journal and which has occupied almost every moment of my time, not required in the duties of the Academy for the last two weeks."[18] After so many months of inaction, why did he race to make the July issue of Silliman's *Journal?* He gave the answer as he continued the introduction to his article on electricity's creation from magnetism: "In the mean time, it has been announced . . . that the result so much sought after has at length been found by Mr. Faraday of the Royal Institution."

Henry stressed that the first announcement—the first notice that he saw of Faraday's achievement—was cursory and indefinite. The blurb appeared as an addendum to an installment on electromagnetism in the London series the *Library of Useful Knowledge*—the same installment, completed in December 1831, whose main text included a summary of Henry's Albany and Yale multicoil electromagnets. The brief announcement neglected the basic elements of Faraday's Royal Society paper in favor of sketching his use of the inductive principle to reinterpret a previously puzzling effect. According to the blurb, to explain why a revolving copper disk deflected a magnetized needle suspended over it, Faraday invoked "a most important principle": when a "piece of metal" and a "magnetic pole" are in "relative motion," electric currents traveling perpendicular to the motion are "induced" in the metal. Not until the April issue of *Philosophical Magazine* (still referred to by Henry as the *Annals of Philosophy*) had completed its journey across the Atlantic and up the Hudson did he locate a slightly fuller account. The account was a two-paragraph summary of comments that the British scientist made at the Royal Institution in February. Distinguishing between "Volta-electric induction" and "magneto-electric induction," it devoted the first paragraph to the former effect: Faraday's ancillary finding that a changing galvanic current in one wire momentarily induces a current in a nearby, parallel wire. In what was for Henry the more critical paragraph on "magneto-electric induction," the account did not mention Faraday's inceptive experiment with an electromagnet; rather, it offered only a succinct description of his follow-up investigation involving the insertion and withdrawal of a regular bar magnet into a helix of wire connected to a galvanometer. The account described this as a prototypical configuration of a magnet, helix, and galvanometer that had frustrated many previous investigators because they failed to realize the necessity of putting the magnet in motion.[19]

From the account of Faraday having used only a regular bar magnet, Henry apparently took hope that his approach involving an electromagnet was novel and still might interest the scientific community. Again, a British researcher had anticipated him, but this time in a major finding; again, he was scurrying to salvage credit for at least his own twist on the finding. "Before having any knowledge of the method given in the above account," he explained, "I had succeeded in producing electrical effects in the following manner, which differs from that employed by Mr. Faraday, and which appears to me to develop some new and interesting facts." Henry apparently meant that, before he saw the more detailed summary in the *Philosophical Magazine*—but after he had noticed the blurb in the *Library of Useful Knowledge*—he had succeeded with his alternate approach involving an electromagnet. Said differently, the more detailed summary must have come to his attention in the midst of his research during the last half of June—intensified research that he had begun following the *Library* announcement.

Henry induced his electrical current using the versatile electromagnet described in his first article in Silliman's *Journal*—the horseshoe with nine switchable coils that could support more than 700 pounds. Apparently, he arranged the device to form a strong quantity magnet, connecting the nine coils in parallel to a small galvanic cell (which produced a sizable current). He modified the original instrument by tightly winding about 30 feet of insulated copper wire around the 7-pound iron armature and rigidly attaching the armature to the ends of the 21-pound horseshoe. Next, he connected this new coil to a galvanometer, which he positioned about 40 feet away to guard against the electromagnet inadvertently influencing the meter's needle. Finally, at the instant an assistant activated the battery, Henry observed the galvanometer's needle swing in "a single impulse"; later, when the assistant disconnected the battery, he was "much surprised" when the needle again deflected. He concluded that he had witnessed the conversion of magnetism into electricity. Unknowingly, he had come close to duplicating Faraday's first experiment. The only significant difference was that Faraday configured his multicoil electromagnet to create an intensity magnet, whereas Henry opted for a quantity magnet; both, however, were strong magnets. Also, Faraday tethered his primary and secondary circuits to two halves of a ring electromagnet while Henry tied them to a horseshoe yoked with a fixed armature. Finally, Henry placed his detector at a much greater distance from the battery and magnet, and he had the good fortune—or foresight, perhaps fostered by the announcement in the *Library of Useful Knowledge*—to station himself near the galvanometer at the precise moment an assistant energized the battery.

In his article in Silliman's *Journal,* Henry reported that after seeing the notice in *Philosophical Magazine* of Faraday's findings using a regular bar magnet, he expanded his own investigation beyond his basic experiment with the horseshoe electromagnet. Still unaware of Faraday's initiative with a ring electromagnet, he now moved even closer to duplicating the ring geometry. Feeling his way through various experiments, he attempted to increase the induced current in the secondary circuit by returning to his trusted technique of using multiple coils of wire connected in parallel; he added the extra coils to new rods linking the armature to the ends of the horseshoe and joined the coils in parallel with the galvanometer. Following these promising tests, he happened on a slightly different arrangement that produced "the maximum of electric development from a given magnetic power." In this optimum arrangement, "the poles of the magnet were connected by a single rod of iron, bent into the form of a horse-shoe . . . ; around the middle of the arch of this horse-shoe, two strands of copper wire were tightly coiled one over the other." That is, by mating a horseshoe electromagnet to a horseshoe armature, he had molded two U-shaped pieces of iron into an elongated ring. Indeed, he essentially had recreated Faraday's distinctive geometry for the instrument. He also echoed Faraday's general results by pointing out that two coils of wire connected in parallel to the galvanometer were much more effective than a single coil in detecting an induced current in the secondary circuit. When he attached both coils, the resulting surge of electricity was so strong that the galvanometer needle spun all the way around. But what most astounded Henry—and further aligned him with Faraday—was that he could draw "a small but vivid spark" when he opened a gap in the external wire of the secondary circuit. Without knowing it, Henry had duplicated all the essential features of Faraday's initial experiment—a ring electromagnet, multiple coils, and induced currents and sparks.

The creation of sparks—direct, visual evidence of electricity—figured prominently in Henry's thinking. As he moved toward publishing his findings on electromagnetic induction, he seemed to take solace that his London rival presumably had not detected sparks. In a letter to a colleague on the last day of June, as he wrapped up his flurry of experiments, he underscored his success with sparks rather than with induction itself. "I have lately succeeded in a most interesting experiment," he wrote, "that of drawing sparks of electricity from a magnet. I have some hopes of fusing platina wire by means of the same principle."[20] In the eventual article, he labeled the spark "the most surprising effect."

Presumably pinning his hopes for recognition on the spark, it must have come as a disappointment when Silliman, upon receiving Henry's letter about his pending publication, wrote back on the last day of June that another British investigator had beat him to the finding. Silliman cited James Forbes's preliminary announcement that he had recently generated a spark using a regular magnet. Moreover, the University of Edinburgh natural philosopher described his achievement as an extension of Faraday's discovery and mentioned having heard that an Italian investigator (Nobili working with Antinori) had also detected a spark. Having already included a disclaimer in his paper acknowledging Faraday's priority in announcing the achievement of electromagnetic induction, Henry now was forced to add a last-minute disclaimer recognizing Forbes's priority on the spark. "It appears from the May No. of the *Annals of Philosophy,*" he wrote in the final paragraph of the main body of his article, "that I have been anticipated in this experiment of drawing sparks from the magnet by Mr. James D. Forbes of Edinburgh, who obtained a spark on the 30th of March; my experiments being made during the last two weeks of June." Noting that the Forbes announcement contained few details, he added that "my result is therefore entirely independent of his and was undoubtedly obtained by a different process."[21]

When Henry received the July issue of Silliman's *Journal* containing his article, he perhaps was further chagrined to discover that he was not even the only American claiming to have detected the spark by a different process. Though his hurried article appeared again as an appendix at the end of the issue, another late arriving report followed his. Joseph Saxton, a Philadelphia inventor who had recently relocated to London, submitted letters dated 14 April and 11 May that recounted his successful attempts using a regular horseshoe magnet to reproduce "Faraday's curious discovery."[22] Henry was left with only the hope that his use of an electromagnet set him apart from Faraday and the other experimenters.

6 Staking a Claim

By dating his electromagnetic induction experiments to the closing weeks of June 1832, Henry matter-of-factly acknowledged that his actual, tangible results trailed Faraday's announcement of similar findings.[1] All the same, Henry maintained that he had been working toward the experiments since at least the previous August—a contention that his letters support. Also, not suspecting the full scope of Faraday's investigation, Henry believed that his June results using an electromagnet provided a distinctive complement to Faraday's findings using a bar magnet. Whatever the exact sequence of their particular thoughts and actions, however, it remains true that Faraday and Henry—independently and about the same time—used strong electromagnets to isolate electromagnetic induction. How did it happen that individual researchers in London and Albany achieved the same breakthrough in the short period from mid-1831 to mid-1832?

To answer this question, it is useful to distinguish between scientific theory, experiment, and instrumentation, though the three are closely if not inextricably interwoven in actual scientific practice. On first glance, Faraday and Henry's duplicate findings might appear to result merely from shared theoretical orientations—that is, separate scientists working concurrently but unilaterally on the same idea or within the same conceptual framework. In particular, both men were avowedly seeking a supposed complement to Oersted's effect. This goal of using magnetism to produce electricity stemmed from a longstanding theoretical opinion in the research community on the close relationship between the two domains. As their papers on induction show, Faraday and Henry were particularly sympathetic to Ampère's fashionable theory that magnetism was due to underlying, circular electrical currents. Both men initially alluded to this prevailing theory to justify a connection between magnetism and electricity. Faraday later returned to the theory's main points as he wrestled to interpret the peculiarities of induction and Henry subsequently pressed "the beautiful Theory of Ampère" as he lectured on electromagnetism at Princeton. Curiously, their mutual interest in a complement to Oersted's finding and in Ampère's theory would likely have led the two researchers to seek a steady rather than transient current in a wire near a magnet. This is because, in analogy with Oersted's finding that an uninterrupted current in a wire sustains a nearby magnetized needle in a deflected position, the two men would have assumed that a magnet would produce a steady current in a neighboring wire. Similarly, Ampère's theory of underlying, continuous currents would likely have led them to seek a steady rather than transient current in a wire near a magnet.[2]

Closer scrutiny of their theoretical orientations, moreover, shows key divergences in their outlooks. Except for sharing a broad commitment to finding a complement to Oersted's effect and a general affinity for Ampère's theory, each man favored a different theoretical model of the detailed operation of electricity and magnetism. These disparate models might just as likely have led the two practitioners in divergent rather than parallel research paths. On the one hand, Faraday, though hesitant to discuss theory publicly, toyed with the hypothesis of the wave character of electricity. Indeed, this speculation—along with speculations related to his other areas of interest such as acoustics—may have predisposed him to look for a transient rather than a steady inductive effect. Regardless, after recognizing the effect, he preferred to package his views on the new phenomenon in the language of "electro-tonic states"—an original conjecture about underlying particles that, unlike Ampère's basic theory, accounted for the fleeting nature of the inductive current. On the other hand, Henry, in his initial probe of induction, seemingly avoided finer theoretical details altogether. In his earlier work on electromagnets, however, he had opted for an electrical hypothesis with North American roots. Specifically, he favored a hypothesis based on Hare's projectile current, a type of imponderable fluid akin to Benjamin Franklin's "electrical fluid"; according to Franklin, an excess or deficiency of this single, invisible fluid resulted in a normally neutral body becoming charged "positively" or "negatively."[3]

With Faraday and Henry's commitment to the same general, but not specific, theoretical framework adding only a degree of plausibility to their shared achievement, the question remains: How did they arrive at the same result in such a short period? Distinguishing between the realms of theory and experiment, it is evident that not only were Faraday and Henry working in the same broad theoretical context but also they were initially attempting similar experiments. Likely through the common catalyst of Shepard's investigation with the Yale electromagnet, both researchers began their inquiries by trying to elicit currents in comparable ways. That is, Henry likely decided to look for a complement to Oersted's effect, and Faraday at least bolstered his earlier commitment to the search, through common exposure to Shepard's prescient note on attempting to decompose water with Henry's strong Yale electromagnet. But again, in line with Shepard's experiment and those of his predecessors, Faraday and Henry probably would have been conducting their initial experiments with the expectation of exhibiting a steady rather than transient current. In light of this expectation, therefore, even the realization that the two men were initially attempting similar experiments does not fully explain their shared achievement. An even more subtle distinction needs to be drawn: between experiment and instrumentation. An experiment can be thought of as subsuming the research instrument or apparatus itself as well as a complex web of spoken and unspoken concepts and guidelines for making and interpreting observations. In turn, the detecting instrument or apparatus entails not only physical hardware but also an associated set of explicit or implicit techniques or skills for arranging or manipulating the hardware. Though Faraday and Henry executed analogous experiments in pursuit of the same general theoretical goal, their mutual success reflects primarily a shared *instrument:* the multiple coil electromagnet that Henry pioneered; that Silliman, Shepard, Cleaveland, and other Americans immediately took up; that Moll tried to recreate; and that Faraday replicated in mid to late 1831. Faraday and Henry shared the instrument in that they relied not only on similar assemblies of iron cores, wires, insulation, chemical batteries, and other materials but also on corresponding procedures and perceptions for managing the device.[4]

Specifically, what the two men had in common was a strong magnet possessing an inherent physical trait not present in traditional, regular magnets: magnetism that sharply rose and fell in response to activating or deactivating a chemical battery. And change in magnetic strength provided the long-sought key to inducing currents. In Faraday's case, although particular theoretical notions may have predisposed him to look for a transient rather than a steady inductive effect, it seems more plausible that his detection of the fleeting current was fortuitous rather than the outcome of a conscious search for a transient effect. Similarly, Henry does not seem to have been intentionally seeking a transient effect. For both investigators, their detection of the fugitive current was inadvertent to the extent that it resulted from the electromagnet's distinctive property of relying on a chemical battery—a depletable device that needed to be activated or deactivated. Thus, their near simultaneous detection of electromagnetic induction was primarily instrument driven. Working with theories and experiments predicated for the most part on the assumption that a steady magnetic presence would produce a steady current, they succeeded in spite of their theories and experiments.

That they built nearly identical electromagnets and manipulated them in analogous ways reflects more than keeping abreast of current developments reported in the main journals, including even Shepard's insightful remarks in the April issue of Silliman's *American Journal of Science.* In fact, the likeness of their electromagnets suggests a personal cross-fertilization of instrument designs and uses. On the one hand, around 1827, Henry cut his teeth on Faraday's anonymous "Historical Sketch." This comprehensive article featured the laboratory instruments and techniques that investigators were using in the new field of electromagnetism, including a mention of Schweigger's multiplier. He also encountered Faraday's views in the excerpts from the Londoner's 1827 student manual on performing chemical experiments that Green included in his 1827 guidebook, in which Green also likely borrowed from Faraday's "Historical Sketch." On the other hand and more obviously, in mid to late 1831, Faraday drew inspiration for strong electromagnets and their operation from the plans of Henry and Moll; he also seemingly patterned his multicoil arrangement, and his handling of the arrangement, on Henry's design as detailed in his three articles. Thus, given the awareness of each other's general research endeavors, their near simultaneous breakthroughs were independent only in the narrow sense that neither man was aware of the other's specific efforts in using a strong electromagnet to detect an induced current.

Although Faraday and Henry shared an instrument and used it in similar ways to isolate the same effect, they diverged in their perspectives toward the device and its product. The division was not sharp, but Faraday tended to view the electromagnet more as a research tool and Henry thought of it more as a pedagogic prop. These contrasting perceptions mirror the contrasting backgrounds and experiences of the two men. Both had humble origins and erratic early educations, but Faraday, who was six years older than Henry, enjoyed an expeditious entry into the mainstream of London science. In 1812 Sir Humphry Davy at the Royal Institution enlisted him as an assistant after the eminent chemist briefly blinded himself. Founded in 1799 by American expatriate Benjamin Thompson (better known as Count Rumford), the Royal Institution of Great Britain provided a scientific forum and sponsored its own research while offering popular lectures with the intent of "diffusing the knowledge." Faraday certainly devoted much attention to the institution's public offerings—and to any scientific instrument's pedagogic potential. Working under Davy, however, he became well connected to the leading scientific practitioners particularly in Great Britain and on the Continent, and he enjoyed easy access to established scientific societies and publications. Becoming director of the laboratory of the institution in 1825, he had the means, time, and mind-set to initiate a research program in electromagnetic induction, carry it to fruition, and report his findings expeditiously. That is, as he expressed it in one of his favorite maxims, he was committed to the regimen of "work, finish, publish."[5]

In contrast, Henry trained under Beck at the Albany Academy and found his main scientific interactions through the Albany Institute—two comparatively provincial organizations whose members endorsed an image of science that tended toward popular education or public demonstration. Indeed, during his early years as an academy professor making presentations to the institute, he emphasized display rather than inquiry, exhibiting eyecatching devices ranging from a "Thaumatrope" (a "newly invented philosophical Instrument" that used spinning pictures to simulate motion) to a "Whirling Table" (modified "to show a class of interesting experiments on the whirling of bodies suspended by a string").[6] As a professor he had little free time, scanty resources, minimal technical support, and spotty outside encouragement to do original research. Accustomed to the values and practices of his colleagues at the academy and the institute, he lacked external incentive to sustain a run of experiments and write up the results for Silliman's *Journal,* his only viable outlet for reasonably wide publication. He was only beginning to think and act in line with the broader American scientific community represented by Silliman, let alone in line with the European community. Just as his large electromagnets, telegraph, and reciprocating engine were to some degree "philosophical toys," so too was the new arrangement for demonstrating the induction of currents and sparks. Of course, this pedagogic orientation also had benefits; behind all his breakthroughs in electromagnetism lay his original motivation to build larger instruments dependent on smaller batteries—a set of instruments suitable for the lecture hall.

Henry's pedagogic mind-set was not the sole reason for his delay in completing the 101-pound magnet, investigating the "identity" of electricity and magnetism, and publishing his findings. He also faced the inherent difficulty of recognizing which of the subtle currents and transient sparks that he was encountering were scientifically significant—that is, of interest to the broader community or relevant to existing research agendas. In the unsettled state of early nineteenth-century science, electrical and magnetic effects often appeared intractable or confusing. For example, after Faraday announced his results, Ampère was chagrined to realize that in an experiment with De La Rive a decade earlier he had achieved a form of electromagnetic induction but failed to appreciate its full import. Even Faraday in the early stage of his work on induction doubted that he had hit on a meaningful result. A month before presenting his findings to the Royal Society, he wrote to the friend and editor who a decade earlier had solicited his "Historical Sketch of Electromagnetism" for the *Annals of Philosophy:* "I am busy just now again on Electro-Magnetism and think I have got hold of a good thing but can't say; it may be a weed instead of a fish that after all my labour I may at last pull up."[7] With less experience than Faraday in the broader community of scientists and fewer opportunities for interchanging views and obtaining reactions, Henry probably experienced even greater difficulty in discerning significant effects. That is, it was not only his penchant for building philosophic toys but also his difficulty in distinguishing fish from weeds that probably hampered his investigation of electromagnetic induction.

But while angling for different currents and sparks during 1831 and 1832, Henry did land one new fish. At the end of his article on producing electricity from magnetism, in a paragraph-long addendum, he distinguished what became known as self-induction. "I have made several other experiments in relation to the same subject," he began, "but which more important duties will not permit me to verify in time for this paper. I may however mention one fact which I have not seen noticed in any work, and which appears to me to belong to the same class of phenomena as those before described." What he noticed was that he could produce a "vivid spark" by disconnecting one of the ends of a wire running between the terminals of a battery. Moreover, he realized that he could strengthen the spark not only by increasing the intensity of the battery but also by substituting a longer wire or, best of all, a wire coiled into a helix. "I can account for these phenomena," he concluded, "only by supposing the long wire to become charged with electricity, which by its re-action on itself projects a spark when the connection is broken." His supposition turned out to be correct.

Scientists later agreed that Henry had isolated an elusive but fundamental instance of electromagnetic induction. In particular, breaking the connection causes the current in the helix to wane and thus also its associated magnetic field; this changing magnetic field induces in the same helix a momentary current flowing opposite to the original, dying current. Said differently, the fading magnetic field produces a potential difference (voltage) across the helix that partially offsets any further decrease in the original current. Thus, a net potential difference briefly appears between the end of the wire and the battery terminal resulting in a spark. Scientists came to refer to this new effect as *self*-induction, which they distinguished from Faraday's and Henry's previous discovery of *mutual* induction. Self-induction involves a single circuit, whereas mutual induction involves two circuits such as the primary and secondary circuits in Faraday's and Henry's original magnets.

As the notebooks of his Princeton students reveal, Henry repeatedly told his classes through the 1840s that he stumbled on the self-inductive effect through his reciprocating engine and, afterward, duplicated the effect with "a long wire around a room."[8] That is, he first began to pay attention to the peculiar sparks while operating his "Electro-magnetic Engine," the Albany invention that he described in the July 1831 issue of Silliman's *Journal*. The sparks would be evident when the engine's two sets of wires alternately and repeatedly broke their contacts with the two batteries. Alert to the sparks, he then further observed them in long wires—specifically, while arranging his long-distance telegraph at the Albany Academy. The telegraph's lengthy wire and tight coil (on the terminal electromagnet) would have enhanced production of the telltale sparks.

Sometime before completing his addendum to his mid-1832 article, Henry ran rudimentary tests in which he compared sparks from wires of different lengths, thicknesses, and helical shapes. He likely began to suspect the significance of the sparks, however, only when he began to grasp the subtleties of electromagnetic induction. This seemingly occurred following Faraday's portentous announcement of "magneto-electric induction" and "Volta-electric induction"—an announcement that Henry quoted in his mid-1832 article. Thus, in the addendum to his article, Henry tentatively associated the new sparking effect with induction—with, that is, "the same class of phenomena as those before described." What he evidently had in mind was not "magneto-electric induction," the main topic of the article, but "Volta-electric induction"—Faraday's ancillary finding that a changing galvanic current in one wire momentarily induces a current in a nearby, parallel wire. As detailed in the announcement that Henry quoted in his mid-1832 article, the induced current in the nearby wire flows in the opposite direction to the principal current in the main wire (with the directions reversing as the principal current first rises and then falls in response to the battery being connected or disconnected). Apparently adapting this finding from the case of two wires to one wire, Henry speculated that when a single wire is disconnected from a battery, a spark results because the electricity in the wire momentarily undergoes a "re-action on itself."

For Henry, Forbes's announcement of generating a mutual-inductive spark from a regular magnet had added insult to the injury inflicted by Faraday's notice of induction. These two setbacks compounded the "mortification" of earlier being anticipated by Moll on large electromagnets. Now more prudent, Henry tacked the extra paragraph to his mid-1832 article staking a claim for self-induction. In an ironic reversal of roles, however, in late 1834, Faraday independently distinguished self-induction. True to form, he quickly published a brief summary of his findings. Seemingly, he was unaware of or had forgotten Henry's earlier but inconspicuous announcement (the wide-ranging chemist and natural philosopher, after isolating mutual induction, had soon shifted his interest to other topics). A short time later, about mid-December 1834, Henry encountered Faraday's notice. Apparently apprehensive that he would lose credit for even a finding that he had disclosed promptly and publicly, he immediately took countermeasures to protect his claim of priority. By then a professor at Princeton, he had recently won election to the venerable American Philosophical Society (APS), Benjamin Franklin's establishment in nearby Philadelphia, and was scheduled to present a paper there on a new galvanic battery. Spotting Faraday's announcement, he promptly arranged to supplement his scheduled paper on the battery with an update of his research on self-induction—research that he had, coincidentally, resurrected briefly a few months earlier, in the late summer. Writing to an influential friend and University of Pennsylvania professor who was affiliated with the APS, Alexander Dallas Bache, he explained about Faraday's duplication of his Albany finding and added: "I am now anxious to publish my observations on this point as soon as possible and do not know a more ready method than to append them or to incorporate them with my description of the battery. Fortunately I published a notice of the fact soon after I discovered it. . . . Mr F has probably not seen my paper or has forgotten the fact."[9]

Bache, a savvy and well-connected booster of the American scientific enterprise, responded with zeal. After Henry spoke to the APS in January 1835, Bache sent an extract of his friend's key points on self-induction along with a cover note to the *Journal of the Franklin Institute,* in Philadelphia, and the *American Journal of Science.* In the cover note, intended to accompany the extract in print, Bache urged immediate publication of the extract even though Henry had since prepared a longer, formal memoir on the topic. "Mr. Faraday having recently entered upon a similar train of observations," Bache observed, "the immediate publication of the accompanying is important, that the prior claims of our fellow-countryman may not be overlooked." In the formal memoir, submitted to the APS and later issued in their *Transactions,* Henry described Faraday's communication and quoted his own, earlier, 1832 announcement; he also remarked that since leaving Albany he had not had time to expand his inquiry until during the past year, 1834, when he added some new investigations. "These, though not as complete as I could wish, are now presented to the Society," he remarked, "with the belief that they will be interesting at this time on account of the recent publication of Mr. Faraday on the subject."[10]

Soon after submitting the memoir, Henry wrote to Bache for reassurance on his handling of the delicate issue of priority. "Are my allusions to Mr Faraday in good taste and proper?" he asked. He also expressed regret that members of the APS, in response to his account but contrary to his intention, took umbrage presumably at Faraday's seeming usurpation of Henry's discovery. And a few months later, after he had seen the first published versions of Bache's extract and accompanying note in the *Journal of the Franklin Institute,* Henry suggested to Bache that the pending, parallel publication in Silliman's *Journal* omit Bache's explicit reference to priority. He worried that the reference placed him "in rather a pugilistic position with regard to Mr Faraday." "Perhaps I am over squeamish on this point," he continued, "and have been rendered somewhat more so by a remark of Prof Green which amounted to this that there was much cry and little wool.'" Although apprehensive about overreacting—and thus seeming to resemble the proverbial pig that produces a lot of noise but, unlike a sheep, no wool—Henry ended this portion of the letter on an indecisive note. He asked the more experienced Bache to decide whether to allow stand the potentially "pugilistic" sentence in the cover note. When Silliman ran the extract and note in the May issue of his *Journal,* he retained Bache's questionable sentence about securing Henry's priority. Henry spoke more assertively in private communications, to Silliman and brother-in-law Stephen Alexander; following the appearance of an updated summary of Faraday's findings, he argued for his own precedence in even the recent round of his and Faraday's announcements. He did this by reconstructing the announcements' chronology in his own favor.

Henry's defense of his priority in uncloaking *self*-induction, though tempered by solicitude for Faraday's sentiments, contrasts sharply with his disregard of his stake in uncovering *mutual* induction. From his Albany years on, he never pressed, let alone publicly mentioned, any claim as a codiscoverer of mutual induction. Only in the intimacy of his family, friends, and students did he even allude to his role. Indeed, on the rare occasions in his Princeton course on natural philosophy when he broached the topic, he claimed merely either to have been "the first to repeat these experiments in this country" or, at most, to have been scooped in simultaneous experiments when Faraday placed "the result of his labors before the public." Usually in his Princeton course, Henry simply adhered to his outline for his lecture on "Magneto-Electricity," which begins: "Production of electricity by means of magnetism. Discovered by Faraday in 1832." Similarly, a colleague's diary from 1848 records a private conversation in which Henry circumspectly recalled that, although his "apparatus had been prepared a year or more . . . he was prevented from making his experiment by interfering circumstances until he heard the indefinite report that Faraday had produced electricity from magnetism."[11] Henry's public silence on the topic reflected a combination of factors. It reflected, of course, his adherence to the norms of a scientific community that valued timely announcements of tangible experimental demonstrations. It also reflected a mixture of his awe toward Europeans—particularly the eminent Faraday—and his own modesty or insecurity. Perhaps also contributing to his silence was an apprehension that Faraday might protest as he had in the misunderstanding involving Nobili and Antinori or, to a lesser degree, Ampère; in these various exchanges, Faraday had vehemently defended his independence in the discovery of mutual induction.

But Henry had successfully established his precedence in identifying *self*-induction and Faraday did not contest the matter. In fact, by late summer of 1835, Henry received word through an APS officer that a contact in London had heard from Faraday himself on the matter. "Had Dr Faraday known that Professor Henry had been making the experiments published by your Soc. in the Franklin Journal," the contact reported, "he would have noticed them in his last essay. He considers them as Very Interesting." As part of this ostensibly first personal communication between these British and American scientists, Faraday also had the contact forward a copy of his article to Henry. For Faraday, the detection of self-induction was simply a further progression in his rapidly advancing program of electromagnetic researches; between 1832 and 1840, he would publish in the Royal Society's *Philosophical Transactions* a series of seventeen "Experimental Researches in Electricity" encompassing a wide scope of topics.[12] In contrast, for Henry, the isolation of self-induction became not only a chief object of pride but also the fount of an ongoing research program.

Normally self-effacing in public forums, Henry later let slip his pride in being the first to identify self-induction, pronouncing it a "remarkable phenomena." This occurred in 1857, at age 60, when he felt obligated to clarify his research record because of his clash with Morse over the telegraph's invention. Reconstructing his personal history slightly, Henry even invoked self-induction to justify his otherwise vexing inaction in fully developing the telegraph. Insinuating that original scientific research takes precedence over the mere practical application of research, he claimed that he discovered self-induction while experimenting with the long-distance telegraph and that this "remarkable phenomena" caused him to suspend his telegraphic studies. In fact, as we have seen, Henry *first* began to pay attention to the peculiar sparks while puttering with his reciprocating engine; only after becoming aware of the sparks, did he further observe them while tinkering with his long-distance telegraph. Also contrary to his 1857 contention, he did not interrupt his telegraphic trials nor any other investigations to concentrate on the sparks, at least not immediately. Although he ran rudimentary tests on the sparks, which he alluded to in the addendum to his mid-1832 article, he left no record of an early sustained investigation. In fact, as we have also seen, he likely only began to suspect the relevance of the sparks when he began to fathom the nuances of electromagnetic induction following Faraday's announcement. Nor would Henry immediately concentrate on the curious sparks after tentatively linking them to electromagnetic induction in his mid-1832 addendum. Two years would elapse before he finally returned to the subject, in the late summer of 1834. As he himself explained in his first paper devoted to self-induction—the memoir submitted to the APS in early 1835 and later issued in their *Transactions*—the cause of the two-year gap lay in the disruptive "new duties" associated with his move in the fall of 1832 from Albany to Princeton. To be sure, Henry did preliminary experiments on the sparks toward the end of his tenure in Albany, announced his suspicion of a connection between the sparks and electromagnetic induction in mid-1832, and briefly returned to the experiments in the late summer of 1834 after settling in at Princeton. However, he would press his investigation of the effect only at the close of 1834. This was after encountering Faraday's disquieting announcement of his similar research. Just as Faraday had kindled Henry's intensified investigation of what eventually became known as mutual induction, so too did the London luminary spark Henry's expanded inquiry into what eventually became known as self-induction.[13]

In this same 1857 statement, soon published in the Smithsonian Regents's *Annual Report* as part of the record of the Morse-Henry controversy, Henry went on to insist that other of his Albany contributions were firsts. "I was the first," he kept repeating as he enumerated his accomplishments in refining electromagnets, in matching intensity batteries and intensity magnets for applications over distances, and in devising the rudiments of the electromagnetic telegraph. Not wanting to seem prideful, he attached a disclaimer to this list of firsts. He emphasized that all these discoveries came early in his career and that he now considered them relatively unimportant. He added, "had I not been called to give my testimony in regard to them [in the Morse proceedings], I would have suffered them to remain without calling public attention to them, a part of the history of science to be judged of by scientific men who are the best qualified to pronounce upon their merits."

In fact, Henry neither considered these early accomplishments to be minor nor resigned himself to wait silently for history to adjudicate them. Ten years earlier, in 1847, he had published a comprehensive review of recent international contributions to electricity and magnetism in which he elevated his own discoveries to a foremost position. He had taken this self-aggrandizing stance in an unsigned article that filled fifteen double-column pages of fine print in a special volume updating the *Encyclopædia Americana*. Admittedly, the editor had encouraged Henry to do "justice" to himself by including coverage of his own contributions. "Do not let any modesty prevent you from doing so," the editor had explained, "for you will bear in mind that it is only I myself who will be responsible to the public for the accuracy of the statements made."[14] Even allowing for the license the editor gave Henry, the resulting article reveals that the publicly modest scientist privately held an immodest opinion of his Albany researches. It also reveals that he believed that educated Americans—that is, readers of this popular encyclopedia—should be apprised of what he personally considered the high value of his accomplishments. Nowhere specifying authorship, the article opens with a statement of its goal: to update all the entries on electricity and magnetism in the last full edition of the encyclopedia, issued a dozen or so years earlier. "No branch of physical science, during this period," the text begins, "has been enriched with a greater number of important discoveries." The article then reviews recent developments in seven subcategories of research. The fourth subcategory, "Electro-Magnetism," starts with an overview: "The general laws of this branch of electricity, as expressed by the ingenious theory of Ampère, have been given with sufficient detail in an article in a preceding volume of this encyclopædia. The additions since made relate principally to the development of great magnetic power in soft iron, the most important results of which were obtained in our own country by Professor Henry, of Princeton." Next comes a detailed account of Henry's early investigations, including devising a telegraph and an electro-magnetic machine. "The invention of the first machine of this kind is due to Professor Henry," the article insists as it goes on to mention later "modifications" of the machine by Ritchie and others around the world.

When the article arrives at the sixth subcategory, "Magneto-Electricity and *Galvanic Induction,*" full credit redounds to Faraday for his "first discoveries in this division of our subject." In a matter-of-fact summary, the article distinguishes, as Faraday had in 1831–32, between two closely related phenomena: "magneto-electricity" (the transitory current arising from magnetic induction) and "galvanic induction" (the momentary current induced in a wire by a changing current in a parallel wire). Having introduced Faraday's exposition of the latter, electrically induced currents, the article immediately jumps to a related contribution: "After the discovery by Dr. Faraday of the foregoing principles of galvanic induction, the most important additions to this branch of electricity have been made by Prof. Henry." These "most important additions" spring from, the article declares, Henry's identification of and research on self-induction, the subject of the remainder of the sixth category. While comfortable extolling his work on self-induction, Henry remains true to his lifelong pattern and avoids even mentioning that he paralleled Faraday's findings on mutual induction.

Henry was using this anonymous article to associate his Albany researches with those of the eminent Ampère and Faraday along with a handful of other acclaimed European researchers featured in the text. Indeed, except for his extensive coverage of Faraday's multifaceted investigations in nearly every subcategory, Henry devoted the most ink to his own researches. And, while fair to Faraday, he used the opportunity to deliver one rebuke. He disagreed with Faraday's theory, by then gaining popularity in Great Britain, of "the specific inductive capacity of different bodies"—a theory relating electrostatic induction to the behavior of the underlying particles of intervening materials such as air. "Although we place the highest value on the experiments of Dr. Faraday, and are disposed to treat with the greatest respect whatever opinions may be advanced by him," Henry wrote under the cover of anonymity, "yet in the present case we are forced to believe, from an attentive study of the experiments, that the results do not warrant the conclusion." Henry preferred applying a refined version of the traditional theory of countryman Benjamin Franklin in which induction occurs through the inherent force of a self-repulsive electrical fluid associated with different bodies—a force that acts over distances independently of the intervening space. In the confines of his own lecture hall, Henry had been even blunter, expressing regret that Faraday's theory obscured phenomena that Franklin's theory explained simply.[15]

This self-interested encyclopedia article appeared a decade before Henry's 1857 public avowal to await history's judgment on his early researches. Fourteen years after the declaration, he was still anonymously extolling the researches. He continued this tack in a broadly reprinted biographical sketch—actually an autobiographical sketch published anonymously. He drafted the sketch originally for a special retrospective and indexing volume of the widely read *Princeton Review,* but it was reissued in national periodicals such as *Popular Science Monthly* and the *Eclectic Magazine.*[16] In the sketch, Franklin's revered mid-eighteenth-century electrical work serves as the referent for a flattering comparison with Henry's early work. The sketch recounts that when Henry was a professor at the Albany Academy, "he commenced a series of original investigations on electricity and magnetism, the first regular series on Natural Philosophy which had been prosecuted in this country since the days of Franklin." "These researches made him favourably known, not only in this country but also in Europe," the sketch proceeds. After tracing later phases of Henry's career, the sketch provides a list of his twenty-two most significant research projects and discoveries. His early studies of electricity and magnetism figure prominently in the chronological list: the first electromagnets powerful enough to lift over a ton despite using small batteries; the first continuously operating machine based on an electromagnet; the development of the concept and apparatus essential for electromagnetic telegraphy; and the discovery of self-induction. As in Henry's previous self-appraisals, however, the list does not even allude to mutual induction.

The Princeton students who, year after year, had taken Henry's mandatory but popular course in natural philosophy would have recognized most of the later public contentions. Beginning with the encyclopedia article, which Henry drafted in 1846, he drew heavily on portions of his lectures from preceding years. His pupils had dutifully copied into their notebooks not only his basic, pedagogic narrative but also his often personal, unguarded commentary. The surviving notebooks, most of which date from the 1840s, suggest that Henry portrayed himself to classes in earlier years as the sole inventor of large electromagnets. He eventually began to acknowledge Sturgeon's contribution, but he still asserted the originality of his own design and, on at least one occasion, complained that he was "in part deprived of that credit which he deserved" merely by a "European philosopher" (Moll) publishing first. He attributed his breakthrough in devising large electromagnets to being ahead of all of his American colleagues and most English researchers in appreciating Ampère's theory. A notebook from the winter term of 1840–41 registers the further assertion that, though Europeans initially overlooked Henry's Albany-era electromagnet, the instrument "has now passed into the history of the science." "Professor Henry has been the most successful experimenter in electromagnetism," a later student similarly recorded as he logged Henry's lecture on powerful electromagnets.[17]

The students' notebooks consistently list Henry as "the first" to build a machine using an electromagnet and routinely explain why this "Sheep's Tail" would remain into the foreseeable future "a mere philosophical toy." Moreover, the similar machines of Ritchie and other Europeans, along with the devices of Americans such as Davenport and Page, amount to nothing more than mere elaborations of Henry's principle that took shape following publication of his original article. The notebooks also consistently depict Henry as the catalyst of the main telegraphic systems that were beginning to operate in Europe and the United States. In particular, by refuting Barlow's pessimistic prognosis and showing that galvanic currents could travel long distances, Henry singlehandedly had established the "practicability" of the telegraph. And the word quickly spread following publication of his finding in Silliman's *Journal;* one notebook specifies that "his article was translated into all the languages of Europe and the project of a telegraph was again revived." Although always assigning full credit to Faraday for having disclosed mutual induction, Henry did purport to have acted simultaneously with Antinori in detecting a mutual-inductive spark. As for self-induction, the student notebooks abound with accounts of Henry being the first to distinguish the effect and with detailed descriptions of his many follow-up investigations.[18]

The student notebooks from the 1840s, the 1847 encyclopedia article, the 1857 statement of record in the dispute with Morse, and the 1871 biographical sketch all suggest that Henry placed high stock in his Albany investigations. They also suggest that, contrary to his public avowal of being disinterested, he sought to portray his investigations as not merely competent and credible but coequal with those of Franklin, Ampère, Faraday, and other luminaries of electricity and magnetism. He pressed this self-aggrandizing portrayal either surreptitiously through anonymous writings or privately in the confines of his classroom. Personal letters, such as those about deserving credit for the first electromagnetic machine that he sent to Page in 1838 and to De La Rive in 1841 (and again, with renewed emphasis, in 1846), also provided vehicles for discreetly pressing his claims of priority. Indeed, about when first exhorting De La Rive, he also included self-assertive letters in packets of his latest publications that he sent to the Danish fountainhead of electromagnetic science, Hans Christian Oersted, and to the venerated Swedish chemist, Jöns Jacob Berzelius. For example, in 1841, on the pretext of introducing himself to Oersted, he turned to his Albany accomplishments and portrayed himself as not only having discovered self-induction but also having originated (with Philip Ten Eyck) strong electromagnets "before Dr Moll's experiments on the same subject were published" and having invented the first electromagnetic machine but receiving "no credit for it in Europe." "I mention these results of my earlier labours," Henry repeated to Oersted, "inorder that you may identify the person who addresses you and because I think I have not received for them the credit in Europe which is my due."[19]

A few years after writing these letters, while corresponding with another acquaintance, Henry could unblushingly contend that, although tempted, he had always refrained from asserting his priority in discoveries that were rightfully his but that others usurped. Indeed, he seemed sincerely to perceive himself a man of great self-restraint, explaining that "in the long run I have never found cause to regret my forbearance." Similarly, he continued to project a public image of selflessness, and his many lay admirers in later years considered him to be beyond pride in past achievements and above any need for recognition. Thus, shortly after the 1871 biographical sketch appeared, the editor of the *Washington Evening Star* happened to comment on recent statements by two of Morse's associates crediting the self-effacing Henry rather than Morse with the crucial discoveries behind the electromagnetic telegraph. "Prof. Henry has been too modest to advance his claim to the leading place," remarked the editor of this largest circulating paper in the District of Columbia, "but it is now fully accorded to him by those who should know. . . . It is curiously refreshing in these days, when every man acts on the principle that he that bloweth not his own horn, the same shall not be blowed,' to see an act of voluntary justice of this sort done towards a quiet, unassuming man of science like Prof. Henry."[20]

Henry's covert tendency to applaud himself probably reflected a variety of factors, extending from psychological to cultural. Of these factors, the most immediate likely included the deflating experience of being scooped by Moll, Faraday, and Forbes in announcing critical findings and the distressing ordeal, in other cases, of perceiving priority to be misassigned to more noticeable innovators such as Ritchie, Faraday, and Morse. By quietly assigning top billing to his own performances, he was discreetly protecting against the reproach of being merely a provincial actor on an international stage. This furtive, defensive stance also perhaps reflected a more latent factor: the "lasting impression" of his father's alcoholism and related death. Likely harboring the self-doubts and insecurities that had arisen during his traumatic childhood, he still apparently was following the misbegotten pattern of publicly cloaking anger and skirting confrontation while, at the same time, surreptitiously contending for personal affirmation. Whatever his inner drives, Henry took singular pride in his 1832 isolation of self-induction and built around it an ongoing program of research. But he unfailingly kept silent on mutual induction, never publicly claiming even partial credit as an independent discoverer. Indeed, his silence on mutual induction provides a counterexample that illustrates the effectiveness of his anonymous self-appraisals and other behind-the-scenes self-evaluations in raising peer and public awareness of his other research contributions. Many later associates and commentators would first become aware of his specific accomplishments in mutual induction only following his death in 1878. This was when they started to look back and reappraise his scientific career—a process aided by an 1886, two-volume republication of his scattered articles and by an ongoing campaign by daughter Mary to promote public appreciation of his achievements. Ironically, these reappraisals—continuing through the twentieth century and increasingly offering comparisons with Faraday—would eventually mark Henry's probing of mutual induction as his most salient research achievement.[21] His apologists seemed determined to remake him into the American Faraday.

In 1893 his work on self-induction *and* mutual induction apparently figured prominently in the decision of a multinational group of scientists and engineers to assign his name to a fundamental unit of electrical measure. The occasion was the International Congress of Electricians, held in Chicago as part of the World's Columbian Exposition. Delegates from Great Britain, Germany, France, Italy, the United States, and other nations honored pioneers of the field when they formalized the international units of electrical measure—the *ampère, volt, coulomb, farad, joule, watt,* and *henry.* Because their country was hosting the Congress, the Americans likely enjoyed an improved chance of placing a native son in the short, final list of otherwise European honorees. Whatever their motives, whereas the delegates designated the *farad* the unit of capacitance (for gauging the storage of electrical charge), they reserved the *henry* for the unit of inductance.[22]

7 Joseph and Harriet

In his first year of teaching at the Albany Academy, Henry received a letter congratulating him on his inaugural address. "May your Professorship be as splendid as its commencement was brilliant," the well-wisher wrote after reading a newspaper extract from the address. He closed the letter, however, with a bit of amiable advice. "Let me exhort you not to give yourself so exclusively to the dry bones of diagrams but consider that you are partly made for your friends and social intercourse."

Although the new professor devoted most of his waking hours to the dry bones, he found some time for social interaction. He made friends easily and his base of earlier Albany relationships expanded rapidly as he became more active in groups such as the Albany Institute. His deepest relationships, however, existed among family members, not only his mother, sister, and brother but also his many aunts, uncles, and cousins—all bound together, on both his maternal and paternal sides, by their Scottish heritage. In fact, the person writing to congratulate him on his inaugural talk was a cousin, the son of Henry's uncle John Alexander. Uncle John, twin brother of Henry's mother, still lived in Galway where he had lent a hand in raising young Henry when the boy was under the care of his grandmother. While close to his Uncle John's family, Henry had grown even more intimate with the family of another of his mother's Scottish born siblings, Alexander Alexander. Having settled in Schenectady, where he became a respected and prosperous merchant and landowner, "Sandy" Alexander died in 1809 at age 44. His large estate, though gradually eroded through the executors' mismanagement, ensured that his widow, Maria, could provide amply for their three-year-old son, Stephen, and their one-year-old daughter, Harriet. Stephen and Harriet grew up in a refined household, reading, drawing, playing piano duets and, as Harriet later recalled, passing together "years of undervalued bliss." By the time Henry became a professor at Albany Academy, Stephen had graduated from Union College and was teaching at Yates Polytechny, a small school in Chittenango, west of Albany. Meanwhile, Harriet continued to live with her mother, an aunt, and two close African

American servants (who Henry, following contemporary usage, called "the two Darkies"). Henry was a recurrent guest in the imposing stone home of the Alexanders, taking advantage of the frequent stagecoaches plying the 14-mile turnpike between Albany and Schenectady. He took time to visit, for example, even as he returned in December 1825 from the exhausting road survey; in a letter describing that particular stopover to Stephen at Yates Polytechny, Joseph affectionately addressed him "My Dear little man," an allusion to his cousin's short height and slight build, traits also shared by Harriet. In the next few years Joseph's relationship with his younger cousins would strengthen, with Stephen almost becoming another brother. As for Harriet, she and he would wed in 1830.[1]

A year and a half earlier, in mid-November 1828, Joseph had sent Harriet a letter laced with banter. He was about to turn 31, but still mistakenly thought he was at least two years younger; she was only a little beyond 20. What triggered his zestful letter was a request from Harriet to purchase some cloth for her in Albany. "If, at the time when I present the account," he deadpanned after tracing his efforts to fulfill the request, "it should not be convenient for you to make immediate payment, I will accept as security of the debt, a bond and mortgage on your real or *personal* property." "On the subject of promises," he coyly continued, "when am I to have the pleasure of performing the one, which depends on your coming to Albany?" Whether they furthered their budding relationship with a meeting in Albany, the couple did travel together a few months later to visit family and friends in Galway. By then, every week, "Cousin Joe" was also sending his "Dear Cous." a copy of an Albany literary gazette.[2]

As his relationship deepened with Harriet, it also expanded with her brother and mother. In one chatty letter, he offered his Aunt Maria advice on travel and finances before telling her about a day spent with Stephen and a colleague from Yates Polytechny who were passing through Albany. Along with dinner presumably at Joseph's mother's home, a trip to the barber, and tea with a family friend, the trio visited a local artist for minor alterations of a miniature portrait of Henry. About the portrait, which seemingly he was having painted for Harriet and his aunt, he commented: "Mother knew it immediately and has taken the charge of it until Harriet and yourself come down." The highlight of Stephen's stopover was an evening visit to a traveling show that not only featured a "grand exhibition of the conflagration of Moscow" but also displayed mechanical figures such as an "Automaton Rope Dancer." "At half past nine," Joseph told his aunt, "I bid good-by to the two Gentlemen at the Eagle Tavern where they had taken lodging in order to start with the stage at one o'clock the next morning." He closed by voicing a wish that his aunt and Harriet would soon visit Albany.

By November 1829, one year since jesting with Harriet about buying cloth, Joseph was sending his letters in the daily "penny post" with playful salutations such as "*A Mon Cher Cous.*" He was also still sending gifts of reading material. But time for personal trysts had been at a premium that fall. In early December, he told Stephen: "my labours have been more arduous during the last three months than at almost any other period of my life." Not only had he been contending with new courses at the academy and an expanded enrollment of more than 200 students, but he also had been rushing to complete his analysis of New York topography for publication in the *Atlas* and the Albany Institute's *Transactions.* Despite all the bustle, however, he found time to become engaged to Harriet. In fact, the reason for his early December letter to Stephen was to seek approval of the pending marriage.

Joseph broached the topic gingerly with Stephen, apologizing for his "unmanly" reluctance to speak in person about the "delicate subject" during a recent time together. A letter allowed him to express his feelings more candidly, he explained. He began by revealing anxieties about the marriage's effect on others:

> Your mother and your aunt are both devotedly attached to their children and would be rendered miserable by any misfortune that should befal either you or your sister. Harriet has had offers which certainly in pecuniary and perhapse in other points of view would be much more advantageous than the present. These considerations have made me more anxious about the result of the connexion than under less peculiar circumstances I might perhapse have been. With regard to myself I am fully confident that I shall be both happier and better by the union. My only source of anxiety is that Harriet may not be as happy as she deserves or that she may not be as fully so as I wishe she may be. Connubial happiness in some degree depends upon worldly success and this is not always within the reach of personal exertions for while much depends on individual labours much also proceeds from the dispensations of that providence over which human power has no controll.
>
> I hope from the first that I have been fully impressed with the responsibilities and importance of this connexion. I have certainly endeavored to act in such a manner that Harriet might be as little as possible under the influence of any feelings which might bias her judgment in making her decision. And even now although I am confident that my future happiness depends principally on this event were she to repent in the slightest degree of having made the engagement or were any serious objection raised on your part or on that of the family I would instantly relinquish all claim and suffer her again to decide a new.

He closed by asking for Stephen's frank reaction.

Joseph would wait more than three months for Stephen's reply. Meanwhile, Christmas day arrived and he wrote to Harriet about a disappointing change of plans. "For the last three weeks I have been constantly looking forward to this day with the liveliest anticipations of hope," he began, "and in truth no Christmas since the days of my boyhood has ever appeared so pleasant in prospect as this." Unfortunately, the academy wanted him to transact some business in New York City, and he needed to embark that very day. A greater disappointment came a month later, when the couple had a spat. Apparently, an encounter in Schenectady had culminated with Joseph departing abruptly in a sullen mood.

Joseph took the lead in breaking the resulting silence, writing Harriet a long, introspective, conciliatory letter. Trusting that they retained warm feelings for each other, he expressed the hope that those feelings had not been irreparably damaged by what he conceded to be his "unmanly or improper reserve." This concession led to a broader self-indictment of past behavior:

> In reviewing our past intercourse, I find much to sensure and to regret, in my conduct; it has often been careless and inconsiderate: At first I treated you with much apparent coldness and reserve; and several times since I have acted with rudeness, highly improper, towards the woman whom I love and respect. But this has not arrisen, from any want of a proper estimation of your worth, or from the slightest intention to manifest indifference, or disrespect; on the contrary. I have from the first regarded you with increasing esteem and affection; and have never intentionally or willingly given you either pain or offence.

If not disrespect, what was the source of his misconstrued aloofness? The source lay, he contended, in a fundamental difference between "the Character of female attachment" and the "attachments of men." The more "tender" or "sensitive" a woman is, the more likely that "every harsh word" or even "the slightest breath of suspition, or unkindness" will upset her; only with time does a woman gain enough confidence in a relationship to remain unrattled by "a quick reply, a trifling error, or even one unkind expression." In contrast, men "are not so permenantly affected by those little annoyances which sometimes ruffle the temper, but leaves undisturbe the heart." Henry insisted that, whereas this difference between men and women explains his past behavior, it does not excuse it. "I should have guarded each word, and each action, so as not to have given, if possible the slightest cause of disquiet; and that too more particularly at the commencement of our intercourse." He ended by admitting that, though men are thicker skinned than women, he remained vulnerable to her reproach. "I must acknowledge that you do possess the power, in a very considerable degree, of giving me pain; and when I am smarting under the lash of your displeasure, I may show for an instant a spirit of rebellion but one kind word, or one smile of reconciliation will I am sure never fail to call me back to my allegiance." On that note, he implored her not to test her power to make him suffer, and pleaded for even "two lines" of response to his letter. One full week after Henry penned this soul-searching letter, Harriet responded: "I received yours on Saturday last and would have answered it immediately, had not a sore finger disabled me."

Although she had delayed her reply—for a seemingly suspect reason—once at the task she matched Joseph in level of self-reproach. She told him that his letter brought her relief from worries about "the impression my, at best, foolish and inconsistant conduct, may have made on you at your last visit; particularly, as your grave phiz and cold air at parting bespoke no little dissatisfaction and which I will acknowledge, for a moment, heightened my displeasure." "But a little reflection," she continued, "soon convinced me that *I* was the offender. I can scarcely tell, what at first displeased me, some mere trifle which ought to have passed unheeded; impute it not to any act of yours but rather to that inability of temper which so often involves me in difficulty." Not as verbose as Joseph, she quickly closed the subject with a mischievous twist: "I regret the inquietude, which I fear, it has occasioned you, but fear I should be half inclined to repeat the offence if I could be assured of it its being followed by so pleasing a communication." In his reply, Joseph extended this thought as he expressed his own heartfelt thanks and relief. "If after every storm comes such a calm," he confessed, "I care not dear girl how often you ruffle the more irritable parts of my temper." Cheered by Harriet's response, he followed her example and closed with wit. He offered a mock scholarly discourse on Harriet's apology for the "shortness" of her letter and her parenthetical query if the word *shortness* properly conveyed her intended meaning of brevity. Paraphrasing Shakespeare's Macbeth on shortness of delay, Joseph argued for the moral propriety of taking quick action once there is agreement on an important plan—a veiled appeal for Harriet to set an early date for their wedding. For months to come, plays on the word *shortness* would provide private jokes in letters between the couple.

Joseph had hoped to thank Harriet for her conciliatory letter in person, but was unable because, on even the weekend after receiving the reply, he was "occupied too much with lectures and our semi-annual examination." Shortly after, he pointed out to Harriet that his official duties made it difficult merely to find leisure to pen a letter, but that an even greater distraction was keeping him from his personal research.

> This is the third time I have attempted to write to you since last week and since I have been interrupted, you must not think my employments or pursuits have for an instant displaced the thoughts of you or of the tender relation in which we now stand to each other. On the contrary I may say in truth that you occupy at present more of my thoughts than is compatible with a very rapid advance in my studies. The thoughts of you have proved powerful rivals to Chemistry and Mathematics, which formerly occupied a much larger share of my contemplations. I now find it almost impossible to confine my attention to an abstract subject for any length of time as it is liable to be diverted by the slightest association which savours of our union.

Coupled with the strains of the classroom, the distractions of courtship further slowed Joseph's personal program of research. And this slackening of research occurred during a crucial period. About one year earlier, during the early spring of 1829 when the relationship between the cousins was blossoming, he had displayed a prototype of his enhanced electromagnet to the Albany Institute. But it would take the "mortification" of Moll's announcement at the end of the current year, 1830, to jar Joseph into compiling his own advancements and hastily reporting them in Silliman's *Journal*.

Meanwhile, in mid-March 1830, Joseph finally received Stephen's reaction to the contemplated marriage. "Your union," Stephen affirmed, "shall not meet with the least objection on my part." He mentioned, however, one qualm. "Although I would not myself marry my own cousin, I still cannot say that such a thing ought not to be done. If therefore your affections are placed upon each other, your happiness shall not be marred by wanting my consent." Marriage between cousins lingered as a grey area in early-nineteenth-century morality and custom. Henry preferred not to call attention to the issue, avoiding mentioning it by name even in his response to Stephen: "Permit me my dear cousin to assure you I duly appreciate your generous consent to my union with your beloved sister—to a union the result of which I know is intimately connected with your own happiness and which in one particular at least is at variance with your opinions of a proper connection." Harriet reported to her brother that Joseph was pleased with the letter of consent. She also informed him that Joseph and her mother had finally persuaded her to set an early date for the wedding. "From his importunity seconded by the arguments of my mother," she announced, "I have at length consented that the matter which has so long occupied our thoughts shall be consummated as early as the first of May."

As May approached Joseph still had to contend with duties extending beyond his normal classroom assignment, including delivering the final lecture of a public course and overseeing the academy library. When he could not travel to Schenectady to visit his betrothed, he continued to rely on the speedy "penny post." In one mailing he included a new novel for Harriet's recreational reading. "Although I do not approve of much reading of the novel kind," he instructed, "yet I think under present circumstances to prevent too serious thought on a certain subject you may safely indulge a little." On another occasion he picked up in Albany a bonnet that Harriet had ordered; before having a chance to hand-deliver it to her, he rhapsodized: "It is certainly beautiful and I cannot refrain from adding, that I am sure it will exhibit in all its native sweetness the face which in my eye is interesting beyond all others." One Monday evening, he managed to fit in a long horseback ride that left him physically refreshed and in exceptionally good spirits, without his normal anxiety about the pending wedding. "With such a state of mind and the fact of my enjoying better health then at any other period during the last six months, it is not surprising that I feel inclined to look at the sunny side of the *present* and to cherish high hopes of enjoying in the future as much domestic happiness as is consistent with our present state of existence." Harriet and Joseph wed, in Schenectady at the Reformed Dutch Church, on Monday, 3 May 1830. Although two days earlier Harriet had turned twenty-two, the wedding certificate lists her age as still twenty-one; moreover, it lists Henry's as twenty-seven, an underreckoning of five years—the same error that Principal Beck had made earlier.

After the ceremony, they traveled down the Hudson River to New York City. Early the next morning, they boarded another boat bound for New Haven, Connecticut, where they stayed "in a little room in a tavern." Mixing business with pleasure, the newlyweds spent the next few days touring the facilities at Yale College, sightseeing, and socializing with Yale faculty members and other friends. Denison Olmsted, natural philosopher and astronomer, showed them the physics apparatus in Yale's "philosophical hall." Harriet described him as "a very scientific and exceedingly pleasant man." Benjamin Silliman, in this first meeting with Joseph, guided the couple around the "chemical lecture room," which included one of Hare's large deflagrators. On a following morning, the famous chemist and his daughter called on the Henrys and offered to escort them to New Haven sites of interest. In the evening, the newlyweds returned the call by visiting Silliman's home; Harriet observed, "I am delighted with him." They left New Haven Saturday and backtracked to New York City, where they spent a couple of days enjoying various art collections, and lastly visited the National Academy of Design. Samuel Morse presided over this professional society of artists, though, apparently, Henry did not then encounter this painter and inventor who was later to complicate his life.[3]

Arriving home in Schenectady on Wednesday, Mr. and Mrs. Henry received well-wishers for the next few days, as was the custom. Harriet commented to her brother that "there seemed no end to the calls." She added: "The events of the last fortnight seem more like a confused dream than reality." When Monday arrived, because the couple had yet to establish their own household, Harriet remained in her mother's home while Joseph returned to Albany for the week's normal duties. The brief separation stirred in Henry a "sense of lonliness, a feeling of anxiety." These emotions led him to probe the nature of their relationship. He wrote to assure Harriet that they would not likely disappoint each other because they entered the marriage with realistic expectations. "Our fancies have not been sickened with visions of happiness human nature can never hope to enjoy," he suggested, "nor have we imagined perfections in each other which belong only to the beings of fiction and romance." Continuing in this rationalistic vein—but with language reminiscent of that he first encountered while acting Shakespeare's plays—he reminded Harriet of a principle on which they might ground their relationship:

> While my eyes, which find to please so much more in you than in any other woman, are not blinded entirely to the fact that you may possess some of the weakness of your sex, I love you the more because you are not perfect. And although the defects of my character have been too glaringly displayed to escape even your notice yet I hope, nay I am sure you look upon my faults with an eye of extenuation. Let us perpetuate our present feeling by a rigid adherence to our resolution of implicit confidence and by considering our marriage as a bond of perfect friendship, in which the faults and weakness of each are to be met on the part of the other by a spirit of gentle reproof and compassion.

A slightly later separation also stirred passions. As the first month of their marriage drew to a close, Joseph implored Harriet to follow through with her plan to visit him in Albany: "Do not disappoint me for you know that I possess but little of the virtue called patience and two weeks is by far too long an absence for a person of less ardent feelings than myself." Apparently, on occasion, Joseph also stayed in Schenectady and relied on a morning stagecoach to get to the academy by the required time of 9:00 A.M. One morning the coach failed to pick him up and Joseph, pleading the urgency of his predicament, persuaded another coachman to make a special trip to overtake the errant vehicle. Traveling rapidly, the driver successfully caught the coach and transferred Joseph, an unusual action that surprised the other passengers. Joseph's embarrassment was compounded a little later, however, when yet another speeding driver caught and hailed the coach, announcing that he had a warrant for the arrest of one of the passengers. All eyes turned to Joseph, but the fugitive turned out to be a man sleeping next to him.

Before the marriage, Harriet had agonized about breaking up the household that she shared with her mother and aunt, "our dear little family." A solution presented itself, however, when Harriet's mother and aunt agreed to move to Albany. Not only Joseph but also his own mother and sister, Nancy, encouraged the move. Both still lived in Albany; although Ann Henry continued to take in boarders to provide income, Joseph had been redirecting half of his academy salary to her and Nancy's support. By August the newcomers had located a house and Nancy took charge of readying it. Because the academy trustees, perhaps feeling financially pinched, did not provide the newlyweds a house of their own—a perquisite that the other professors enjoyed and that Joseph had expected—they would be living under the same roof with Harriet's mother and aunt. The family received an unexpected boost when Stephen decided not to return to Yates Polytechny but, instead, to relocate to Albany. Joseph reacted to the news of this additional housemate by designating a room as a combination study and library in which he and Stephen, a scientist who was already a corresponding member of the Albany Institute, could work together. With Joseph's brother, James, also in Albany working at a bookstore, the newlyweds enjoyed daily interactions with their closest relatives—an unusually interwoven clan in which the two mothers-in-law were also aunts. Writing to Harriet while on a short trip, Joseph closed by bidding her: "Give my love to all the members of both families."

A year later Harriet's "dear little family" underwent even a more profound change when she and Joseph celebrated the birth of their first child. William Alexander Henry was born in October 1831. Joseph probably shared at least some of the high hopes that a friend, in a congratulatory letter, sportively listed for the new baby: "The mathematical world and the votaries of natural science ought to watch the rise of the new luminary with great care and anxiety. His name may possibly cast those of Leibnitz, Euler, Newton, Laplace, Volta, Lavoisier &c. in the shade of oblivion. On him American Science may have to rest its hopes and that in all probability with reasonable grounds of having them gratified."

Stephen's relocation to Albany meant that Joseph could include him in his projects, thus combining research and friendship. Although Stephen's main interest was astronomy, particularly the relationship between solar eclipses and the latitude and longitude of observational sites such as Albany, Joseph enlisted him to help with his newest area of serious study—terrestrial magnetism. The mapping of the earth's magnetic patterns, a longstanding international quest, offered practical benefits for practitioners in fields such as navigation and intellectual rewards for scientists in areas ranging from electricity and magnetism to meteorology. Indeed, in meteorology, persons envisioned ties between weather, climate, and the earth's magnetism. For decades, investigators in various countries had been painstakingly observing delicately balanced magnetic needles to contribute local information on three key variables: magnetic "dip" or "inclination," the direction of the magnetic pull in the vertical plane; "variation," the compass orientation in the horizontal plane, expressed as a deviation from true geographic north; and "intensity," the location's relative magnetic strength. Joseph joined a distinguished American lineage of students of terrestrial magnetism when, in 1825, he briefly assisted Simeon DeWitt, surveyor-general of New York and anchor of the Albany Institute; DeWitt was updating an ongoing survey of magnetic variation begun in 1672 by Harvard natural philosopher John Winthrop. Joseph also joined a distinguished international network when, in the late summer of 1830, he agreed to participate in Columbia College professor James Renwick's survey of magnetic intensity; Renwick had acquired two, sensitive magnetic needles from English geophysicist Edward Sabine, who, in turn, had obtained one of the needles from Norwegian geomagnetic specialist Christopher Hansteen. He had only recently met Renwick, but Joseph had examined Hansteen's views on terrestrial magnetism three years earlier when trying to explain the aurora borealis that ominously appeared after a hanging. Now, as the summer of 1830 ended and the Henry-Alexander clan settled into their new home, Joseph made observations with Renwick in New York City and then borrowed the two needles for his and Stephen's own studies in Albany.[4]

When teaching duties relaxed each day at the noon hour or after five o'clock, Joseph and Stephen would delicately mount one of the needles in a specially crafted mahogany box and place it on a fixed post on the academy grounds. After emptying their pockets of keys, knives, and other metal objects and waiting for the apparatus to reach the same temperature as its surroundings, they would set the needle in slow oscillation and use a well-regulated "quarter-second watch" to time how long it took to complete 300 oscillations—about sixteen minutes. Stephen later recalled the joint effort with Joseph: "He noted every 10th oscillation of the needle calling it out distinctly, while I noted the times by chronometer in seconds and portions." Using the arithmetic mean of the incremental readings to calculate the total elapsed time to the nearest hundredth of a second and using a Hansteen formula to compensate for the effect of different temperatures, they soon generated a set of relative magnetic intensities that they specified to five decimal places. The first run of observations, from late September through late November, established a mean value of the intensity useful for comparisons with past and future figures. In April 1831, with the resumption of warm weather, they returned to their post each day to make such a comparison. Following a procedure that had become routine, they were finding that the new tests duplicated the original results. But the regularity vanished on 19 April when they detected "a remarkable anomaly." Whereas the noon observation on that day showed nothing unusual, a six o'clock repetition indicated "a great increase in the magnetic intensity of the earth." Moreover, that evening, an unusually spectacular aurora filled Albany's sky. "The idea for the first time now occurred to me," Joseph reported, "that this uncommonly brilliant appearance of the aurora might possibly be connected with the magnetic disturbance observed at 6 o'clock." To test the speculation he and Stephen returned to the academy grounds during the peak of the event and remeasured the magnetic intensity. They were chagrined to find that the value was no longer elevated, as they had anticipated, but had slipped significantly lower than normal.

Within the week Joseph described to the Albany Institute how the needle's oscillations had been "strangely altered without an apparent cause." He also began a systematic search for an explanation in scientific publications, including his standbys, the *Edinburgh Philosophical Journal* and *Philosophical Magazine.* Furthermore, the curious scientist wrote to Benjamin Joslin, William Campbell, and other colleagues living in nearby towns asking if the aurora had been visible in their localities. He waited nine months, however, until late January 1832, to update his findings for his institute associates. Not only did he inform them that colleagues in neighboring towns had seen the aurora but, more important, he reported that Hansteen had already published an article documenting an increase in magnetic intensity before and decrease during an aurora's appearance. (Joseph had examined the article three years earlier, but he apparently overlooked Hansteen's insight or else had forgotten it. Similarly, he probably overlooked Green's more general characterization of auroras in his 1827 guidebook on electromagnetism.) He also noted that even Prussian polymath Alexander von Humboldt, probably the most internationally visible scientist of the early nineteenth century, had described the auroral fluctuation of magnetic intensity.

By March, Joseph had learned even more. After examining the meteorological records annually compiled by the state, he realized that the aurora had extended beyond Albany to the furthest reaches of the state. Moreover, reading the *Journal of the Royal Institution of Great Britain,* he discovered that Samuel H. Christie, professor at the Royal Military Academy, in Woolwich, near London, had witnessed the identical aurora and similarly measured an irregularity in magnetic intensity. (What Henry had no way of knowing was that Faraday, who supplemented his position at the Royal Institution with a chemistry professorship at the Royal Military Academy, also had a hand in the British sighting. Faraday recorded in his diary for 19 April that he noticed the aurora at ten o'clock, called Christie's attention to it, and remained with Christie as he ran his needle measurements.)[5] Perhaps again provoked into action by seeing results similar to his own appear in a British journal—as with Moll's electromagnet announcement—Joseph finally published his findings on the April 1831 aurora in the April 1832 issue of Silliman's *Journal* and concurrently in the state's annual report containing its meteorological records. He added his own twist to the findings, stating: "I am not aware that a simultaneous disturbance of terrestrial magnetism, in connexion with an aurora, has ever before been noted at two places so distant from each other." He explained that he subsequently searched records and linked previous auroras that had been visible at the same times in England and New York. Also, he tendered a brief theoretical explanation, derived from Hansteen, that posited "the beams of the aurora to consist of cylindrical portions of some kind of matter, which becomes luminous as it passes into the higher regions of the atmosphere." In concluding the article, which he offered as a corroboration of similar international findings and as a call for further American investigations, he emphasized that an aurora "cannot be classed among the ordinary local meteorological phenomena, but that it must be referred to some cause connected with the general physical principles of the globe."

In submitting his aurora paper to Silliman during the early months of 1832, Joseph stressed that he had devoted "considerable attention" to terrestrial magnetism. He mentioned that, along with carrying out the aurora study, he and Stephen had begun testing for a shift in magnetic intensity between the base and peak of a high hill. Moreover, he intended to use his aurora publication to persuade state officials to purchase new apparatus for use on a major expedition. Conjuring a plan reminiscent of his 1825 road survey, he hoped to travel along the border between New York and Pennsylvania checking for changes in magnetic variation since the line was first surveyed fifty years earlier.

While Joseph was showing a heightened interest in conducting the exacting, quantitative, field measurements of terrestrial magnetism, he was shying away from his more qualitative, laboratory studies of large electromagnets. Indeed, it was in a letter concerning the aurora paper that he told Silliman that he had suspended his work on electromagnets. Although he had "not abandoned entirely" the work, "circumstances," including the loss of a suitable room at the academy, had forced him to interrupt his "extensive plan." Except for completing Cleaveland's large magnet and exploring the electromagnetic separation of iron ore, he had "little to communicate at the present time." In contrast, he and Stephen continued to study auroras into the spring of 1832. However, Joseph soon would return full time to his electromagnets. During the last two weeks of June, he would be scurrying to salvage some credit for the isolation of electromagnetic induction, recently announced by Faraday.

Many factors contributed to Henry's delay in completing his latest round of electromagnetic studies—the round that would contain results pertinent to mutual induction and self-induction. Along with conceptual and empirical obstacles connected with the research, there were obstacles arising from limitations of time. Duties at the academy and institute along with obligations to his double family all constrained his laboratory investigations. Moreover, he had to parcel the remaining time between competing research interests. As if he did not have a full research agenda with topography, electromagnetism, and terrestrial magnetism, he had yet another major investigative interest, meteorology. Henry's meteorological work overlapped his involvements in topography and terrestrial magnetism, particularly as they pertained to New York State. Indeed, he held a unified perspective toward meteorology, topography, and geomagnetism—an integrated perspective that mirrored not only that of his elder colleague at the Albany Institute, DeWitt, but also that of the eminent Humboldt. Through his writings, research, and world travels—

including a short visit to the United States early in the century—Humboldt symbolized an international trend to ally studies of climatic patterns with mappings of the earth's contours and magnetic features.[6]

In 1825, at DeWitt's urging, government officials in New York mandated that the scholastic academies around the state compile systematic meteorological records. Data began to accumulate from an increasing number of observation sites, and in 1829 the state published the first annual "Abstract of the Returns of Meteorological Observations." Henry had a hand in all phases of this project. Initially, he oversaw the construction of the state's instruments for measuring temperature and rainfall. With Beck, he also oversaw the local collection of data at the Albany Academy, and together they compiled the statewide figures and prepared the first five "Abstracts," extending through 1833. This collaboration paid off early with their 1829 report in the *Edinburgh Journal of Science* on mean temperatures around New York State—Henry's first overseas publication. Issuing his "Topographical Sketch" of the state at the time of the *Edinburgh* report, Henry could appreciate the sketch's potential ties to the state weather data; he told Stephen, "I hope the article will be in some degree usefiıl both as it regards the subject of internal improvement and the study of the climate of the State, particularly in connexion with the meteorological observations made by the several academies." Henry also published together with the state "Abstract" for 1830 a report on the latitudes, longitudes, and elevations of the state's forty-two weather stations and, with the "Abstract" for 1831, a copy of his aurora paper. Writing to a colleague early in the summer of 1832, he summarized his commitment to meteorology: "In connection with Dr. T. R. Beck I have the principal direction of the Meteorological observations made by the different academies of the State. . . . In this work I am considerably interested and have hoped at some future time to deduce many facts from it of importance to the science of meteorology."

8 The Call to Princeton

Back in 1825, as Henry was finishing the road survey, he had thanked T. Romeyn Beck for befriending and supporting him. "To you sir alone," he attested, "I owe what little reputation I may possess."[1] Ironically, the man whom Henry had credited with sparking his initial reputation was the one whom Henry subsequently would censure for becoming jealous of his rising reputation.

With Henry's appointment as an academy professor, he had at first drawn even closer to Beck as they collaborated on projects such as the annual meteorological "Abstracts." However, Henry's relationship with Beck and a few others at the academy began to deteriorate. In later decades, Smithsonian clerk Rhees listened as Henry reconstructed the problem. "As long as he worked for others," Rhees recorded, "they assisted & praised him. As soon as his own papers appeared & they gave him reputation his old tutors became jealous." Particularly disturbing was Beck's decision to remove Henry's name as partner in a pending state meteorological report. When Henry objected, Beck apparently responded by telling him "if he wanted to go alone he could do so." Unfortunately, the ill feelings also involved Lewis Beck, the principal's younger brother, and Philip Ten Eyck, Henry's frequent collaborator. Perhaps exacerbating the purported resentment was a stark difference in academic training between the upstart Henry, a mere graduate of Albany Academy, and the Beck brothers, both graduates of Union College who had gone on to become physicians; Ten Eyck also was a trained medical doctor.[2]

The falling out with Lewis Beck revolved around whether Henry had given him sufficient credit for his help in building, testing, and refining electromagnets. Silliman innocently sowed the seeds of the dispute when he asked Henry to clarify not Beck's but Ten Eyck's role in the investigations reported in the seminal January 1831 paper on electromagnets. Ten Eyck had assisted Henry as had, to a lesser degree, Lewis Beck; though Henry had acknowledged their contributions in the paper, Silliman wondered if as editor he had made a blunder by publishing the paper without adding Dr.

Ten Eyck's name as a coauthor. Henry assured him that only his name should have appeared on the title page, and he supported this contention by briefly contrasting his own fundamental contribution with Ten Eyck's supportive work. Without Henry's approval or knowledge, Silliman summarized Henry's candid overview and included it in an extended editor's footnote that he appended to Henry's latest publication, his and Ten Eyck's April 1831 account of the Yale electromagnet. Silliman intended the footnote not only to call attention to Henry having constructed the world's strongest magnets but also, retrospectively, to do "exact justice" to both of the contributors in the January experiments, Henry and Ten Eyck. The footnote upset Henry's fellow investigators. Whereas Ten Eyck most likely found disturbing what he took to be an underestimation of his contribution, Beck probably found distressing what he took to be his complete neglect. The younger Beck was just beginning to solidify a national reputation; he not only published a *Manual of Chemistry* in 1831 but followed it in 1833 with his *Botany of the Northern and Middle States.*[3]

By July, Silliman heard through a mutual friend that not only was Ten Eyck troubled by the footnote but also Henry had never intended his candid words for publication. Silliman apologized to Henry and offered, depending on Henry's wishes, to try "to set the matter right." Seven months passed before Henry replied. In a letter about his pending aurora paper, he added that his delay in responding on the Ten Eyck issue did not reflect ill feelings toward Silliman who had displayed throughout the episode only the highest motives. "The truth is," he elaborated, "your last letter refered to a most disagreeable affair and I was not at the time of its receipt in a state of mind to answer it properly. I have been very delicatly and unplesantly situated in reference to my magnetic experiments & my feelings have been so deeply wounded that even at this time I find I cannot refer to the subject without most unpleasant associations." Although he regretted that the experiments had led to "unpleasant feelings" between himself and the other persons associated with the project, he again insisted that the original article included an accurate accounting of the division of labor. Moreover, Silliman's later footnote, the unintended cause of "unjust censure" toward himself, stated the truth and merited no retraction or apology.

"These experiments cost me much time and thought," Henry further informed Silliman, shifting to a reflective tone, "and instead of the results producing pleasure they have been a source of much anoyance and many wounded feelings." He pointed out that he was not complaining about Ten Eyck; though he and Ten Eyck had experienced rough moments, they had both come to regret the misunderstanding and had since mended their friendship. But there remained another person, Henry added, "of whom I think I have cause of complaint." Propriety prevented him from naming the person, but apparently it was his only other collaborator on the experiments, Lewis Beck. Indeed, in his later reminiscences with Rhees about his falling out with T. Romeyn Beck, Henry said that he had also experienced "difficulty" with Lewis Beck "about proper credit for labors &c." Henry further told Silliman that this "other person" finally seemed to be shifting to a "conciliatory course." "I am now willing to forget & forgive," Henry acknowledged in his closing words, "but at the same time shall be somewhat tenacious of my rights in reference to magnetism." Four months later, he would be putting this prophetic caveat into practice as he asserted his rights—or, at least, what remained of his perceived rights—in the face of Faraday's and Forbes's recent announcements on electromagnetic induction.

Friction with the Becks brought to the surface Henry's ambivalence toward the Albany Academy. In spite of being a dedicated and hardworking member of the faculty, he initially had been reluctant to take the job and subsequently never displayed unbounded loyalty to the academy—at least not the loyalty befitting a supposedly indebted graduate and employee. Ten months after assuming his post, he was already seeking an appointment as professor of chemistry at the U.S. Military Academy. Indeed, his visit to West Point three months before beginning to teach in Albany—for which he cut short the geological tour of the Erie Canal—perhaps also involved a possible chemistry position. Of course, for an aspiring scientist—who was particularly adept in engineering—a West Point appointment offered opportunities and resources that an Albany professorship could not match; also, having almost obtained a post with the Army's corps of civil engineers, he likely realized the benefits of a military career. In July 1827 he felt "great anxiety" as he awaited a decision from West Point. Though he again visited the academy, directed a letter to the Secretary of War, and enlisted supporters to write on his behalf, the chemistry job fell through and he remained in Albany.[4]

Charles Lanman, an author and one of Henry's Washington friends, later reported that the scientist also told him about his "difficulty" with Lewis Beck, how the problem rose out of "a spirit of rivalry," and how "that circumstance made him very unhappy." "At this particular time," Lanman continued, "while out walking, he met his warm personal friend, George Clinton, who inquired the cause of his apparent dejection, and on being informed of the cause made this remark: 'Henry, it is your duty to leave Albany; as you know, a prophet is not without honor, save in his own country.'

But the time for his departure had not quite arrived." Though Lanman's telling of the story smacks of embellishment, Henry did relate a similar version to Rhees.[5] Whatever the anecdote's accuracy, Henry's time for leaving the Albany Academy did come shortly after he confided to Silliman about his discord with Albany coworkers. But it was not a departure for West Point. During the two most pivotal weeks of his young career—the final two weeks of June 1832—he not only hurriedly readied his paper on electromagnetic induction but also received a feeler for a professorship at Princeton.

John Maclean, vice president of the College of New Jersey—that is, Princeton—extended the feeler. The chair of natural philosophy likely would come vacant in a few months, and Maclean was confident that the trustees would endorse Henry's appointment. Henry delayed answering. When he finally responded, he explained to Maclean that, for the past two weeks, he had been too busy to reply; he had been devoting nearly every moment not consumed in teaching to experiments he was preparing to announce in the next issue of Silliman's *Journal*. As to Maclean's offer, Henry began his response circumspectly: "I answer that my only views at present are to secure a comfortable support for my family and next to establish and to deserve for myself the reputation of a man of science. I have determined to confine my attention principally to a course of study and investigation intermediate to pure Mathematics on the one hand and the more detailed parts of Chemistry on the other. Any honourable situation in which it would be a part of my duty to teach those branches and which would afford me superior advantages to those I now possess for prosecuting them will be acceptable." Having demarcated the breadth of his interests in physical science—presenting as outermost boundaries the same two interests, mathematics and chemistry, that during courtship he had found Harriet to rival—he summarized the advantages and perquisites he enjoyed in his Albany post. These included access as Albany Institute librarian to a rich collection of scientific literature and involvement as codirector with Principal Beck in the state's meteorological program. Regarding the drawbacks of his present assignment, he revealed, "my duties in the Academy are not well suited to my taste. I am engaged on an average seven hours in a day, one half of the time in teaching the higher classes in Mathematics, and the other half in the drudgery of instructing a class of sixty boys in the elements of Arithmetic." Finally, observing that he would be "more pleasantly situated in Princeton than I am at present in Albany," he informed Maclean that he would accept the chair if offered.[6]

Henry closed his letter with a list of references and a disclaimer. For colleagues who could judge his scientific ability, he named Silliman and Renwick. For friends who could attest to his background and character, he listed Stephen Van Rensselaer, DeWitt, and Jacob Green. The disclaimer centered on his lack of a college degree. He asked if Maclean realized that he was "principally self educated" and if this deficiency would trigger objections to his appointment. At best, he mentioned, his formal credentials included an honorary degree from Union College and recent election as a corresponding member of Edinburgh's Royal Physical Society.

Maclean could extend the tentative offer to Henry, sight unseen, because he and his Princeton cohorts had already received unequivocal assurances of Henry's abilities from two scientific colleagues closely associated with Princeton. Without even knowing that the chair might come open, Green had praised Henry. And when the chair became a possibility, John Torrey, who had left West Point for joint teaching positions in New York City and Princeton, encouraged the appointment. Maclean told Henry that he and the Princeton faculty did not need to obtain letters of reference because they already recognized Henry's scientific accomplishments. However, for the trustees's sake, he would contact a few men from Henry's list. The letters were slow to arrive during the summer months, but they were all enthusiastic in their support. Renwick responded that in "the purely physical branches of Natural Philosophy" Henry was "possessed of knowledge probably unequaled in this country." "Of his qualification to teach well the branches of Nat. Phil. & Mechanics usually taught," Green replied, "no one acquainted with these subjects and with Professor Henry can doubt." And Torrey, in his response, looked to the future: "I have no doubt, should his health be spared for a few years, but he will stand among the first philosophers of the country."

When Silliman heard from Henry of the pending appointment, he also quickly dispatched an endorsement. He first listed reasons for his belief that "as a physical philosopher" Henry had "no superior in our country," adding that Henry enjoyed "the important advantage of being an excellent practical mechanic." And though Henry apparently had the flaw of having "not cultivated classical literature," his speech and writing were still "pleasing, perspicuous & correct." All told, Silliman was confident of Henry's "intellectual power, his scientific attainments and his moral excellence." In later months, confidentially, he would encourage Princeton to provide time and money for Henry to continue his private research. He also would write to colleagues advising them of Henry's appointment and arranging introductions. "He is a most promising man," Silliman informed Robert Hare, "of a fine tone of character & with modest & winning manners."

The Princeton professorship offered a light teaching assignment that contrasted dramatically with his onerous Albany classroom duties. "On an average," Maclean explained, "you will not be occupied more than from one to two hours of the day." Maclean also pledged "social and friendly" interchange with the faculty. The position further carried with it an initial salary of $1,000, which matched his Albany pay, and the use of a house. The guarantee of his own house also contrasted with his Albany situation. In later reporting Henry's memories of Albany, Rhees conveyed Henry's bitterness toward the academy on this issue, noting that the trustees had reneged on their promise to provide a house if he got married.

Probably because of his strained relationship with T. Romeyn Beck, Henry neither included him as a reference nor informed him of the Princeton invitation. In late August, with the October opening of Princeton's fall session fast approaching, Maclean urged Henry to reveal the pending appointment to Beck. When the Princeton trustees finally formalized Henry's appointment in late September, Beck and the Albany trustees reacted by electing Ten Eyck to fill their own vacancy. With time, Henry and Beck would reestablish a civil relationship; on a later visit back to Albany, Harriet would make a point of telling Joseph that Beck had been "very polite and friendly."

Not receiving official word on his appointment until September, Henry persisted through the late summer with his normal routines. During these months of waiting, life for Henry probably remained under a cloud because of the tensions with the Becks. Meanwhile, a much more serious cloud had descended over all of Albany: a cholera epidemic that was sweeping American cities. Joseph took advantage of the break between the academy's summer and fall quarters to seek refuge for his family from the virulent and often fatal intestinal disease—from what he called "the pestiferous atmosphere of Albany." Uncle John Alexander's home in Galway provided a haven for him, Harriet, son William, and Harriet's mother. "The pure air of the country has had an almost magical effect in restoring our prostrated strength," he wrote back to Stephen who had remained in Albany with Harriet's aunt. He worried, though, about his own mother and Nancy who had both shown signs of illness, and he urged his brother, James, to be sure to move them from Albany if the cholera "should come into your street." Meanwhile in Galway, enjoying visits to "the whole clan of relatives" along with casual studies of local geology and an aurora, Joseph hoped that the academy would delay its fall opening until the epidemic ebbed. To his dismay, the school reopened on schedule, although the enrollment was "very slim." Returning alone to Albany, he wrote frequently to "My Dear little Wife," and grew melancholy. He told Harriet that "all things appeared gloomy after enjoying myself so much in the country." "I wandered involuntary from room to room," he lamented, "as if in expectation of meeting you and the *boy*. In the front room up stairs the cradle with its top counterpain neatly arranged, I suppose by Aunt, probably reminded me of its little owner." "Stay in Galway," he advised her, recalling the cholera, "as long as possible."[7]

Melancholy also enveloped Henry as he pensively reflected on the reality of leaving Albany for Princeton. An incident occurred that brought these feelings to a head. He recalled it in later years when he informally addressed a meeting of the Albany Institute, where New York physician and historian Franklin Hough heard it; Henry also repeated the anecdote to Lanman, Rhees, daughter Mary, and other close associates. Probably in early October 1832, he was returning by steamboat on the Hudson River after a preliminary visit to Princeton, where he would soon begin teaching. He was experiencing "mental anxiety," Hough recorded, "least he might prove unequal to the responsibilities to which he had been invited." Not only did he doubt "his ability to justify the good opinion of his merits which the invitation implied" but also he was "sad beyond the power of language to express, at leaving the associations of past life, for the uncertainties of a future career." Similarly, speaking to friend Charles Lanman about the boat trip, Henry repeated that "one of the greatest trials of his life was his departure from Albany." "He felt," Lanman remembered, "as if he could not possibly sever the ties which bound him to his early home, and the future was so uncertain." Rhees also recorded that Henry was "timid & afraid to go to a College so well known & where there were so many eminent & learned men." Apparently, the new Princeton appointee was sitting alone at the rear of the cabin, with his head buried in his hands, absorbed in his thoughts, and looking distraught. He felt a reassuring hand on his shoulder and looked up to discover an acquaintance, William Dunlap, a 66-year-old playwright and painter. Dunlap asked what troubled Henry and received a frank admission of his insecurity. The playwright responded to his fellow traveler's "doubt and self distrust" with words of comfort that touched Henry, giving him new courage and boosting his resolve. In Lanman's, Rhees's, and Mary Henry's telling of the anecdote, Dunlap additionally responded to Henry's disillusionment from his treatment by Albany colleagues. "Don't be depressed, my good fellow," Dunlap reassuringly urged; "the time will come when Albany will be proud to claim you as her son!"[8]

Although the anecdote took on apocryphal tones in the various renderings, it does have a documented basis: Dunlap, in a diary entry for 8 October, had mentioned encountering Henry on the steamboat. After using part of the entry to praise the new Princeton appointee, he predicted: "His name will be enroll'd with those of Franklin, Silliman, Rittenhouse & other Americans who have transmitted light from the West to the East, and from the region to which light has been travelling for ages to that whence it emanated."[9]

9 Settling In

During his first winter in the town of Princeton, with its modest population of eleven-hundred, Joseph reported to his brother, James: "We are all well and going on as usual without meeting with anything to change the monotany of a village residence"; he facetiously added that the tedium broke only when "the *expresses* employed by the Editors of New-York pass here on the run four times aday." Severed from their relations and routines in the bustling state capital of Albany, Joseph and his family probably did find some diversion in the horseback couriers rushing in relay to supply New York newspapers with the latest dispatches from Philadelphia. The family's two-story frame house, set along the back perimeter of the college's modestly sized quad or green, provided a view through the intervening elm and poplar trees to the village's main street. Nassau Street not only fronted the quad and separated it from the shops, taverns, and houses on the north side of the broad avenue but also provided the midpoint of the well-traveled, hundred-mile road wending between Philadelphia and New York.[1]

What village life lacked in excitement, the college more than offset with its student body of 150 and staff of thirteen. The staff consisted of: the president and vice president, who both also taught; seven professors, including Henry; three tutors; and a steward who oversaw housing, food, and other practical matters. Vice President Maclean, who had hired Henry, shouldered more of the daily operation of the college than President James Carnahan, a well-meaning but lackluster Presbyterian cleric. By spring Henry could report to Silliman: "I am much pleased with my situation and thus far have no cause to regret my exchange." Both age and character distinguished the College of New Jersey. Chartered by the governor in 1746 and settled in its permanent Princeton home in 1756, it was the nation's fourth oldest institution of higher education. Moreover, it boasted a distinguished lineage of presidents, faculty members, and graduates. Henry aptly described the college's legacy, writing "that she has educated a large proportion of the men who have directed the affairs of the general government since its first establishment and that from the first she has been the steady friend of true

Religion and has furnished the church in this country with a long list of eminent Divines who have earned for her the appellation of 'the school of the Prophets'."[2]

For government leaders, Henry probably had in mind a distinguished group of administrators, professors, and graduates that included two signers of the Declaration of Independence, one-sixth of the delegates to the Constitutional Congress, and the nation's fourth president, James Madison. By religious divines and prophets, he likely meant the college's impressive yield of Presbyterian clergymen and theologians. The college, though formally unaffiliated with any denomination and open to students of all faiths, was a creation of Presbyterians caught up in the Great Awakening, the mid-eighteenth-century religious fervor that swept the colonies. And Presbyterians—often with ties to Scotland, a wellhead of their embodiment of Reformed or Calvinist religion—continued their influence into the nineteenth century. The opening of the Princeton Theological Seminary in 1812 reinforced this ongoing, informal leverage. Officially, this institute for training Presbyterian ministers functioned independently of the college. The seminary and the college, however, drew from a common stock for their faculty, board members, and students (graduates of the college often went on to the seminary) and also shared a Calvinistic and Scottish legacy. When Henry arrived in Princeton, the venerable Archibald Alexander and the ascendant Charles Hodge were only the most recent Presbyterian divines to dominate the Princeton scholarly community. Henry was likely aware that his own Presbyterian and Scottish roots probably contributed to the decision to invite him into the community. And brother-in-law Stephen, who accompanied the Henrys to Princeton, realized that he could further his ambition of obtaining his own professorship with the college by enrolling straightaway in the seminary; by the next spring, about the time he gave his first sermon and consented to being "incased in a frock coat of the Princeton cut," he received an appointment as a college tutor—a position that quickly led to a lifelong tenure as a professor specializing in astronomy.[3]

That college officials intended for Henry to be an integral member of the Princeton scholarly community was evident in the location of the house they provided for him and his family. The house stood near the southwestern corner of the smallish, quadrangular campus between two main buildings, the "College" and the "Library." The College, later known as Nassau Hall, was a stately landmark extending along the rear perimeter of the quad, facing Nassau Street. Originally it had been the only academic building and—after weathering occupation by British and American troops during the Revolutionary War—it continued as the hub of student life. There, students slept under the supervision of resident tutors, woke at 5 A.M. to the blast of a horn blown by the "rouser," attended morning prayers in the building's chapel, presented recitations to their tutors and professors, attended vesper services at 5 P.M. in the chapel, and met the 9 P.M. curfew. The Library stood on the other side of Henry's house, along the quad's edge running perpendicular to Nassau Street. Besides holding a modest collection of books on the second or main floor, it contained recitation rooms on the ground level and housed the student's two academic societies in its upper rooms. The American Whig Society and the Cliosophical Society underpinned all aspects of college life. Nearly every student routinely belonged to one of these secret literary societies, where they found academic support, fellowship, and a competitive motive to win the college's scholarly prizes. Soon after Henry's arrival, the Whig Society elected him an honorary member.[4]

Across the quad from the Library, along the other perpendicular running to Nassau Street, stood Henry's main haunt, "Philosophical Hall." As Joseph explained to James, "the third story is the Museum and the chemical and Philosophical rooms in which most of my time is ocupied." Students also called the building the "Refectory" because they ate their meals there, served by "negro lads," at the cost of $2.50 per week; the dining room occupied the main floor while the kitchen was on the ground level. The remaining buildings on the quad—that is, the remaining buildings at the College of New Jersey in the fall of 1832—were Carnahan's residence (between the Library and Nassau Street), Maclean's house (between Philosophical Hall and the street), and the college steward's home (on the other side of Nassau Hall away from Henry's home). With the arrival of spring, Joseph could rhapsodize to James about his new surroundings: "The College Campus or green has a most beautiful appearance and every thing about Princeton at this season is delightful." And toward the end of spring, he could further relate to James that no less a dignitary than President Andrew Jackson, recently reelected for a second national term, spent a morning visiting the campus. After passing the night at "the hotel almost directly opposite our house," Jackson joined the students and faculty for their daily prayers at 5:30 A.M. in the chapel and then toured the Library and Philosophical Hall, leaving "a favorable impression."

Harriet was pleased with the family's new house, declaring it to be "comfortable and quite convenient." She was particularly happy that the interior had been freshly cleaned, painted, and wallpapered. The downstairs rooms consisted of, on one side of a central hall, a formal parlor with front windows looking out to the green and rear windows opening to a large yard with a garden and, on the other side of the hall, a casual "dwelling room" distinguished by a fireplace with closets bracketing the chimney. To the rear of the dwelling room extended a spacious kitchen, and upstairs were three large bedrooms with attached closets and one small bedroom located over the downstairs hallway. The house adequately accommodated not only Joseph, Harriet, and son, William, but also Harriet's mother, aunt, Stephen, and—through the years, in the small bedroom—a series of college students with whom the family had personal ties or juveniles that helped Joseph in the laboratory. In addition, the family apparently brought from Albany a young British immigrant—William Bannard or "English William"— who served as a domestic and probably slept in the kitchen. They also brought from Albany, by boat and cart, most of the furnishings. These included utilitarian articles such as bedsteads, mirrors, tables, a sideboard, and crockery, along with special items such as a piano, Harriet's music box, and a small painting of Ann Alexander Henry—Joseph's mother and Harriet's aunt. In a letter, Joseph asked James to tell their mother that the painting stands on the lintel over a fireplace directly in front of his and Harriet's bed and "that I never go to sleep without thinking of her."[5]

Initially, Harriet and her mother, Maria Alexander, found themselves overwhelmed trying to keep up with the cooking, laundry, and other household labors. Part of the problem traced to a faulty kitchen stove that Stephen had purchased in New York. But a more significant aggravation sprang from what they considered a lack of household help. English William provided daily assistance, but Maria Alexander apparently was accustomed to more help. She immediately engaged an African American woman presumably to be a live-in domestic, but unfortunately the woman could not begin work for a few weeks. Eventually, the family employed additional African American workers. For example, in the spring they hired an older man to help English William spruce up their courtyard. A year after that, a "black girl" was assisting with child care. And a year after that, in the spring of 1835, Joseph could speak of "English William and the other auxiliaries of our establishment" and add that there were "no fewer than *three* Blacks scrubbing at one time in as many different rooms." While the household probably included at least one live-in domestic throughout these early Princeton years, by 1840 the Henrys had moved into an adjacent, brand-new, larger house and the number of domestics had definitely risen to three, one male and two females.

Although Harriet did her share of washing, ironing, and other tasks, she doubtless increased her appreciation of the hired help during the next seven years with the addition of three more children, all daughters. (She also bore two other sons, but both died in infancy.) During Harriet's second winter in Princeton, when William was about two years old, Harriet gave birth to Mary Anna. The mother soon pronounced the bright baby a "prodigy." About two years later, "Bub" and "Sis" welcomed a new sibling, Helen Louisa. Finally, in 1839, Harriet bore Caroline, a baby "quite smart and well made." By then William was old enough for active, outdoor play with his father. Whether his father was ready is another matter. Henry seriously sprained both wrists when he stumbled over a pile of stones while "attempting to assist Willey in raising his kite." At Christmas the next year, events unfolded more smoothly as Henry got swept up in the excitement of William, Mary, and Helen's expectations. "The three older children are making great calculations," he wrote to James, "relative to the important ceremony of hanging up the stocking. Many conjectures are expressed as to the probable contents of the articles." Adopting St. Nicholas's words from the closing line of an increasingly popular, children's poem that had appeared in New York during the 1820s, Henry ended this letter to his brother with the wish: "A merry Christmass to all and to all a good night."[6]

Earlier, in April 1835, the death of Joseph's mother shattered the calm between the births of Mary and Helen. Ann Henry died at age 74, after being bedridden for a few years with recurrent fevers, intestinal pains, paroxysms of coughing, and shortnesses of breath. Her illness perhaps traced to her longstanding habit of smoking a pipe—a habit so ingrained that even young William, to the amusement of his elders, had learned to "smoke and spit" in imitation of his grandmother. Out of a sense not only of heartfelt duty but also lingering anxiety over having taken his family from Albany—sentiments probably traceable, at least in part, to the traumas of his childhood—Joseph had felt compelled to make frequent and sometimes emotionally "painful" visits to his ailing mother. Though siblings Nancy and James provided daily care, Joseph also had continued to furnish financial support and watch out for his mother's special needs; this proponent of temperance even had arranged for the purchase of half a barrel of a locally brewed alcoholic beverage in case, as he told James, "Mother would like occasionally a glass of ail." In her final months, a menacing fire near the family's Albany home had led Joseph to reflect on her personality: "I hope it did not much alarm Mother. She is however a much better soldier when danger really comes than a person would imagine from her usual timidity." When her final moments came, Joseph later remarked, she expired "with scarcely a struggle and her last prayers were for her children and grandchildren." "The event," he told Stephen, "was a great shock to me. I have come to Albany several times with much anxiety but had become accustomed to find her better, and in this case was not in the least prepared for the circumstance of her departure."[7]

Shortly after moving into their Princeton house, Harriet had begun to appreciate the significance of the large garden out back. She had anticipated raising "vegetables enough for our family." Under Maria Alexander's watchful eye, the garden prospered with, for example, English William planting peas. But Harriet's mother went a step further during the family's first spring in Princeton. Accompanied by President Carnahan's wife and driven in a wagon by English William, she went into the country and bought a cow and calf for $24. Soon selling the calf, the family kept the cow in an enclosed portion of their yard where it produced enough milk for the large clan—especially William, whom Harriet was weaning. In fact, the cow produced a daily excess of four to five quarts of milk, which Maria Alexander sold at a profit to the college steward. In 1844 the family would acquire more livestock when one of Henry's appreciative students gave William a pony, a gift that would make the twelve-year-old, according to his father, "the envy of all the Boys in the village." He would enjoy hitching the pony to a small wagon or sleigh and giving his mother or sisters rides around the campus.[8]

Harriet, her mother, and her aunt initially missed their Albany coterie and felt "dispair of ever finding any in Princeton with whom they can associate on such terms of intimacy and friendship," but they established close bonds with their new neighbors almost as quickly as they set up the new household. "We find the President's family and also Mr Mclean's very pleasant and kind," Joseph soon informed James. Meanwhile, strolling across the green between this lively household and Philosophical Hall, Joseph was launching an approximately fifteen-year professorship that he would later judge the most personally satisfying episode of his life.[9]

The new professor of natural philosophy relished the prospect of a manageable workload. His formal, weekly assignment involved spending only about two hours for each of three days in lectures and recitations plus attending chapel every day at 5 P.M. Also, accustomed to "the noisy boys of the Academy," he initially found himself "almost embarrassed" by the "quiet attention" of his Princeton charges. Silliman assured him that the new position was "a very honorable & advantageous retirement from your severe labors & general treatment in Albany." Henry quickly discovered, however, that his new assignment did require intense, if not "severe," labors. Previously in Albany, even with his advanced courses, he had grown accustomed to a traditional, regimented teaching style of recitation and repetition linked directly to lessons in a textbook; he typically reserved formal lectures, more challenging to prepare and still uncommon in American schools, for only special topics such as electricity and magnetism. Also at Albany, he had gradually assembled a well-stocked cabinet of philosophical instruments to illustrate basic phenomena. But at Princeton he found himself having to devote "undivided attention" to merely two lectures and three recitations each week because of the need to frame new lectures in natural philosophy and to ready apparatus for demonstrations. When he came to the unit on mechanics, for example, he explained: "My duties are some what more arduous on account of my never having before lectured on this subject and also because the apparatus is much out of order and repair."[10]

Mechanics was only one of the basic but complex topics that Henry presented in his course. He also prepared sets of lectures on somatology (the study of matter), hydrostatics and hydrodynamics, pneumatics, heat, sound, electricity and magnetism, light and radiant heat, and meteorology. As in Albany, in presenting topics from natural philosophy, he emphasized not merely abstract theories but also concrete applications such as steam engines. Moreover, continuing his Albany practice of compiling background notes through exhaustive reading and other experiences, he would toil through coming years to refine and update his lectures even on familiar topics. A decade after coming to Princeton, he would still be devoting months to master fully a single subtopic such as capillary interactions. As Henry developed into the college's leading champion of instruction based on the innovation of lectures rather than recitations, the course would grow to eighty-nine lectures presented three or four days each week plus supplementary recitations. Through most of his fifteen years at Princeton, Henry would pack the lectures into the college's winter session, typically running from early November through April.[11] (After giving the students a short vacation in late April, the college continued with a summer session between May and September; commencement followed in late September, along with another vacation in October. In 1844 the college shifted commencement to late June and introduced a revamped schedule anchored by fall and spring terms.)[12]

Part of Henry's challenge in teaching natural philosophy at Princeton was that all seniors had to study the topic though they had only a smattering of training in science. With most "gentlemen" matriculating as sophomores at ages as young as fifteen or sixteen and aiming for careers in the ministry and other "learned professions," students followed a basic, conventional course of instruction before the senior year. Although they studied some applied and basic science, juniors, sophomores, and the few freshmen concentrated on ancient and modern languages, mathematics, mental philosophy (the branch of philosophy stressing psychological processes), evidences of Christianity, natural theology, and, every Sunday, the Bible. Henry's course on natural philosophy fit into this carefully prescribed curriculum in a manner similar to president Carnahan's course on moral philosophy, which was also limited to seniors. Both courses sought not merely to impart factual knowledge but to provide final training of the seniors' higher mental faculties. Building on his Albany commitment to the psychological theory of mental faculties—and the associated mental-discipline approach to teaching—Henry would come to stress that natural philosophy was rightly restricted to seniors because only they had adequate training to benefit from the use of scientific, inductive reasoning to arrive at unifying principles. "When I commenced this course," he reminded a later class in his closing comments at the term's end, "I informed you that my object was not to teach you the mere facts of natural philosophy but general principles—to exercise you in the analysis of facts and the explanation of them by referring them to general laws or in other words my object has been to teach you to think—to philosophise—to arrive at general prin[ci]ples." The resulting mental discipline could transfer, Henry pointed out, to theology, law, medicine, and all other fields.[13]

Also, in line with his own commitment to liberal education and the cultural relevance of science, Henry informed his students of even broader goals for the course. Year after year, the students heard their professor read from the notes of his introductory lecture: "The ultimate tendency of the study of Physical science, is the improvement of the intellectual, moral, and physical condition of our species. It extends the bounds of human thought and human power. It unfolds the magnificence, the order, and the beauty of the material universe. It affords most striking proofs of the beneficence, the wisdom, and the power of the Creator. It habituates the mind to the contemplation of truth. It enables man to control the operations of nature and to subject them to his own use." Students also transcribed their professor's spontaneous elaborations on this prepared statement. Reflecting his own and the college's commitment to natural theology and the harmony of Christianity and natural science, Henry developed the idea that the "immutable" laws of physics provide evidence of the design of the "Divine Intelligence." He also amplified the notion that physical science not merely gives the species control over nature but even improves the species. According to one student's notes, physical science "tends to bind closer together[,] by the strong ties of friendship[,] the whole of the human family—to advance the arts and sciences—and to hasten forward the civilization of the whole earth." And, the notes continue parenthetically, this civilizing tendency of science had a more tangible, literally visceral consequence. "It has been established as a fact that civilized man is larger—his body more developed—and more able to endure fatigue than the savage man."[14]

General laws, expressed with as little mathematics as necessary, capped Henry's lectures on each of his main topics such as heat or sound. At the outset of each topic, however, the lectures revolved around particular, fundamental observations and experiments. That is, he structured his course to mimic what he took to be the historical development of each topic, with the probe of phenomena preceding the inductive articulation of laws. This tidy, linear view of the history meshed with his guiding educational principle—dating back to his Albany concern with mental faculties—that students need to learn the "concrete" before the "abstract." (While in Albany, he had preached the precedence of the "doing faculties" over the "thinking faculties.") Aiming to ground his lectures in the tangible, he illustrated many phenomena by routinely drawing diagrams on either the blackboard or auxiliary black paper and by occasionally projecting images on a wall with a "Magic Lantern." For the lectures on steam and steam engines—part of his coverage of the broader topic of heat—he and his assistant gave the students "quite a treat" by showing them not only sketches of different engines but also a working model.[15] In illustrating most phenomena, however, Henry relied on full-scale demonstrations and experiments using standard philosophical instruments and makeshift apparatus.

Unfortunately, he had to contend with Princeton's inadequate collection of implements. He often spent Saturday afternoons repairing or constructing instruments and would travel to New York City in search of appropriate materials and supplies. In fact, he devoted much of his second year in Princeton to building a large, versatile battery. "It has cost me nearly all my leisure time for a year past," he complained to Silliman in October 1834, shortly after exhibiting the battery at commencement. "As there are no mechanics who work in metal to be procured in this place," he elaborated, "all the soldering has been done by myself and an assistant. This has been a very unprofitable waste of the little time my college duties have left unoccupied." A year later, nevertheless, this frustration had turned to satisfaction after he began to capitalize on the battery in the lecture hall. In an update to Silliman, he exclaimed: "I have shown to my class with my large battery some of the most magnificent experiments on electromagnets which I believe have ever been exhibited."[16]

These extraordinary experiments would have as their centerpiece a new electromagnet that Henry had constructed a little earlier. The *New-York Standard,* reporting on the magnet's unveiling during commencement in September 1833, decreed it a "wonderful instrument" that "far surpasses in power every thing of the kind." For the core of the magnet, Henry probably used the 101-pound iron horseshoe that he had failed to take advantage of in Albany when he first considered experimenting on the identity of electricity and magnetism. When mounted in a large wooden frame and eventually joined to the new battery, the electromagnet would support 3,600 pounds, a new record. One student who had assisted Henry recalled that the professor "fairly leaped from the floor in excitement when he saw his instrument suspending and holding a weight of more than a ton and a half." Never publishing a report of this imposing magnet—a magnet that over a decade latter he could still label "the most powerful yet made"—Henry reserved it primarily for demonstrations in the lecture hall. To impress on his students the strength of the device, informally known as "Big Ben," he enjoyed loading eight bulky young men on a board extending across the large pan of "a hay maker's scale" hanging below the armature, and enlisting a few other students to pull down on an iron lever linked to the armature. Although graduates of Henry's course would vividly recall the brute strength of his giant magnet, they also probably remembered the side effects of the large battery used to power the magnet. As one senior noted in his diary during an 1846 demonstration of the magnet, the great volume of hydrogen gas evolved by the battery caused "sneezing & coughing"—each student's sense of smell "became tortured & wrung until everybody's eyes became red & gave forth water." Similarly, in a run of experiments in 1840, Henry attributed recurrent head pain, stomach upset, and even a momentary, partial loss of vision to the fumes.[17]

Beginning sometime in the mid to late 1830s, Henry introduced a new twist with the electromagnet. By remaining at a distance and using a telegraphic signal to deactivate the magnet, he could cause a large load of weights to crash dramatically to the floor. In particular, he arranged a small, intensity magnet in a long telegraphic circuit to lift and disconnect a movable wire in the local circuit supplying the large, quantity magnet. Again, he did not publish a description of this apparatus—an apparatus that anticipated the electromagnetic "relay," an essential component of commercial telegraphy more doggedly pioneered in London about the same time by Charles Wheatstone. He merely impressed on his students that the procedure presaged the remote control of enormous mechanical effects over large distances, such as the ringing of faraway church bells.[18]

The large battery and electromagnet bolstered Henry's lectures on electricity and magnetism, but there were other topics in the course that still lacked adequate instruments. During the spring and summer of 1837 Henry made a buying trip to Europe to remedy the remaining deficiencies. Nevertheless, by the fall of 1837, even before the newly purchased European instruments had reached Princeton, Henry could assure his seniors that during the course they would witness a strikingly full slate of demonstrations. On the first day of class, he declared: "I pledge myself to give an experimental illustration of every important principle in Natural Philosophy. In order to accomplish this however considerable labour is necessary in preparation as the apparatus is defective and instead of finding an article for each experimental illustration it is necessary to make one article serve many purposes. Notwithstanding this we hope to shew as many as are given in any college in the country making up by industry what we may be deficient [in] instruments." In another first meeting, he informed his new class that "the lectures cost much trouble and labour: 3 hours before, 1 during, 1 after, 2 in putting up [the apparatus]. The class must labour in accord."[19]

Soon after beginning to teach natural philosophy at Princeton, Henry explored the possibility of buttressing the course with a textbook. Many European texts enjoyed high reputations, but few American texts even existed. Henry learned that a colleague at West Point, Edward Courtenay, was about to publish a translation of a respected French textbook on mechanics—a translation that, through added commentary, Courtenay had tailored to his students. Beginning in the 1834–35 session, Henry assigned this new translation of Jean Louis Boucharlat's *Elementary Treatise on Mechanics.* A few months into the session, however, he discovered that the calculus-based text added to his already heavy workload. "My time thus far in the term," he told his brother, "has been very much occupied as I lecture 3 times each week and hear as many recitations in a new work on Mechanics which I have never before read and which requires considerable research among other works on the same subject in order to a full understanding." A year later he turned to another colleague, Alexander Dallas Bache, at the University of Pennsylvania, for his new American edition of a well-regarded British text, David Brewster's *Treatise on Optics.* Bache had enhanced this up-to-date textbook with additional commentary.[20]

Ultimately, Henry found frustrating the dearth of original American textbooks. He attributed the deficiency to the lack of a copyright law that would make it illegal for American publishers to "filch with impunity" existing European texts. Without such a law, publishers had free access to foreign materials, available for only the cost of setting them in type, and so were averse to paying for original works by American authors. Indeed, the reluctance to fund American authors would perhaps contribute to his and Bache's failure to follow through on a plan to write their own two-volume textbook on natural philosophy—a plan ultimately disrupted primarily by Bache's move from Philadelphia to head the U.S. Coast Survey in late 1843. Eventually Henry judged as unsatisfactory the Boucharlat translation that he had been using and other textbooks that he had tried. As a result, he stopped assigning any textbook in his classes. By 1842 he resigned himself to providing his students with merely a detailed outline of his lectures—an outline "containing catch words of my lectures and the principal formulae given in my courses."[21] Normally, at the beginning of each lecture, an advanced student would write the outline on the blackboard for the class to transcribe. About 1844 Henry began to prepare fuller outlines and had them printed for the convenience of his classes. But again, even preparation of this *Syllabus of Lectures in Physics* consumed so much time that Henry managed to publish only the first three units of the course—the introduction and the sections on somatology and mechanics. He discovered "the task a much more difficult one than I had anticipated and when I came to put my propositions on paper many of them required more profound investigation than I had previously given them and such is the habit of my mind that I cannot put on paper what I do not thoroughly understand."[22]

What initially seemed a comfortable teaching assignment at Princeton turned out, year after year, to be a time-consuming succession of lectures and demonstrations. Moreover, natural philosophy was not the only course that Henry taught. During the summer session of his first year, he was also responsible for the senior course in chemistry, the sequel to natural philosophy. (Henry was substituting for the regular chemistry professor, John Torrey, who was overseas.) Henry further spent this first and the next three summer sessions teaching astronomy to seniors, a job that he relinquished to Stephen Alexander in 1837 after his brother-in-law secured a permanent faculty position in astronomy. Over the years he also presented introductory versions of his natural philosophy course to juniors. In addition, juniors made up his main audience for an elective course that he offered in the mid-1830s on civil engineering and architecture, topics that he had taught at Albany; he illustrated his lectures, for which students paid a supplemental fee, with models of buildings and architectural drawings mounted on pasteboard. In 1838 he turned over the architecture course to mathematics professor Albert Dod, pressing him to continue teaching the class using the more innovative lecture method. This was part of Henry's campaign to encourage wider adoption of lectures in a Princeton curriculum still heavily dependent on student recitations; Vice President Maclean endorsed the campaign in 1836. Henry also eventually supplemented his main teaching assignment with an elective course on geology. Having neglected the field of geology since traveling the Erie Canal with Eaton, he consumed a good portion of the summer of 1841 brushing up his knowledge and preparing materials and lectures, including commentary on the compatibility of geological science and Biblical revelation.[23]

Most of Henry's courses—natural philosophy, chemistry, architecture and civil engineering, and geology—involved large enrollments. Natural philosophy averaged about sixty seniors each session throughout Henry's tenure at Princeton. These large rosters meant long hours spent at the end of every week reading the mandatory notebooks in which students, after deciphering their casual classroom jottings, reconstructed Henry's lectures. Judging by the handsome calligraphy and meticulous illustrations in many of the surviving notebooks—along with a strict fidelity to Henry's polished, stock lecture-hall utterances—the students also invested much time. "We the undersigned, do hereby certify," two initiates interjected in their transcription of a hydrodynamics lecture in January 1843, "that we were engaged in *writing lectures* from 10 o'clock in the evening of the 21st until 6 o'clock A.M. of the 22nd." The heavy enrollments also meant much time spent conducting recitations and examinations, both of which involved one-on-one, oral exchanges. After returning from college prayers one evening in the middle of his first week-long examinations in natural philosophy, he told Harriet that he was "fatigued." The long stretches spent in examining the students, he explained, were "very tedious" and "appeared of double the length of any equal space of time since I came to Princeton."[24]

Propelled not merely by necessity but also by conscience and duty—remnants likely of drives acquired in childhood—Henry found that classroom activities alone filled his workdays at the College of New Jersey. But his dedication—coupled with a talent for teaching, sensitivity toward students, and robust physical bearing—apparently won him the admiration of Princeton's juniors and seniors.

Professor Henry 10

Student testimonials to Henry's classroom prowess abound. A typical tribute came from Theodore L. Cuyler, who entered Princeton as a sophomore in autumn 1838. He later recalled that, though the faculty "contained several remarkable men," without question "the pride of the College was *Joseph Henry.* He was then in his splendid prime."[1]

Cuyler and his classmates particularly found Henry's appearance to be striking: "The physical beauty was as attractive as his modest & manly bearing. A clear fresh complexion, rosy cheeks & rich brown hair made his a 'face to dream about'. He sometimes came into his lecture-room with a broad white collar turned back from his neck, & we college-boys used to think that Byron at his best could not have been a handsomer man." William E. Schenck, who assisted Henry during the winter of 1833–34 in building his large battery, went on to become a member of the 1838 senior class. Reminiscing about his old teacher, he also recalled being struck by Henry's appearance: "He was a man of *presence*—He was a handsome man—of fine form & physique, with a full round benignant face, inviting approach & intercourse—No pupil, no child, probably ever feared to address him & this was an important factor in his success as a teacher—He had a frank, open, commanding & dignified *presence* with an unaffected simplicity of manner inviting familiar & unconstrained intercourse." Stephen G. Dodd echoed Cuyler's and Schenck's impressions when he similarly reflected on his experiences as a member of the class of 1846. Musing on Henry's appearance and temperament, Dodd wrote: "Prof. Henry was a most noble type of man. He had large, lustrious, blue eyes, and an expressive and winning countenance. As a lecturer and professor, he was simple and natural." Joseph Halsey—a member of the class of 1842, who took "first honor" in Henry's natural philosophy course—elaborated on the professor's temperament: "Prof. Henry was the rarest Combination of Nervously Excitable Energetic intellectual power, with the most excessive caution, modesty, and gentle tenderness of heart, I ever met. The purest mettle magnificently tempered."

Schenck further saluted Henry's meticulous organization of lectures and his ability to speak with "clearness & precision & force that commanded the attention of his flabbiest pupils." Looking back, Schenck also underscored his old teacher's emphasis on general principles and his patience with students:

> He hated unnecessary words—No pupil could satisfy him by memorizing a string of words in lieu of mastering the thought or principle or explanation & presenting this in his own language—and thus he was a capital teacher—clear, sharp, distinct, impressive, always securing attention, knowing what to withhold as well as what to present, successful in experimental illustration & remarkable for his patient repetition & elucidation after lecture, of what may not have been previously fully understood—I have often thought he appeared to best advantage as a teacher when, after lecture, with the usual crowd around his table, a question was asked of a matter not previously fully comprehended—He would thus go over & over the simplest truths or explanations with the dullest pupil, saying at intervals, do you see? until failure to comprehend was no longer possible—The simplicity of the question & the answer made no difference.

Dodd also found memorable the degree to which Henry was "fond of being questioned by us in our recitations, and teaching us to answer our own questions, by making application of the principles and laws he taught us, to facts and conditions." "He was an enthusiast of true science, preeminently an educator," Dodd continued, "all the students having for him an unbounded admiration and affection. He was the culmination of the College. I have heard many College graduates talk by the hour of Professor Henry, of their affection for him, and their very great indebtedness to him."

While Henry strove to ensure that his students understood the substance of his lectures, he also "endeared himself greatly" to the students with the insertion of personal anecdotes. Halsey remembered that the students appreciated "his cordial, frank expression of little incidents happening to himself, & which for a time seemed to affect his delicately sensitive nature." For example, it was to Halsey's class that Henry recounted the disappointment he felt after his first public lecture on galvanism, when an Albany merchant implied that the flashier presentation of an itinerant but glib lecturer had more substance.

Most Princeton graduates probably would have considered it an understatement when Schenck lauded Henry for being "successful in experimental illustration." Student notebooks reveal that the professor not only figuratively but also literally galvanized the aspiring scholars.[2] "Prof. Henry gave the whole class a very respectable shock," one senior recorded, only to add a week later, "Prof. Henry galvanized an unfortunate chicken which thus fell a victim to science." Decades afterward, graduates still were enthusiastic about Henry's lecture-hall spectacles. Along with dramatic displays of his giant electromagnet, demonstrations of his telegraph ranked high among the graduates' most vivid memories. Though he had covered the topic of telegraphy every year that he taught at Princeton, he introduced a particularly striking illustration beginning in the early 1840s. He strung a single wire from his third floor lecture room in Philosophical Hall to the study in his new house (which the family had occupied in late 1838 and which was next to the site of the original, now torn-down house). To complete this long electrical circuit, he dispensed with a return wire and, instead, adopted the innovative technique of using the earth. That is, he "grounded" the ends of the wire by immersing them in deep wells adjoining the hall and his house; although a German in 1837 had devised this technique that would prove to be crucial in commercial telegraphy, Henry had been the first to apply it in North America.[3] M. B. Grier, a student at the Theological Seminary whom Henry had invited to sit in on his class during the winter of 1842–43, recalled that Henry had announced in advance the telegraphy demonstration and, on the appointed day, the students and guests gathered "in eager expectation of the wonder to be wrought." What Henry had arranged was a two-way transmission, sending an electromagnetic signal to Harriet at the house and having her respond by electromagnetically ringing a bell-like triangle in the lecture hall. "We waited in earnest expectancy, for three or four minutes" Grier remembered, "when the little triangle rang out the news that the Professor's signal had been received at the other end of the wire, and a response returned. As the rapid sounds came from the triangle, the demonstration seemed to be so complete, that the class broke out into applause." A fellow witness added that "Prof. Henry stated with evident and great satisfaction that a wire for the return current was not needed." Henry would continue to impress students with this display for the remainder of his Princeton tenure.

Another favorite and long-remembered demonstration in acoustics involved transmitting the sound of one violin to another, distant violin. This demonstration grew out of a similar, well-received display that he had derived from Wheatstone and presented in 1839 as part of a heavily attended lecture series for the Mercantile Library Association in New York City. In preparation for the Princeton demonstration, part of his lectures on sound, Henry ran a long pole from the basement of Philosophical Hall to the ceiling of his third-floor lecture room. Attached to the bottom of the pole was a violin while screwed to the top was "a rude imitation of a fiddle." When the contraption was ready, Henry stationed his assistant, Sam Parker, in the basement. In their reminiscences, Princeton graduates stamped Parker as Henry's "mulatto man 'Sam'" or "negro Sam"; the college had hired him in 1840, at an annual salary of $48, to help Henry with his apparatus and experiments in teaching and research. As for the violin experiment, P. C. Van Wyck of the class of 1845 recalled: "On a bell signal from the Professor, Sam would saw away with his bow in the cellar, the Professor calling the attention of the class to the weird music his fiddle discoursed in the lecture-room." "On these occasions," Van Wyck added, "Professor Henry always remarked that the function of the philosopher ceased when he demonstrated the principles of nature in his discoveries: that it then fell to the share of the inventor, by ingenious devices, to subordinate them to the uses of man."[4]

The lapsed decades between these students' college courses and their later reminiscences might have led them to romanticize their portraits of their old Princeton professor. But contemporary letters of appreciation written by recent graduates confirm that students already revered Henry. "The period which I passed under your eye, was certainly the serenest and happiest of a life, that for its short duration, has been somewhat checquered," wrote a grateful Georgia lawyer two years after graduating in 1844. He thanked Henry especially for developing in him the mental habit that was the most useful, from all his basic education, for a lawyer—the "habit of generalization," specifically the facility "to reduce particular cases under the great general principles of the Law." A few years later, another student who had become a lawyer commented: "Until I came under your direction I cannot recollect of ever having possessed an idea—This is no exaggeration, but my serious opinion—I mean that I had never once thought for myself on any one subject."[5]

A diary kept by John R. Buhler, a senior in 1846, offers further substantiation of the students' admiration for Henry. Buhler, a loquacious Louisianian and would-be poet, first recorded a detailed description of his teacher after a chance conversation in a bookstore—a conversation that ranged over physics and metaphysics and touched on Kant, Keats, and Humboldt. Seventeen-year-old Buhler wrote of his professor, then in his late forties: "He is a splendid old fellow to talk with—he affects no superiority over the smallest, is free & familiar & social—but 'for a' that', there is that about him—an indescribable *Je ne sais quoi*—which excites within one, an emotion akin to Awe! I always feel as if I were in the presence of a Superior Being—infinitely far above *me* or my hopes of future being." After Buhler completed his final examination in natural philosophy and was awaiting commencement, Henry invited him, in the interim, to attend his shorter course of lectures on natural philosophy for juniors. Flattered, Buhler raised his level of elocution in his diary: "The old Fellow is a Walking Awe to me—a Peripatetic Superstition! I dont know the littleness of my own being until I am thrown into his company & then the *Lilliput* of my Mind crouches before the *Brobdignag* of His! I feel as though I were in the Presence of awful disembodied Intellect & my puny Insignificancy fairly writhes before the mighty nervousness of the Presence!" He ended this paean—after first disparaging Princeton's other professors—by wishing, may "a Star ever shine on the head of Nassau's Nobleman—the Great & Wise HENRY!!" A few days later, the star continued to shine and even brightened as Buhler audited the junior lectures. Buhler explains:

> He grows upon you. There's no popping up & then popping out about him. He isnt a kind of Water Pot that thro' a number of small holes allows his Instructions to filter through & drizzle out on the Understandings of his Pupils—but he is a Great Canal—laugh not at the Simile—like a Canal in its Constancy[,] Uniformity, Depth & Majesty of Flow. Or like his own *Galvanism*—a Strong & Constant & Powerful Current—not possessing the momentary pungency & the Rapid Brilliancy of *Electricity* is true, but having that which *it* has not, a Continuity & a Deep Power in it that lasts & lasts with strong *Effect*.[6]

Although awestruck by "Nassau's Nobleman," Buhler could use cutting language to describe Henry—more precisely, Henry's physique—when the professor was acting to enforce discipline. A week after their bookstore conversation, the young diarist watched in chapel one day as Henry reacted to boisterous freshmen and sophomores who were stamping their feet; Henry warned the perpetrators that, if caught, they would be punished. "Several fellows thereupon began a laugh," Buhler further records, "& the potgutted old Prof. ordered them personally & by name to leave the CHAPEL which they had to do." Edward Hall, a student who graduated two years after Buhler, similarly recalled not only the middle-aged professor's "tendency to corpulency," but also his penchant for maintaining firm discipline. According to Hall, Henry's normally dignified deportment concealed an agitated, authoritarian manner:

In the lecture room, there was much in his bearing that showed that he was a man of power. In speaking there was often an implication of self restraint, that he was holding himself in, that he was using the curb. There was a nervous tension, which showed itself in his quick glances, and in the tones of his voice, although these were always conversational. And when the attention of any student flagged, he brought down the rattan cane that he used in the lecture room with an impatient whack on the high table before him, and with a vigor that showed that there was plenty more nervous energy behind the arm that wielded the cane. If the whack came, when the nerves were tense while writing, it was like an electric shock.

That Henry shared in the era's belief in corporeal punishment shows in an offhand comment that he included in a letter to Harriet when she was out of town and he was helping his mother-in-law care for toddler Mary. "Mary has behaved remarkably well," he remarked, "and is very obedient to me. I have not whiped her and have only on one occasion given her a scoalding."[7]

In recalling Henry's authoritarian side, Hall added that he had heard that the professor had a sluggish sense of humor. Specifically, if someone in a group made a funny remark, then Henry laughed only after others had started to laugh, as if he needed time to grasp the joke. Entries in Buhler's diary, however, suggest the contrary: that Henry had a quick sense of humor. But it was a droll jocularity, perhaps a product of his Scottish heritage. When Buhler, for example, related a story of a recent Princeton graduate becoming certified as a surgeon in the South after taking merely one medical course, Henry laughed so heartily that he "shook up his Bass Viol Belly into a perfect Jelly of Mirth!" Indeed, Henry usually was the party spinning the tales and causing his students to laugh. For example, he ended his bookstore conversation with Buhler by relating an amusing anecdote, presumably with New York origins, about the frustration of a "Dutch Savan" trying to converse with a "Yankee Numskull." Moreover, he preempted an entire recitation in his senior natural philosophy course to whimsically debunk mesmerists' claims about animal magnetism and religious zealots' testimonials about their bodies supposedly being seized with "The Jerks," "The Jumps," or "The Barks." "HENRY'S manner of telling the story," Buhler wrote, "was irresistibly humorous & kept the Class in a constant Roar." Even during the final oral examinations of his students, with somber trustees present, Henry retained a sense of levity. The day of the examinations in 1846 happened to be rainy and, according to Buhler, "HENRY took advantage of the circumstance to perpetrate a little bit of facetiousness." In particular, he comically advised the trustees that if the student then being examined did not do as well as expected it was because the human head "is a kind of Barometer & is affected correspondently with the Conditon of the Atmosphere."

Henry's personal letters also reinforce the impression that he had a droll sense of humor. Writing to Harriet from the snowy outlands of upstate New York—where he was traveling between sites trying to ameliorate earlier mismanagement of landholdings in the Alexander estate—Joseph offered a jestfully self-deprecating caricature of himself. Making this "pilgrimage mounted on a nagg" with a bulky carpetbag and horse blanket strapped behind the saddle, Henry implied that he already looked ridiculous; but, he added, the "large rope halter and short stirrups completed an equipment which rivaled that of Icabod Crane." Looking back at this period in his life, Henry also poked fun at himself for being too intellectually distracted to correct a persistent, mundane problem with his undergarments; in her diary for 1862, daughter Mary recorded a banter in which her father remarked that for years "he suffered from cold in a certain part of his nether limbs in consequence of both drawers & socks being rather short." Finally, after sixteen years of torment, this "philosopher with his thoughts in Heaven" realized that "by lengthening the former the difficulty might be obviated." Henry's lacing of bits of Scottish vernacular into his letters and conversations also reflected his lighthearted side. Instead of simply saying that he hoped to have a long talk with a colleague, he wrote: "I am anxious to have a long crack with him as the Scotch men say." He also occasionally spiced his interchanges with excerpts from poems, including facetiously closing a letter with a paraphrase of Scottish poet Robert Burns: "My page is full and the slowly moving finger of time 'points to the west of the key stone of nights black arch.'" Similarly, he cited a satirical couplet by English poet Lord Byron to make the point that he had no desire to author a book: "I know 'that it is plesant to see ones name in Print/ A Book's a Book although theres nothing in't.'"[8]

Reflecting on his Princeton years, Henry appraised his own teaching in the widely reprinted but anonymous autobiographical sketch that first appeared in 1871. At the end of the sketch, he commented about himself: "He was successful as a teacher, and never failed to impart to his students a portion of his own enthusiasm. His object was not merely to impart a knowledge of facts, but mainly to give clear expositions of principles—to teach the use of generalizations—the method of arriving at laws by the process of induction, and the inference from these of facts by logical deduction." A few years earlier, in his less formal reminiscences to Smithsonian clerk William Rhees, he offered a simpler evaluation of his Princeton teaching. He told Rhees that he "commenced a new order of things, collected all the old apparatus, put it in order, gave lectures for the first time & showed experiments & gained the confidence & attention of his class." In hindsight, the many testimonials of admiring students confirm that Henry did successfully impart his enthusiasm for natural philosophy to his classes and gain, at the least, their confidence and attention. However, because he structured his course not simply as a conduit for technical information but as a vehicle for cultural enrichment and mental development—and because Princeton was grooming its student body for occupations primarily in fields such as the ministry, law, and medicine—few of the graduates of his course settled on careers in science. Indeed, of the scores of students whom Professor Henry stirred during the 1830s and 1840s, only two went on to achieve moderate national notoriety as scientific researchers.[9]

Besides being dedicated to his courses, Henry also served with his usual diligence in a wide range of assignments well beyond the classroom. For example, during his first summer, the college's officers took advantage of his technical skills and delegated him to supervise the Fourth of July fireworks. Princeton celebrated Independence Day with a procession, orations by representatives of the Whig and Cliosophic societies, evening illumination of Nassau Hall, and a campus display of fireballs, rockets, and stars. Fireballs were globs of turpentine-soaked cotton wick that the students lit and threw back and forth across the green, having almost a hundred in the air at the most dazzling moments. Fearing an accident, Henry remarked that "the eavening was rather an axsious one to me." A few years later, as a member of the commencement committee in late September 1836, Henry became anxious about another student tradition involving a ball—the students' commencement dance. Because of "the evils" associated with past balls and the temporary unavailability of a hall, the college trustees were considering dropping this popular, dressy dance. Henry intervened on the students' behalf, however, and suggested a party at Philosophic Hall as a substitute for the ball. The students not only got use of Henry's building but also reinstated the ball. "The Philosophical Hall is under a process of great change," Henry commented two days before commencement. "The old dining room is to be used as a ball room. Dr Torrey's room is to be used as the supper room for the Ladies and my room upstairs to be converted into a Ladies dressing room." Apparently, commencement went smoothly, with a large crowd attending the exercises held in the flower bedecked Presbyterian Church that adjoined the campus. A highlight of the day was a speech on the Library steps by William Henry Harrison, the Whig candidate for president (who would lose the upcoming election to Martin Van Buren). Although Princeton's Whig Society had no political connections to the national Whig party, Henry helped greet Harrison on behalf of the student literary society. Incidentally, Henry himself had no overt connections to any political party. As he told James in 1842: "Politics in our country, and under any circumstances, is a dangerous and unstable business, and the less we are connected with it the better. When once a man becomes imbued with the spirit of party politics, he becomes unfit for the sober duties of life."[10]

By all reports Henry also freely devoted much time to helping students in their personal lives. During his fifteen years with the college, his compassionate behavior outside the lecture hall complemented his solicitous demeanor inside, thus heightening even further the students' affection and respect. One of his devotees was John Newland, a bright, young, Scottish emigrant who had studied under Henry at the Albany Academy and entered Princeton with advanced standing in 1835. Years later Newland wrote to Harriet Henry recalling the moment of his arrival in Princeton. He described "how, when I, a timid boy of fifteen reached Princeton at dusk & felt I was in a land of strangers, your husband met me at the stage coach door and almost taking me in his arms, saved my drooping spirit by his general loving manner and soon conducted me to your house where I was again reassured and cheered by your kind & motherly interest in me." Correspondingly, when Joseph's old friend and New York surveyor general Simon DeWitt died in late 1834 at age 78, the Henrys took in DeWitt's teen-age son; before his death, DeWitt had requested that Joseph take the young man as "a private pupil" and "an inmate of your family." And the Henrys' wards were not limited to children of family friends; in 1842 Joseph and Harriet took in a student "from the far South" who was thought to be within a few days of dying from typhus and helped restore him to health.[11]

Not all of Joseph's (and Harriet's) personal interventions were so benign; enjoying the trust of the students, Joseph also could defuse squabbles and fights that flared between the undergraduates themselves and sometimes involved young men in the village. One morning a decade into his Princeton career, Henry received an ominous note: "As you are the friend to one of the parties that are going to be engaged in a duel, I think myself bound to inform you of it. They are Petigru & M^c^Whorter. If you wish to be of any service, act speedily." Later, Henry matter-of-factly scribbled on the bottom of the note, "attented to this matter [and] had the difficulty settled." Though sensitive to the plights and passions of students, Henry believed in strict discipline—as Buhler learned when he witnessed Henry evict unruly worshipers from the chapel. He shared his colleagues' aversion to student vices such as card playing and drinking, two activities with unpleasant associations from his own youth. Indeed, in 1836 he took the initiative to call a meeting of the faculty to rectify "the moral state of the College."[12]

Henry ensured the lasting affection and respect of his students by continuing to show personal concern for them after they left Princeton. When an 1842 graduate and aspiring lawyer invoked the "excuse" of reporting a scientific curiosity to justify writing to his old teacher—"one whom I have so much reason to revere, and whom I hope never to forget"—Henry responded with encouraging words about the compatibility of science and the law. "I hope your friends do not think that your attachment to physical science has rendered you a less profound jurist. 'He who can travel,' says Dr. Johnson[,] 'ten miles in one direction can travel the same distance in another[,]' and this is particularly true when the roads in the two directions are of the same general character. The written laws of Nature and of Nations are both expressions of general principles, and he who has sufficient strength of mind to grasp the one, can also by proper application master the other." To another recent graduate and novice lawyer who also reported a curious scientific phenomenon to his old teacher, Henry again responded, as was his wont, with aphorisms reminiscent of those that Samuel Johnson—or Benjamin Franklin—had used to edify and uplift. "Besides knowledge and skill in your profession," Henry advised this 1843 graduate, "aim at *deserving* a character for candor[,] honosty[,] and truth. Strive to be *worthy* of the confidence and patronage of the Public and you will ultimately secure them. Do not be discouraged should your rise be not as rapid as you could wish or might have reason to expect. The tree of slowest growth strikes the deepest root." And to a graduate who was training for a career in engineering rather than the more populated fields of law and medicine, Henry reassured him by revealing his own preference for this profession that he once followed. "I think it requires more originality of mind to make a good practical engineer than a passable Lawyer or Doctor. It requires but little mind to attain a knowledge of the ordinary forms of law pleadings and the mistakes of the Physician are often buried with his patient while the labours of the engineer are seen and appreciated or at least criticised by many."[13]

During his third winter in Princeton, Joseph informed James that "my college duties now occupy so much of my time that I have scarcely leisure for anything else." Nevertheless, with an already taxing commitment to teaching and assisting students, Henry still made time for the college's ambitious fundraising and building programs. Henry came to Princeton at the beginning of an era of aspiration and growth. In fact, his hiring was part of an effort to reinvigorate the college after its near collapse during the mid to late 1820s. By the time of his arrival, enrollment was on the upswing, having already reached 150; though fluctuations would occur from session to session, the number would crest as high as 270 in the next two decades. Neither a national financial crisis (precipitated partly by President Jackson's monetary policies) nor a national schism in the Presbyterian Church caused lasting harm to the college's recovery. Drawing students mainly from New Jersey but also attracting a smattering from adjacent states and a sizable block from the South, the college was prospering. In particular, the growth in student body went hand in hand with ventures to erect additional buildings and garner greater financial backing.[14]

Henry realized that Princeton still lagged behind the handful of top American colleges. He concurred with others, however, that only a lack of money kept the institution from ranking "among the first in the country." Thus, he had high hopes for a campaign that college officials mounted in the mid-1830s to raise $100,000 among alumni. Joining in the effort, he successfully solicited, for example, a $500 pledge from Stephen Van Rensselaer, his old, Albany champion who had briefly attended Princeton. Unfortunately, the alumni campaign faltered, collecting only a much smaller sum, a few hundred dollars of which went to the chemistry and natural philosophy programs. Future efforts to increase Princeton's monetary endowment would be more successful,[15] but Henry's main contribution to Princeton's material growth would come not in raising money but in raising buildings.

Drawing on his knowledge of architecture and civil engineering, he devised a master plan for the placement of buildings on the college grounds. The trustees adopted the plan, rectilinear in design and emphasizing geometric symmetries. Henry paid particular attention to the layout of a new quad developed in the mid-1830s behind the "College" (that is, behind Nassau Hall) and flanked by two new dormitories, "East College" and "West College." Anchoring the rear perimeter of the new quad was a pair of buildings that Henry promoted vigorously: new quarters, in the Neoclassic style, for the Whig and Cliosophic societies. In an undertaking of more personal significance, Henry also pressed a plan for a major expansion and renovation of Philosophical Hall. "This plan will render the building one of the most convenient and pleasant edifices for Philosophical purposes in the country," he argued. With the trustees' approval, but only after various delays lasting into late 1835, carpenters refurbished the old building and constructed a three-story annex on the back. Torrey moved his classes onto the expanded second floor, which contained not only a lecture room but also a chemical laboratory, zoological cabinet, and workshop. Henry designed the enlarged top floor to house philosophical apparatus and to accommodate his lecture hall along with a laboratory, workshop, and office.[16]

Henry apparently was as fastidious about storing articles in their assigned locations in the renovated facility as he was about positioning buildings on the campus grounds. A sign in his laboratory displayed an illustration of a long-lashed whip below the warning: "A place for everything and everything in its place." In line with this injunction, he had labeled in chalk the proper locations of even common tools, such as "files and punches" and "saws and hatchet." To acquire articles of philosophical apparatus to fill their designated berths in the glass cases ringing the lecture hall, however, Henry would first have to journey in 1837 to London and Paris. Finally, in 1844, he could state: "After 11 years labour I have succeeded in getting the apparatus[,] the lecture room[,] & the Laboratory all arranged to my satisfaction and conveniently fitted both for my labours in the line of instruction and in that of research."[17] However, as Harriet Henry had come to realize over the years, Joseph's unflagging labor toward this goal had brought with it occasional setbacks in health.

During their second winter in Princeton—though Joseph was busy with lectures, recitations, and other responsibilities—Harriet optimistically believed that "it all appears to agree with him." "He has grown very fleshy," she wrote back to Albany, "and looks uncommonly well, at least so I think." Although Joseph was driving himself, Harriet apparently accepted his relentless exertions as routine. On a Monday later that winter—while still confined to her upstairs bedroom following the birth of Mary, her second child—she could dispassionately note: "Joseph is as usual much occupied in preparing for the class." By summer session, Joseph himself could observe in a letter to a colleague that "I am at present so much engaged in my course that I can find no leisure for any thoughts but those which relate to my college duties." A year later, during winter in 1835, he similarly would declare "that the present session has been one of the most laborious periods of my life." The following fall found him devoting from "8 in the morning until 10 at night" to "college duties." By winter session of 1836, the refrain had become familiar: "This session has been one of more labor to me than any other since my connection with the college." Finally, during that winter, the escalating workload took its toil by contributing to a protracted illness. "I have been afflicted for several weeks," he wrote to Silliman, "with a slow feaver and pain in the head which although not confining me to my room have prevented my applica[tion] to any thing of a mental kind. Want of exercise during the long continuance of snow and too close attention to the business of my lectures have produced the attack." A few weeks later, he told Torrey: "I have never worked as hard as I have done this winter, have been in very bad health and worse spirits." When he finally delivered his last lecture in the spring—not long after Harriet delivered their second daughter, Helen—he exclaimed that he felt "as if a heavy pressure were suddenly removed."[18]

Nevertheless, as at the Albany Academy, a term's end never meant for Henry a period of leisure and rest. He used the intervals between college sessions for concentrated research, continuing to push himself unremittingly. Vice President Maclean worried that the intensity with which Henry riveted himself to his research would further jeopardize his health. During the summer after Henry's illness, Maclean told him: "I am satisfied, that it would be a sacrifice of not only health but of life of itself, for you to apply yourself so closely for months in succession. For your health, I feel deeply interested, not merely from feelings of friendship, but in a very great degree from a regard to the College, whose reputation is becoming more & more identified with your own." Silliman had similarly recognized the broader effect of Henry's illness, projecting not merely local but national ramifications. "I grieve to hear that you have been ill but trust in God that you are restored," Silliman wrote, "& I pray that your important life may be long preserved, as I think you are destined to a brilliant career in the science of this country."[19]

Despite Maclean's counsel, Henry persisted in his research, though overwhelmed by lecture room obligations, student commitments, and administrative assignments. As a teacher, he was exceptionally dedicated and compassionate. As a researcher, he was extraordinarily resolute and meticulous. Indeed, this Princeton natural philosopher was still displaying patterns of personality and behavior that likely traced to having grown up the oldest son in his family and, moreover, the child of an alcoholic. Not surprisingly, this inveterate overachiever somehow found time to expand not only his experimental investigations but also his circle of scientific contacts. What investigations was he squeezing in at Princeton?

11 New Experiments

Stages of growth and cycles in time were implicit themes of Joseph's letter to James in late May 1834. Joseph recounted that Harriet was still nursing baby Mary, who was gaining weight seemingly as fast as Harriet was losing it, and that William was growing taller and thinner. Meanwhile at the college, Joseph was returning to his teaching duties with the opening of the summer term. Also, the students were awaiting with curiosity the predicted reappearance of so-called seventeen-year locusts, technically cicadas. Joseph amused his brother with a local explanation of the cicadas' long stay underground as larvae. "Amoung the *Blacks* of this place," he reported, "the story is that these insects are hatched in China and eat their way through the earth to us in the course of 17 years." "Not so bad an explanation!" he added. Joseph himself was about to emerge from his own period of inactivity and begin a new cycle of laboratory research on electricity and magnetism.[1]

As the summer session wound down in mid-August 1834 and commencement approached, Henry found himself with a modicum of spare time. Not only were college demands slacking because of the season but also his large battery was essentially completed, almost ready to take its place alongside the new electromagnet of the previous year. Moreover, Harriet and the children were vacationing in Albany, where they would remain until mid-September. Although lonely, Joseph sequestered himself in Philosophical Hall, about a one-minute walk from his house, and concentrated on his research and teaching. "I have confined myself to my little study and to my pursuits in the Hall," he wrote to Harriet. He grumbled, however, that a lack of resources tempered his enthusiasm for the research: "I have been much interested in some of my late results but meet with constant retardation in my experiments from a want of ready access to materials and from the tardiness as well as want of skill of the mechanics of the place."

With loneliness perhaps provoking self-reflection in this letter to Harriet, Joseph shifted from a preoccupation with work to an apologia for the pattern their marriage had assumed. He confided: "Sometimes I fear my devotion to the Hall and to my study may seem to interfere with my duty to my wife, but I do not believe that such pursuits can ever be incompatible with the warmest feelings towards my family. We should not perhaps forget that I have now the responsible duties and the dignity of a somewhat public character to support and you those of a wife and a mother." Cautioning Harriet that she was "no longer a girl," Joseph offered his vision of a mature relationship between husband and wife in which, "when of one mind and one interest, the labour of each serves only to heighten the love and esteem of the other. If I may have lost in the eyes of my wife some attractions as a ladies man, may I not think I have gained on the other hand in respect since I am endeavoring in the constant exercise of my official duties and with every honorable means in my power to extend the field of my usefulness and to establish for myself and to deserve the reputation of a scientific man." He ended by reminding Harriet of his deepened dependence on her:

> Never has my wife been more to me than now. By my removal to Princeton I have been separated from all my early associates—from the friends interested in my welfare because they have witnessed my early struggles and the many dangers a Divine Providence has enabled me to escape. Here the friendships we have formed are as yet but new, not long tried. I have been more than ever impressed of late with the fact that here at least I stand alone; that I must rise or fall by my own exertions and in this state of mind even when I am actively engaged in my pursuits I would draw my wife even more closely to me and have her yet more intimately a part of myself.

When Joseph mentioned to Harriet, "I have been much interested in some of my late results," he probably had in mind a fresh run of experiments that he was chronicling in a new laboratory notebook. Entries in the notebook show that he had at last returned to experimenting with coils of insulated wire, compass needles, pieces of iron, galvanic batteries, and the other familiar apparatus of electromagnetic research. Of course, he had never stopped thinking about the topic. Right before moving from Albany, he had finally examined a copy of Faraday's 1832 article on mutual induction; after relocating, he continued to keep abreast of latest developments as reported not only in Silliman's *Journal* but also in foreign periodicals. And, though he had suspended his laboratory experiments in electromagnetism, he had never interrupted his field research in a related area of interest: terrestrial magnetism. During his first spring in Princeton, he used the short vacation between terms to complete his measurements in Albany of magnetic dip, variation, and intensity. Afterward he could boast that Albany stood probably as the nation's only site with a complete specification of its geomagnetic characteristics. His ongoing interest in terrestrial magnetism further led him to assemble data in the hope of calculating the height of the aurora borealis, an issue of "considerable discussion among the Meteorologists of Europe." He later explained, however, that he was unable to make new auroral observations in Princeton because forested hills obstructed his view to the north. Also during his first years at Princeton, he had persisted in exploring ancillary phenomena such as "animal electricity." In the late summer of 1833, he rendezvoused with a group of colleagues in Morristown, New Jersey, to undertake a type of galvanic experimentation that he had been disappointed in performing six years earlier—experimentation with a large battery on the corpse of a murderer who had been publicly executed only moments before. In a similar vein he found a more congenial use for his knowledge of animal electricity. Specifically, he arranged for Princeton theologian Charles Hodge to receive galvanic shocks in hopes of lessening chronic, debilitating pain in his hip.[2]

Although he had built a new lecture-hall electromagnet, designed and constructed a versatile battery, carried out geomagnetic measurements, and delved into animal electricity, Henry returned to laboratory experiments involving electromagnetism only in the late summer of 1834. He then adopted a more systematic approach to his research. In particular, he attempted to maintain in his new notebook a dated, running account of his investigations; eventually, this "Record of Experiments" would grow into three volumes, with the first extending into 1839, the second into 1842, and the third into 1863. The initial line in the earliest entry, 15 August 1834, reads: "Exp on the spark from long & short wire with circular battery." That is, after a lapse of two years, Henry had reengaged his study of what became known as self-induction. He was investigating the varying intensities of sparks that resulted from disconnecting a battery from coiled versus straight arrangements of different lengths and gauges of assorted wires and copper ribbons. But four days later, according to his notebook, he had switched to another topic. He would continue through August on this new tack, testing iron's tendency to shield magnetic effects and relating this tendency to his favorite theory of magnetism, Ampère's notion of underlying electrical currents. Moving into September, he switched again, to trials on the conductivity of wire and tests of his new battery. The latter tests, made in association with exhibiting the battery during Princeton's commencement, included occasional self-inductive effects, heightened by the battery's power.[3]

By late September, Henry had ended his brief season of renewed electromagnetic experimentation. Caught in the rush of the new college term, he would log no further entries in his notebook for the next three months. When he again recorded a run of experiments, it would be over the course of two holidays that even a dedicated researcher such as Henry probably was reluctant to devote to science: New Year's Eve and Day. But Henry had an extra incentive to reopen his laboratory during the holidays. He had recently learned of Faraday's independent discovery of self-induction and, in response, was scurrying to substantiate his claim of priority.[4]

Concerned about the effect of Faraday's announcement, Henry had hoped to report his own updated findings on self-induction to the American Philosophical Society before New Year's Day. Specifically, in mid-December, he had informed Bache, a mainstay of the APS in Philadelphia, of his desire to counter quickly Faraday's recent revelation. He assured Bache, "I am now in the humor for working and will give you my paper before New Year," but a crisis in the family of another professor forced Henry to cover the colleague's time-consuming recitations. Able to reopen his investigation only beginning on New Year's Eve, he progressed rapidly, nevertheless, with a series of focused experiments that built on his late-summer probing of "long and short conductors." By the second week of January he could inform Bache that his latest results "conclusively prove" what he had intimated in his 1832 article in Silliman's *Journal:* that self-induction is "intimately connected" with mutual induction. He summarized his findings before the APS a few days later and submitted a full report in early February. The APS published the report in its *Transactions* as the second number in a special, new series of Henry's research articles, "Contributions to Electricity and Magnetism." The series' first number, which appeared in the same APS volume, covered Henry's galvanic battery. Henry likely modeled the series on Faraday's well-established "Experimental Researches in Electricity," which the London researcher had been issuing steadily in the Royal Society's *Philosophical Transactions.* Faraday's first installment, in 1832, had contained his findings on mutual induction while his ninth installment, in 1835, would formally report his conclusions on self-induction.[5]

In Faraday's initiatives Henry found not only a model for his APS series but also motivation for his latest burst of research on self-induction. Writing to Torrey in late February 1835, he recounted how the British researcher had recently duplicated his 1832 results from Silliman's *Journal,* but without acknowledging the results. "This induced me to arrange my experiments for publication," Henry elaborated, "and in so doing many new suggestions were presented to my mind which required imediate testing by direct experiment. The result was that the subject grew very rapidly under my investigations and has opened quite a field of research." A few months later, after seeing the latest summary of Faraday's self-inductive investigations, Henry conceded to Bache that, though his and Faraday's findings were "almost identical," the Londoner perhaps had "given a more definite & analytical exposition of the subject." He pointed out, nevertheless, that Faraday had overlooked the superiority of coiled copper ribbons over coiled wires in producing sparks; indeed, Henry had determined that a "flat spiral" of copper ribbon, even without an iron core, produced prodigious sparks. Faraday had thus failed to achieve, Henry added with his customary fixation on intensifying feeble electromagnetic effects, "results as striking as those which I can now exhibit."[6]

While briefing Bache in mid-January about his latest experiments in self-induction, Henry had interjected: "The subject opens to me a wide field of experiment which I intend to cultivate." One possibility that he would explore in future years had dawned on him a few days earlier. "It occurred to me this morning before rising from bed," he had jotted in his notebook, "that if I could solder two copper handles to the coil used in the foregoing exps a shock might be obtained." In other words, he had chanced on an alternative to the normal procedure of creating a self-inductive spark by breaking one of the connections of a coil with the battery. In particular, he realized that he could grasp copper handles positioned at the extremities of the coil (near the connections to the battery terminals) and experience a shock in his own body when breaking the contact. (In self-induction, recall that breaking the contact creates a momentary potential difference across the coil of wire. Normally the resulting voltage manifests itself as a spark jumping through the air in the gap where the break in the circuit occurred. But the copper handles and the person's intervening arms and torso provide an alternative, less resistive path for the voltage to manifest itself—as an electrical shock.) Teaching responsibilities soon sidetracked his investigation of the effect, but he briefly returned to it at term's end in May. He considered the effect to be a manifestation of "lateral discharge"—an emission of a puzzling "redundant" or "extra" electricity that previous researchers felt showed itself as a small spark from the side of a Leyden jar charged with common electricity. Lacking convenient and reliable instruments for measuring the strength of the discharges from the copper handles, he and an assistant evaluated the strength by estimating how high up their arms the shocks migrated. The strongest discharge, Henry scrawled in his notebook, produced "a peculiar thrilling in both arms." To double check this result, he put one copper handle in his mouth while continuing to grasp the other, thus producing a shock "in the other arm and in the teeth." In coming years, he would expand his studies of lateral discharge, concentrating on sideways emissions of static rather than galvanic electricity.[7]

At the end of May 1835, Henry interrupted his laboratory studies of electromagnetism because of another project he had instigated: renovation and expansion of Philosophical Hall. The drawn-out construction kept him away from significant electromagnetic experiments until March 1836. Meanwhile, he continued with other pursuits, including basic experiments on sound—harbingers of more extensive and exacting experiments in later years. And while in a mood for building, he arranged for the construction of a "magnetic house," a small, iron-free shed to shelter instruments for measuring terrestrial magnetism. A year later, though he had published little on the topic, he could assert with some justification that he was the nation's most active specialist in geomagnetism. "I can say without egotism," he explained to his brother, "that I have paid more attention to that subject than any other persons in the United States." The same was true of his involvement in electricity and magnetism, although he may have let his technical skills get rusty during his forced exclusion from Philosophical Hall. Within a few days of returning to his experiments, he recorded in his notebook a hazardous blunder. He accidentally discharged a sizable Leyden jar through the right side of his body, receiving the largest shock that he had ever experienced. "It produced an involuntary shout," he dispassionately chronicled in his notebook, "the fingers of the right hand which grasped the glass were violently cramped and for the space of half an hour, felt as if they were immersed in cold water. A sensation of fainting came over me, the blood left my face, and the pulse sunk to 38."[8]

Undeterred by this mishap, he persisted with preliminary experiments on the inductive effects of bursts of static electricity—specifically, fleeting discharges of Leyden jars into various circuits. These experiments on "common" or "ordinary" electricity complemented his previous investigations involving "galvanic" currents from batteries. Again, however, distractions—primarily his 1837 trip to Europe—kept him from immediately completing the study. Only in the spring of 1838, after another lapse of about two years, was he able to concentrate on the topic. Working into early autumn when classes reopened, he wrote to American colleagues that he considered his findings "the most important I have ever made" and projected that his eventual report would contain "a greater number of new generic facts than any I have ever published." "I have worked harder on these experiments than on any I ever before engaged in," he similarly informed Torrey, "and have overcome more difficulties and developed more new facts than were dreamed of when I commenced the series."[9]

Although now enjoying a cordial relationship with Faraday after having spent time with him in London (see chapter 14), Henry remained apprehensive that this intrepid researcher—whom Henry considered the world's leading "Experimental Philosopher"—would again scoop him. Henry was aware that Faraday also had shifted his research to static electricity. Writing to Faraday in the fall of 1838, Henry stressed that he was "very anxious" to receive the publications covering the new research, which Faraday had promised to send. Henry additionally offered his frank impression of what he already knew about the Londoner's research: "Your investigations are of a very extraordinary character. They tend to unsettle what was considered some of the best established laws of statical electricity and were the investigations not from yourself I would be inclined to be some what sceptical in reference to their accuracy." In this letter, Henry guarded the priority of his own work on static electricity by mentioning his pending publication, but revealing no details. He similarly withheld the details from Sturgeon, even though he considered him less of a threat, characterizing him to a friend as merely "the head of the second rate philosophers of London." In contrast, to his American colleagues and those European associates not involved in electrical research, he freely revealed that he was probing new effects involving static electricity. Also, to further guard a claim to priority, he rushed to get his preliminary results on record with the APS in Philadelphia. But the American's worry was unfounded. When Faraday finally sent copies of the three latest papers in his series—the eleventh, twelfth, and thirteenth installments—Henry was relieved to discover, as he told Torrey, that his London colleague was "on entirely a different track."[10]

Henry published his full findings on static electricity the next year, in 1839, as the third in his APS "Contributions to Electricity and Magnetism." The lengthy, rambling article, which he distributed to a large circle of colleagues at home and abroad, enjoyed a generally positive reception by a wide readership and would be frequently translated and reprinted. The article even would provide incentive for Faraday and Wheatstone to nominate Henry for the most revered scientific prize in Great Britain, if not the world. Though ultimately unsuccessful, Henry would become one of five nominees in 1839 for the Royal Society's Copley Medal.[11]

Early in his new article, Henry summarized the distinction between two types of induction that Faraday had recognized in 1831; he then justified his commitment to exploring, in not only galvanic but also ordinary electricity, the lesser known type.[12] On the one hand, many researchers were developing what he credited as Faraday's (not his own) discovery of the magnetic induction of currents. These researchers were striving particularly to perfect the associated "magneto-electrical machine"—a popular, embryonic electric generator, nicknamed a "magneto," that produced a series of alternating bursts of current. On the other hand, he was alone in extending "the purely electrical part of Dr. Faraday's admirable discovery." By this "purely electrical part" Henry meant Faraday's ancillary finding that a changing galvanic current in a wire induces an oppositely flowing current in a parallel wire. Henry believed that this "Volta-electric induction" helped explain self-induction, the action of a current on itself rather than solely on a parallel wire. (Recall that he had hinted at this connection in his original announcement of self-inductive sparks, in mid-1832. Subsequently, in the 1835 second paper of his new series, he had theorized—and would continue to theorize in following years—that traditional static-electrical induction could account for Faraday's Volta-electric induction, which, in turn, could account for self-induction in coiled wires and copper ribbons.) Moreover, although he devoted most ink to examples involving galvanic rather than static electricity, he established that Volta-electric induction in galvanic electricity had a counterpart in ordinary electricity. That is, he found that, in analogy to the galvanic case, the discharge of a Leyden jar's ordinary electricity into a wire induced a current in a parallel wire. Since it was common practice to speak of "statical" induction in ordinary electricity and of "dynamic" induction in galvanic currents, Henry would dub this new area of study the "dynamic induction of ordinary electricity." He considered the induction of currents using static electricity an important finding because—as he remarked in his article and then emphasized to his Princeton students—Faraday had cast doubt on the possibility. Indeed, Henry's results caught Faraday's attention; in the fall of 1839, about when he nominated Henry for the Copley Medal, he mentioned in his diary his intention "to work on Henry's late dynamic induction experiments."[13]

Henry also showed the existence of induced currents of third, fourth, and fifth orders. That is, whereas Faraday had established that a changing galvanic current in one wire induces an opposite flowing current in a separate, parallel wire—a second order effect—Henry used successive, adjacent, freestanding coils of metal ribbon and spools of wire to trace the inductive effect through multiple stages. He did this for both galvanic and ordinary electricity, noticing that the direction of the current's flow alternated in each consecutive stage.

After the article appeared, Sturgeon peeved Henry by announcing, in a seemingly amiable tone, that he had earlier not only achieved a third order current but also advanced an electrical theory capable of predicting the alternating flows of current in the consecutive stages. Aware that Sturgeon had previously alienated Faraday, Henry did not hesitate to belittle what he considered Sturgeon's latest, unfounded thrusts. To Bache he rebuffed the Londoner's commentary as "an admirable sample of the would be wise and condesending," and added that "before the publication of my paper Mr Sturgeon knew about as much of the nature of the currents of the different orders as one of the long eared animals so abundant near the spot whence his letter is dated." Although Sturgeon had issued his commentary in his periodical *Annals of Electricity* (after it had reprinted Henry's third article), Henry persuaded Silliman not to republish the commentary in his *American Journal* (alongside another, pending reprint of the third article). Later, in the fourth article in his "Contributions to Electricity and Magnetism," Henry interjected a calmer dissection of Sturgeon's assertions.[14]

While studying arrays of coils for his third article, Henry also found that by varying the makeup of two adjacent coils he could modify the character of the original current. In particular, using a primary and secondary coil, he could employ an initial "intensity" current (a weak current associated with what later became known as high potential difference) to induce a "quantity" current (a strong current associated with a low potential difference), and vice versa. To illustrate the reach of an intensity current created in a particularly large secondary spool, he passed the current through a circle of 56 seniors from his natural philosophy class. Henry's techniques for self-consciously altering currents would contribute to the development of electrical transformers—devices to "step up" and "step down" currents, crucial capabilities in later nineteenth-century science and technology.[15]

Additionally, in his third article, Henry explored how plates of iron, copper, and other electrical conductors, when placed between coils, neutralized the inductive effect. He attributed this result to "an instantaneous current" appearing in the intervening plates. The screening effect of metal plates became one focus of Henry's fourth article in his "Contributions"; this article, which he read to the APS in 1840 but got in print only in 1843, was an elaboration of the earlier article with a concentration on galvanic effects and theoretical generalizations. About a year before reading the new paper, he had recorded in his laboratory notebook the reason for his renewed interest in the screening effect. "I have lately received Dr Faradays 14th series," he jotted concerning Faraday having sent the 1838 installment of his electrical researches, "and find that he has crossed the track of my last paper but in reference to the screening influence draws an intirely oposite conclusion." Faraday, to his own surprise, had found that not even a copper plate screened the induction between coils, an anomaly that he tentatively interpreted through his trusted concept of the "electro-tonic state" of the "intervening particles." Quickly returning to his laboratory to investigate the discrepancy and to search for the hypothetical electrotonic state, Henry reconfirmed his own result, thereby seemingly discrediting Faraday's. "I have thus opened to me a new field of research," he declared to Bache, "and for once have found Mr Faraday tripping." "Mr Faraday for once is entirely in the wrong," he similarly informed Torrey. His satisfaction in besting Faraday, however, was short lived. He soon realized that he needed to temper his initial interpretation of the discrepancy and grant the legitimacy of not only his but also Faraday's result. Eventually, in his new APS article, by arguing that he and Faraday had emphasized different aspects of the inductive process, Henry resigned himself to reconciling his findings on the screening effect with Faraday's apparently contradictory results.[16]

On a lighter note in the third article in his series, Henry also revealed "the means of exhibiting some of the most astonishing experiments, in the line of *physique amusante,* to be found perhaps in the whole course of science." Most engrossing, on placing coils in adjoining rooms on opposite sides of a wall (made of nonconductive materials), he could use the primary coil to induce a current in the other. "The effect is as if by magic," he noted, "without a visible cause." With his usual lecture-room flair, after publication of his third article, he enlarged the apparatus so that the secondary coil alone was four feet in diameter and contained almost a mile of wire. To dramatize the penetrating powers of this enhanced device, he erected a seven-foot barrier of interspersed students between the coils and showed that the furthest student could still perceive a slight shock "in the tongue." Meanwhile, whereas he used galvanic currents to trigger these magical displays, he also reported in his third article that ordinary electricity could operate over much greater distances, up to twelve feet. Such startling effects, Henry concluded, "could scarcely have been anticipated by our previous knowledge of the electrical discharge."[17]

The recognition of static electricity's potency led, as Henry announced in the fifth contribution in his electrical series, to "a remarkable result" of an inductive effect being conveyed over a large distance "by a very small quantity of electricity." He produced the result in May 1842 using an electrostatic friction machine positioned on the upper floor of Philosophical Hall and, on the ground floor below, a secondary circuit containing a special spiral segment that could detect a current by magnetizing an inserted needle. (Recall that years earlier Arago and Ampère had found that a current-carrying wire when wrapped around an insulating glass tube containing steel needles would magnetize the needles.) He found that when a small spark, merely an inch long, was "thrown" from the machine onto "the end of a circuit of wire," the needle in the parallel wire downstairs became magnetized—though two fourteen-inch floors and ceilings blocked the thirty-foot path between the circuits. The result "almost equals clairvoyance," one colleague marveled.[18]

To explain the surprising interplay over such distances, Henry invoked the hypothesis of "an electrical *plenum*"—an invisible medium that permeated all of the space and materials between the two circuits and that a mere spark could readily agitate. Describing the way in which this plenum mediated the electrical disturbance through Philosophical Hall, he noted that the process "is almost comparable with that of a spark from a flint and steel in the case of light." For the past few years, Henry had been leaning toward this popular hypothesis of an electrical plenum and its probable identity with a light-propagating plenum, or as it was commonly known, the "luminiferous ether." In particular, he had broached the possibility in 1838 after confirming that electromagnetic induction, like light, acts through a vacuum; because he assumed, along with most of his contemporaries, that neither electromagnetic induction nor light could act through truly empty space, he postulated the existence of intervening, unseen media. By endorsing the concept of an electrical plenum and its probable identity with the luminiferous ether, Henry was deviating from a strict allegiance to Franklin's theory that electricity was a single, imponderable fluid associated solely with regular matter and not the space between the matter. Instead, adapting various views from European trendsetters such as Ampère and American colleagues such as Hare, he was subscribing to the idea of an intermediary, ethereal agent that permeated not only all matter but all space. Whereas he still favored a refined version of Franklin's single fluid theory when describing the effects of static induction, he now meshed it with the plenum hypothesis when describing the effects of dynamic induction.[19]

A few months after the interfloor result, in early October 1842, he would move the experiment outside and try it on a much larger scale. He would center this new series of trials on a single wire strung from his lecture room in Philosophical Hall to the study in his house—a wire, grounded in two wells, similar to the one that he would use during this period to demonstrate for his class two-way telegraphic communication. For his source of common electricity he either used a large burst from various combinations of Leyden jars, which he had charged with a series of sparks from his electrostatic friction machine, or drew a single spark directly from the machine. Discharging the electricity through the primary wire, he could detect an induced current in a parallel wire on the other side of Nassau Hall at, eventually, the "remarkable" distance of over 220 feet. To register the current, he again used a small spiral in the distant wire to magnetize an inserted needle, typically a common sewing needle. "Each spark sent off from the Electrical Machine in the College Hall sensibly affects," Henry later generalized while lecturing to his class, "the surrounding electricity through the whole village." For his students, Henry also drew a parallel between the electrical spark and candlelight: both probably share the property of being a "wave or vibration of the same medium" and, just as the eye allows detection of the light, the magnetizable needle provides "the organ of a *new sense*" for registering the electrical spark over distance.[20]

Curiously, depending on the distance between the parallel wires and the size of the original discharge, shifts in the north-south polarity of the magnetizable needles showed that the induced current alternated in direction. (Whether a particular end of a needle became a north or south pole depended on the direction of the current's flow in the surrounding spiral of wire.) This strange fluctuation was an effect that Henry had noticed previously in the smaller-scale experiments. He also had tried to explain the fluctuation in his fifth, and what would turn out to be final, contribution to his series on electricity and magnetism—summarized to the APS in mid-June 1842. Specifically, building on an earlier study by Paris investigator Félix Savary, he had found that a Leyden jar discharges not in a single, unidirectional outpouring of some type of "imponderable fluid" but in a rapidly diminishing series of forward and backward strokes. Relating these oscillations to a cascade of self-inductions, he had determined that they continue until electrical equilibrium is reestablished in the Leyden jar. Thus, he had reasoned, because the magnetic effects of the jar's successive, reversing strokes combine differently at different points in the electrical plenum, a particular needle's north-south polarity depends on its distance from the source. Variations in the original size of the oscillating discharge also affect the polarity, he had explained.

Enthusiastic about having identified the Leyden jar's oscillating discharge—an effect that "had never before been recognized"—he had emphasized in his APS summary that he had made the finding only after having magnetized "nearly a thousand needles" and devoted "considerable study" to Savary's previously anomalous results. He similarly impressed on his class during the 1842–43 term that the finding cost him much labor. According to one account in a student notebook—an account that either Henry or the student exaggerated—Henry came to the result only "after making some 5000 experiments" and spending a cumulative total over several years of "8 months incessant thinking." Reminiscing about the finding in an 1868 letter to an American colleague who had asked for an overview of his electrical researches, Henry declared the Leyden jar's oscillating discharge "one of my most important discoveries." The discovery, he elaborated, "opened up for me a wide field of research" that resulted in an additional run of "several thousand experiments." Unfortunately, Henry added, commentators had unwittingly attributed the discovery to other investigators though they had published after him. Henry was possibly referring to independent studies in Europe by Hermann Helmholtz in 1847 and William Thomson in 1853. In fact, whereas Henry had anticipated Helmholtz and Thomson, Savary had anticipated all three to a degree in 1826. In his Princeton investigations, however, Henry had cited and conscientiously repeated Savary's earlier experiments before going on, he believed, to generalize and interpret the French researcher's results.[21]

In June 1842, four months before running the wire from Philosophical Hall, Henry had succeeded with a similar experiment, but even on a grander scale. Having suspected that the far reach of a lightning discharge would trigger a strong inductive effect similar to that arising from a Leyden jar's or frictional electric machine's discharge, he had rigged in his own house (the new one) an apparatus to detect the inductive influence of distant flashes. He soldered a copper wire to the house's tin-plated iron roof and ran the wire into his study before routing it outside again where it anchored in a deep well—the same well and ground connection that he later used for the wire from Philosophical Hall. In the section of wire passing through his study, he spliced a spiral of thinner wire and inserted a detecting needle. He then waited for lightning. After a few days, his hopes rose one evening when, before going to bed, he noticed very faint flashes on the horizon. "At a little before three o'clock," he jotted in his research notebook, "I was awakened by a storm of rain and heard several distant discharges of lightning. I did not rise but in the morning I found the needle strongly magnetic in the direction which indicated a current upwards." As he reported in his fifth contribution to his series—which he summarized to the APS a few days later—lightning many miles away could induce a current in this roof and wire assembly, thus magnetizing the needle. Using a proposed "self-registering electrometer," he further reported, he could conceivably monitor a succession of such currents. He also came to realize that the lightning discharge, similar to an emission from a Leyden jar, consisted of a quickly dwindling series of oscillations.[22]

This study of "atmospheric electricity" held special appeal to Henry in that it merged with his interests in meteorology, a field that he had continued to follow, at least marginally, since leaving Albany. In the late 1830s he had expanded his probe of lateral discharge to include lightning, especially the peculiar sideways sparks emanating from grounded lightning rods, which seemed to compromise traditional rods' effectiveness. Continuing in the footsteps of Franklin, he also had documented the aftereffects of lightning strikes and, eventually, would begin to offer advice on the construction of protective rods. During one month in 1847, he would receive requests for guidance on protecting the dome of the Capitol in Washington and the steeple atop Independence Hall in Philadelphia. At the end of his career, he would unblushingly tell an affiliate: "I now have to say that as far as I know I am the only person who has made a special study of the conduction of frictional electricity in regard to lightning rods." "From all my study of this subject," he would add, "I do not hesitate to say that the plan I have given of lightning rods is the true one."[23]

A young, British military officer with an interest in geomagnetism happened to visit Princeton shortly after Henry had started his lightning tests and while he was still conducting his long-wire experiments. The officer recorded the professor's opinion on the pertinence of these large-scale studies for a research community unduly preoccupied with small-scale, laboratory effects. Henry reportedly pointed out that, "in attending chiefly to cabinet experiments[,] Electricians have wholly lost sight of the facts, as to the distance to which any disturbance of Electricity is propagated." The British visitor also volunteered an evaluation of the professor's disposition. Finding Henry to be an exceptionally "kind and hospitable" host, he commented: "I never met a person who united such simplicity and bonhommie with high talents, I think him one of the most clearheaded men, entirely absorbed in his particular branch of Electricity, and a perfect Enthousiast in it."[24]

In conducting his research over the years, Henry had frequently managed to enlist either acquaintances, students, or other young men for short-term assistance. Although he often personally compensated the students and young men with board and a stipend, he could count on neither their amenability to remain nor their willingness to do certain tasks. When one student assistant became overwhelmed by Henry's requests, the student's father first reassured his son that it was "a privilege to be with an accomplished philosopher, whose name is in high repute among men, familiarly," but then counseled him to extract himself tactfully from the taxing position. Henry himself complained that, because of inadequate help, he had been spending about one-third of his time at Princeton in rigging makeshift apparatus and doing menial tasks such as washing bottles, sweeping floors, and carrying water.[25]

By 1842, however, whether investigating lightning or a Leyden jar, Henry enjoyed the services of a permanent, college-paid helper. Sam Parker, the assistant who helped in the lecture-hall demonstration of sound transmission between two violins, had begun working for Henry in 1840 at the wage of $4 per month. Parker proved essential to Henry. The day after succeeding in magnetizing a needle with lightning, for example, Henry stated in his research notebook that he could not continue with related laboratory tests because of Parker's absence. "My experiments are interrupted," he observed, "on account of the illness of the colored servant of the Laboratory *Sam* Parker."[26]

Whereas Henry depended heavily on Parker for help in building apparatus and conducting experiments, he seemed predisposed to relegate his African American assistant to the rank of a lowly servant. Thus, when a colleague asked Henry how he managed to handle galvanic batteries without suffering bruised and burnt fingers and acid-rent clothes, Henry callously answered that he delegated all disagreeable tasks to Parker. "There is no form of the galvanic battery," Henry began, as he replied in late 1841, "which is not attended with some inconvenience of manipulation and a liability of soiling the clothes with acids and salts. The Trustees of our college have however furnished me with an article which I now find indespensible namely with a coloured servant whom I have taught to manage my batteries and who now relieves me from all the dirty work of the laboratory." Within a few years, however, Henry's trust in and rapport with this "article" had grown to the extent that he was also relying on Parker's vital services in his home. When Harriet was out of town, Joseph called on Parker to help prepare breakfast for a guest (ex-student Eli Whitney, Jr., who was bearing a gift of a model cotton gin from his father). And it was Parker whom twelve-year-old Will repeatedly accompanied to the outskirts of town in 1844 to watch troops gathering in commemoration of the sixty-seventh anniversary of Princeton's Revolutionary War battle. Parker perhaps was filling in for Henry, who viewed the military gathering with disdain. "For my own part," Joseph wrote to James, "I should not be much grieved should the whole be a failure for next to political meetings I have the most sovereign contempt for the business of soldier playing." "*Will* and *Sam,*" he conceded, however, "are very much interested in the matter and have made frequent visits to the camp." The Henry and Alexander clan embraced Parker to such a degree that, a few years later while traveling, Joseph could close a letter to Harriet by asking her to remember him to her mother, Stephen, Stephen's wife, the children, and "Sam & all the other members of the family."[27]

In 1847 Henry paraded his recent findings in his anonymous article in the *Encyclopædia Americana*—the article in which he reviewed latest advances in electricity and magnetism. Giving exclusive billing to his own breakthroughs when summarizing the new field of the "dynamic induction of ordinary electricity," he recounted how "Professor Henry of Princeton" had probed induction over large distances using discharges from Leyden jars and lightning. The distances, he claimed, extended to 300 feet for the jars and 20 miles for the lightning. This summary led him to offer a generalized and nuanced version of his theory of the electrical plenum—a theory that he had incubated in his five "Contributions to Electricity and Magnetism," rehearsed with his students, and elaborated during the previous year in an APS talk. "After a laborious investigation of the phenomena we have described," he wrote about his own research, "Professor Henry has succeeded in referring them all to the hypothesis of the existence of an electrical plenum, through which dynamic induction is transmitted wave fashion; and in showing that, in all cases of the disturbance of the equilibrium, the fluid comes to rest by a series of oscillations." Reasoning in accord with the traditional laws of electrostatic induction, he explained how the discharge of a Leyden jar into a long wire produces within the wire a diminishing series of reversing waves of electricity, "each of which will induce an opposite wave" in an adjacent, parallel wire. In turn, the induced waves of electricity in the secondary wire will produce different polarities in a magnetizable, test needle depending on the distance between the two wires; this is because the successive, diminishing waves combine differently at different points in the electrical plenum. Unlike traditional "statical induction" involving attractions and repulsions of stationary electrical charges, however, "dynamic induction" with its moving electrical currents demands consideration of "*time* as an element of the calculation." Dynamic induction, whether traceable to the discharge of a Leyden jar or frictional machine—or the action of a galvanic battery—"is transmitted in time, wave fashion, through the electrical plenum."[28]

In hindsight, Henry's experiments and interpretations evoke the late nineteenth-century investigations of Heinrich Hertz on electromagnetic waves and Guglielmo Marconi and Ernest Rutherford on radio communication. Unknowingly, in exploring the reach of inductive effects, the Princeton professor had skirted the empirical and conceptual foundations of the transmission and reception of electromagnetic waves.[29]

Reassessments 12

Henry never got past the fifth number in his "Contributions to Electricity and Magnetism." In fact, he never went beyond a preliminary overview of this intended fifth article of the series—an overview that he presented to the APS in June 1842. Although he persisted with his electrical experiments, and occasionally provided oral reports to the APS, he quit publishing formal accounts of the research. In April 1845, he offered an explanation for this inaction to Faraday, who by then had issued eighteen installments in his own electrical research series. "I have not published any thing of any moment for some time," Henry wrote, "although I have on hand a large collection of facts relative to the dynamic induction of ordinary electricity which I have kept back with the hope of finding time to render more complete." He realized, he further told Faraday, that by holding back results he had created opportunities for later researchers to upstage him; but he insisted that this did not bother him. "I am however not very anxious about scientific reputation," he explained, "and I can truly say that I have received much more pleasure from my investigations than from any credit I may have received on account of my labours."[1]

Henry likely believed these lofty words—or at least found them to provide a satisfying rationalization—but there were other, familiar reasons for his failure to polish and publish his ongoing electrical investigations. Though he could rely on Parker's assistance, he continued to shoulder a heavy load of college duties and to lack ready access to laboratory materials—obstacles that disrupted the continuity of his investigations. Also, having decided to pursue the branch of electrical science involving the induction of currents using static electricity—rather than the more popular branch involving galvanic electricity and inventions such as the magneto—he perhaps suspected that he was floundering in the accelerating science's wake. Additionally, during the early through mid-1840s, his longstanding tendency to delve into a wide range of scientific topics carried him repeatedly beyond the normal bounds of electrical inquiry, whether static or galvanic. He explored, on the one hand, the application of electrical techniques to other fields, such as measuring the velocity of a ball shot from a large gun (by having the projectile break two successive circuits) or estimating the relative temperature of sunspots (using, with Stephen Alexander's help, a thermoelectric sighting device). He even brought his experimental skills to bear on the fashionable topic of mesmerism. After entering into a dispassionate study of a "young Negro" reputed by an itinerant lecturer on mesmerism to possess extraordinary magnetic and electrical attributes, Henry emerged incredulous but open to further evidence. (He also emerged amused, with a stock of "laughable" anecdotes that he later shared with his students.) On the other hand, he investigated completely nonelectrical topics, such as phosphorescence and colorblindness. Also, after being distracted one day by the "beauty of the colours" in soap bubbles that some children happened to be blowing, he began a concentrated study of similar bubbles in an effort to shed light on surface tension and its intermolecular causes. On a deeper, personal level, moreover, he continued to have difficulty culling his many research initiatives, and identifying and following through on truly significant leads. In other words—to modify Faraday's old metaphor—whereas Henry did not lack the ability to hook fish among the many weeds, he remained somewhat incapable of reeling them in. Thus in 1852, when an old acquaintance, Angelo Secchi, built on Henry's sunspot research to announce in Rome more sweeping solar studies, Henry conceded to Bache that he himself should have pursued the research more vigorously. Lamentably, he told Bache, his pattern is to grow so accustomed to his own results that he underrates them *until* another investigator "brings them forward and develops them in a more popular or a more definite manner."[2] Indeed, in late 1845 the pattern that Henry had displayed throughout his earlier explorations of electricity and magnetism reappeared: his tendency to grasp the full import of an experimental trail, and follow it to completion, only after another researcher had blazed the way. Again, that other researcher was Faraday.

In December 1845 Henry received preliminary word that Faraday had succeeded in showing that magnetism could affect light. Specifically, Faraday had announced to the Royal Institution that by directing a beam of polarized light past a strong electromagnet he could alter the degree of polarization—that is, he could rotate the beam's plane of polarization. "I consider this next to the connection of Electricity and Magnetism by Oersted," Henry wrote to a former student a few weeks later, "the greatest descovery of the present century." Since the early 1820s, Faraday and other researchers had been searching for an interaction between magnetism and light, often motivated by a belief in the "correlation of forces," in not merely the interconvertibility but the unity of the powers associated with electricity, magnetism, heat, and light. The 1845 announcement of what became known as the "Faraday effect" would influence the course of nineteenth-century physics, especially through the follow-up analyses of William Thomson and James Clerk Maxwell. Earlier, in April 1838, when Henry revived his laboratory studies after returning from Europe, he had momentarily joined the search for magnetism's influence on light. Perhaps motivated by Faraday's 1834 report, in the eighth installment of his series, of unsuccessful polarization experiments in search of the electrotonic state, Henry had transmitted a polarized beam of light through an iron tube that he had magnetized with an electric coil. But the experiment failed. A variation in 1841 also failed. When word reached Princeton in mid-December 1845 of Faraday's success, however, Henry could quickly resurrect his old apparatus, modify it to conform to Faraday's specifications, and confirm the British result. "Sam was engaged all day yesterday," he wrote in his notebook shortly before detecting the effect, "in making me a coil for the purpose [of] repeating Mr Faraday's experiment."[3]

On the day he reproduced the result, 30 December 1845, Henry described to a Harvard colleague the circumstances of the confirmation and his feelings toward it. While surmising that his own, earlier experiment had miscarried only because of "a want of sufficient galvanic power," he insisted that he was not seeking credit for anticipating Faraday's breakthrough. "I do not mention this fact to claim any of the merit of Dr Faradays descovery," he wrote, "but merely to show you that my mind was in a fit state to readily conceive the nature of the expermt and to reproduce the phenomenon." Henry would have agreed with the contemporary judgment of another researcher who had earlier failed in the polarizing experiment; in also granting priority to Faraday, British physical scientist John Herschel had acknowledged, "He who proves, discovers."[4] After detailing his confirming experiment, Henry told his Harvard colleague that he was choosing not to extend the investigation since he might simply be duplicating Faraday's still unpublished results. Moreover, he felt that further work would violate scientific protocol: "I do not think it perfectly inaccordance with scientific etiquette to enter on a new field of investigation before the author of the primary discovery has had time to give his own results to the world." It would be anything but a breach of etiquette, however, to enhance the apparatus for use in lectures, Henry's forte. He traced for his Harvard colleague the procedures for doing this.

Henry closed this late December letter with an assessment of his general research program. In the process, he revealed feelings of resignation similar to those he had disclosed to Faraday earlier in the year:

> I am now engaged during the college vacation in repeating some of my older experiments and in preparing for the press the results I have obtained on the subject of induction during the four or five years past. My college duties are such that I can do nothing in the way of investigation during the term and at the end of the vacation I often find my self in the midst of an interesting course of experiments which I am obliged to put aside and before I can return to them my mind has become occupied with other objects. In this way I have gone on accumulating almost a volume of new results which will require much time and labour to prepare for publication. Indeed I find so much pleasure in the prosecution of these researches that the publication of them in comparison becomes a task. I have become careless of reputation and have suffered a number of my results to be rediscovered abroad merely from the reluctance I feel to the trouble of preparing them for the press.

If this self-appraisal seemed to suggest an ambivalence between frustration and forbearance, then that ambivalence probably reflected Henry's desire to put the best face on his publishing inactivity for his Harvard colleague. In the privacy of his own research notebook, however, when he first summarized Faraday's successful polarization experiment, he inserted a parenthetical remark that betrayed not only frustration but also perhaps self-deprecation. "I have long saught some action of this kind," he began, before adding rather dejectedly, "and have planned a number of experimts on the subject which[,] like many other plans[,] I have not realized." In the intimacy of his own natural philosophy class, he also dropped his guard and revealed the depth of his personal frustration, particularly regarding Faraday's latest achievement. Senior John Buhler discerned this frustration during the final meeting of the course in late April 1846 when Henry demonstrated the Faraday effect using his own enhanced apparatus. Buhler, normally a doting student, depicted in his diary a bitter Henry: "Exhibits *Farriday's* recent discoveries tending to show the *Connexion between Light & Electricity.* I think he is rather jealous of the merit thereof. Seems not to relish this outstripping of himself in his own Honors—modifies the importance & brilliancy of the Discovery by styling merely—'an Interesting Fact!'"[5]

This begrudging tribute, delivered in a private setting to students with whom he had spent the academic year, contrasted sharply with the accolade that Henry bestowed on Faraday two months earlier in a published letter to the editors of the *New York Evening Post.* With justified indignation, Henry was responding to the newspaper's report of Faraday's discovery in which the editors repeated an admittedly unsubstantiated rumor claiming that Faraday had appropriated the discovery from Henry. Correcting the record and denying any credit for himself other than "being the first to verify" the discovery, Henry emphasized the embarrassment that the newspaper report portended for himself and Faraday. "It places me, to say the least, in a very unpleasant position," he insisted, "and involves a charge against the moral character of a gentleman who deserves the gratitude of the whole civilized world for the many additions he has made to the sum of human knowledge, and who commands the respect as well as the admiration of all who are acquainted with the history of the progress of physical science during the last twenty years." If Henry were privately resentful toward the latest discovery of this "gentleman who deserves the gratitude of the whole civilized world," then that resentment probably masked a deeper frustration with his own inability to carry his experiments to fruition and communicate his results. Thus it is ironic that Faraday, Henry's supposed nemesis, also voiced frustration with the dearth of personal communications from Henry. Toward the end of 1846, Faraday entrusted with a mutual friend who was returning from England to the United States a packet for Henry. It contained, for the Princeton experimenter's use, a piece of the special glass with which Faraday had detected the polarizing effect. The mutual friend also conveyed a terse account of Faraday's frustration with Henry. "He complained bitterly," the friend reported, "that he could get no letters from you. I told him you were overwhelmed with duties and that must account for the delays."[6] Henry's delays in passing on findings in electricity and magnetism would only lengthen as his responsibilities became even greater.

As at the Albany Academy, though Henry's teaching often impeded his research, at times it contributed. A dedicated professor, he had committed himself to displaying through demonstrations the effects and principles covered in lectures; this activity provided opportunities to try new experiments. He also had committed himself to achieving a high level of mastery of all the course topics; this endeavor provided opportunities to explore new ideas. During the winter of 1836, for example, he told Silliman that, "while lecturing on the subjects of electricity and galvanism[,] I was able to devote all my leisure time to the repetition of some of the more unusual experiments in these branches." "Some of these," he remarked, "have interested me much and have opened new subjects of investigation."[7]

Henry's rigorous regime of background preparations and lecture-hall demonstrations led him to conduct original investigations in a wide range of topics beyond electricity and magnetism. A particular impetus for exploring new topics came from his struggles in the mid-1840s to prepare his *Syllabus of Lectures in Physics.* In drafting these comprehensive outlines for his students, he confessed to being stymied by phenomena such as capillary action, "which I had never fully understood and which I had previously taught in the popular way in which it is found in the ordinary books." After devoting almost an entire college term to studying capillarity, he realized that he had gained command of the topic. "I found my mind in a proper state to advance the subject," he wrote to a colleague, "and accordingly I commenced a series of experiments the results of which have found considerable favour with the scientific world abroad." Indeed, main British and Continental journals reprinted the findings that he had originally reported to the APS. Incidentally, in his APS paper "On the Capillarity of Metals," Henry seemed to draw on his experience in Albany as a silversmith's apprentice, at one point citing a fact about gold-plated copper that was "well known to the jeweller."[8]

Henry's determination to produce a reliable *Syllabus* even led him to delve into the philosophical and conceptual foundations of natural philosophy. Wanting to introduce his course with a clear account of epistemological and methodological issues, he spent two years in the mid-1840s slighting standard physics topics to study "the doctrine of the origin of our knowledge" and that branch of philosophy concerned with "psychological" or mental processes. Because of the national popularity of what he variously called "German metaphysics," "the a priori philosophy of modern Germany," or "German transcendentalism," he felt particularly obliged to prepare himself to counter the premise that knowledge of reality originates in intuitive ideas rather than sense experiences. After much study and reflection, he judged transcendentalism to be "an ocean of unsubstantial truth." His distrust of transcendentalism or, more broadly, idealism allied him with many of his Princeton colleagues—particularly those perpetuating the college's traditional alignment with the philosophy of Scottish Realism, which stressed "common sense" trust in sensory experiences. Henry told an associate in 1846 that, after having familiarized himself with the views of commentators partial to transcendentalism such as Immanuel Kant and William Whewell, he had "concluded to keep pretty close to *philosophy positive* during the remainder of my life." In alluding approvingly to Auguste Comte's *Cours de philosophie positive*—and elsewhere mentioning favorably the notion of "positive science"—Henry was not offering a blanket endorsement of the contemporary French thinker's multifaceted, positivistic perspective on scientific inquiry. Rather, he was expressing his general sympathy for Comte's overall antimetaphysical outlook. Promoting a variant of empiricism, Comte advocated reliance on observation and reason in searching for invariable natural laws of phenomena as revealed in relations of succession and similarity.[9]

In forging his *Syllabus,* Henry also found it necessary to untangle the fundamental concepts of mechanics. He explained to a colleague that, after moving beyond the general obstacle of philosophical foundations and the particular obstacle of capillarity, "I next grounded on the first principles of mechanics and devoted nearly a session to the study of the proper method of establishing the laws of force and motion." Controversy still surrounded "the first principles," as the scientific community was only beginning to articulate the modern concept of energy and its conservation. Henry labored to enunciate a coherent view of traditional concepts of mechanics such as *"vis viva," "vis mortua,"* and *"power."* The upshot was a paper that he read before the APS in late 1844, "Classification and Sources of Mechanical Power." "It has found favour with the younger men of science to whom I have communicated it," he remarked about the analysis, "but met with considerable opposition from some of the older members of the society."[10]

The more speculative passages of the paper on "Mechanical Power" might have struck some APS old-liners as a departure for the Princeton experimenter. By his own admission, he was exceeding the limit where "the true spirit of inductive philosophy" would normally compel strict empiricists to halt. In earlier years, however, he consistently and openly had been exceeding this limit by relying on tentative hypotheses as guides to further his investigations. Thus, at the conclusion of the second article in his series on electricity and magnetism, the 1835 paper on self-induction, he reminded his readers that the theoretical explanations or principles that he had incorporated throughout the analysis fulfilled significant heuristic roles. "The foregoing views are not presumed to be given as exhibiting the actual operation of nature in producing the phenomena described," he wrote, "but rather as the hypotheses which have served as the basis of my investigations, and which may further serve as formulæ from which to deduce new consequences to be established or disproved by experiment." He offered a similar perspective to Bache later in 1835 while discussing explanations of meteors. Hypotheses, when used with proper caution, "may serve to direct attention to certain points of observation and thus become important aids in the accumulation of facts." Only after "a sufficient number of well attested facts have been collected," is a researcher entitled to deduce "even an approximation to a rational theory" of the phenomenon. By 1840, when he presented the fourth in his series of papers, he felt that he was approaching at least a "wider generalization" about electromagnetic induction. Having repeatedly refined his tentative hypotheses to bring them into "exact accordance with all the facts," he could assert that, "if the explanation I now venture to give be not absolutely true, it is so at least in approximation, and will therefore be of some importance in the way of suggesting new forms of experiment, or as a first step towards a more perfect generalization." A few years later, he reexpressed this point, saying: "My communications are not a mere collection of facts without an attempt at classification or explanation; on the contrary in almost every section an attempt is made to refer the facts to some law of action."[11]

Although Henry prided himself on following "the true spirit of inductive philosophy," clearly, he deviated from the popular method of inductive inquiry commonly attributed to Francis Bacon. Henry likely had encountered Baconian philosophy as early as his late teenage years when first enthralled by George Gregory's *Popular Lectures.* Now, however, he shied from the Baconian notion that a purely empirical accumulation of particular facts could directly lead to comprehensive laws; instead, he favored the use of admittedly provisional hypotheses or tentative theories in eliciting or comprehending new facts. As he expressed it in 1837, "No working man of science advocates Bacons method." In distancing himself from the stringencies of Baconian method, he reflected the various views of not only contemporary British thinkers such as John Herschel, David Brewster, and Baden Powell but also the earlier Scottish common-sense philosophers such as Thomas Reid. While he justifiably labeled his approach "inductive," in that he believed that all scientific generalizations ultimately needed to be grounded in experience, his particular approach increasingly became known as "hypothetico-deductive." That is, in his reports to the APS, he was advocating that researchers formulate provisional hypotheses, develop the hypotheses into tentative theories, and use the theories to logically deduce particular, testable effects. Throughout his tenure at Princeton, he also was pressing this message in his course on natural philosophy. His heavily annotated notes for the introductory section of his *Syllabus*—the section concerned with epistemological and methodological issues—reveal that he was exhorting his students to follow the precepts of hypthetico-deductive inquiry. Also, aware that the terms "hypothesis" and "theory" had fallen into disrepute through misuse, he was supplying his students with myriad examples of the appropriate applications of hypothetico-deductive inquiry. Often the examples were personal, as one student documented when recording an assertion about Henry's Albany dependence on Ampère's theory: "If he has done anything for science, it has ever been by the *proper* use of a *Theory*."[12]

Henry's commitment to hypothetico-deductive inquiry colored a further foundational probe stemming from his *Syllabus of Lectures in Physics.* While preparing these detailed course outlines, he had taken time to reevaluate the widespread hypothesis that the ultimate constituents of matter are minute corpuscles or atoms. Having thoroughly reexamined the issues surrounding the corpuscular hypothesis—and having gone on to weave coverage of the hypothesis into his course—he was primed to respond to the issues when they happened to come up at an APS meeting in 1846. In opening his response, later published in the APS *Proceedings,* he justified the use of such a glaringly speculative hypothesis by restating his hypothetico-deductive credo: "Though simple insulated facts may occasionally be stumbled upon by a lucky accident, the discovery of a series of facts or of a general scientific principle is in almost all cases the result of deductions from a rational antecedent hypothesis, the product of the imagination—founded it is true on a clear analogy with modes of physical action the truth of which have been established by previous investigation." The rational, antecedent hypothesis that he preferred was a variation of the common supposition that all matter consisted of indivisible, indestructible atoms that obeyed the principles of Newtonian mechanics. These atoms, according to Henry, serve two basic functions: they combine in various ways to make up "common matter," which can exist in solid, liquid, and gaseous forms; and the same atoms pervade all space in the form of an "ætherial medium." This dichotomy allowed Henry to explain, in mechanical terms, the behavior of the so-called imponderables—electricity, magnetism, light, and heat. In particular, the actions of any of these imponderables result from "the motions of the atoms of the ætherial medium combined in some cases with the motion of the atoms of the body." In analogy to the ponderable agent of sound traveling through air, the imponderables travel through the ether as "undulations" or "vibrations" and require time to complete their passages. For example, drawing on earlier research done by Wheatstone, Henry pointed out that electricity travels through copper wire "with about the velocity of light." By referring electricity, magnetism, light, and heat to "one kind of matter," Henry was implying a potential unity of the imponderables. He had been more explicit about this possibility in his lectures and, for example, in an earlier letter to a friend. Reminding the friend of recent progress in connecting the phenomena of, on one hand, electricity and magnetism and, on the other, light and heat, he predicted that, in the near future, "the four imponderable agents of nature as we now call them will be reduced to one."[13]

To close this 1846 APS presentation, Henry returned to the subject of hypothesis and issued a warning that encapsulated his view of scientific inquiry:

> In conclusion it should be remembered that the legitimate use of speculations of this kind is not to furnish plausible explanations of known phenomena, or to present old knowledge in a new and more imposing dress, but to serve the higher purpose of suggesting new experiments and new phenomena, and thus to assist in enlarging the bounds of science and extending the power of mind over matter; and unless the hypothesis can be employed in this way, however much ingenuity may have been expended in its construction, it can only be considered as a scientific romance worse than useless, since it tends to satisfy the mind with the semblance of truth, and thus to render truth itself less an object of desire.

For Princeton's professor of natural philosophy, scientific inquiry was a humble but righteous quest.

Philadelphia Connections 13

Living in Princeton along the north-south corridor between New York and Philadelphia, Henry could reach these large cities easily and quickly. During his initial years, he relied on the stagecoaches plying the New Brunswick Pike, arriving at either destination in as little as three or four hours during good weather. By 1834, if he were not in a hurry, he could drift along the just completed Delaware and Raritan Canal, though this waterway mainly carried goods such as coal from Philadelphia to New York City. And by 1839 he could take advantage of a new branch of the speedier Camden & Amboy Railroad. Eventually, the railway would whisk him to either city, on forty-five miles of track, in about two and a half hours.

For Henry, New York held many attractions and he had several colleagues who lived there. With James Renwick of Columbia College, for example, he could enjoy the city's cultural offerings, such as an exhibition of artist John James Audubon's images of birds. For an enthusiast of science, however, the destination of choice was Philadelphia, the nation's second largest city. Thus in 1839, drawing stark contrasts between the two cities, he would write: "I almost always return from New York dispirited in the way of science. I am there thrown as it appears to me amoung all the Quacks and Jimcrackers of the land—I am disgusted with their pretentions and anoyed by their communications. How different is my feelings on a return from the [city] of Brotherly love! ! ! There is there jealousy and rivalry but also science and intelligence, and speculation and money not the only things which occupy the mind."

In Philadelphia, at the popular museum founded by the late Charles Willson Peale, Henry could savor not only fine art but also an outstanding collection of natural history artifacts and philosophical instruments. And surprises abounded. One time at the museum, in spring 1834, he even glimpsed a national folk hero—Colonel Davy Crockett, the frontiersman and politician who, two years later, would become a martyr at the Alamo. "Amoung the curiosities," Henry wrote to James about this visit to the museum, "I had a peek at Col. Crocket, and was somewhat disappointed in his appearance. He reminded me of a Methodist Minister." The next spring, during an afternoon before again visiting Peale's Museum, Henry was anything but disappointed by what he witnessed: a spectacular ascent by an "aeronaut" in a hot air balloon. Allowed as a member of "the philosophical corps" to bypass the crowds and witness the launching up close, Henry found the balloonist's erratic but ultimately triumphant "ascension" to be not merely stimulating but also edifying. "The whole exhibition was to me one of the most exciting I have ever witnessed," he told Harriet, "more however in reference to the moral than the physical effect."[1]

A greater allure than the sight of an aerial voyager, a fabled frontiersman, or novelties at Peale's Museum, however, drew Henry to this historic city bracketed by the Delaware and Schuylkill Rivers. The primary attraction was Philadelphia's prodigious community of scientific practitioners and devotees. Partly through the legacy of Benjamin Franklin—who animated not only the city's political but also scientific life until his death in 1790—many of Philadelphia's leading citizens still embraced an Enlightenment belief in human progress through scientific and technological advancement. Although the belief had been leavened with other cultural currents—for example, the progress-minded republicanism and Protestantism of the early nineteenth century—it flourished among the city's upper classes. These idealistic citizens expressed their commitment to Franklin's vision through their ongoing support of the formal institutions that he had founded: the American Philosophical Society and the University of Pennsylvania. But this venerable learned society and respected center of learning were only the two most visible examples of Franklin's enduring influence. The spirit that Franklin had engendered in his salon—a spirit of earnest intellectual exchange and unfeigned fellowship, fueled by ample food and drink—lived on through "Wistar Parties" in the private homes of a network of eager Philadelphians. Named after the late, fourth president of the APS—Casper Wistar, who had followed Franklin, David Rittenhouse, and Thomas Jefferson in the office—these regular gatherings brought together two groups: the Wistar core members, drawn from the APS's leading lights, and talented guests of all backgrounds who showed promise in scientific, technological, or other areas of cultural or social enrichment. During the 1830s the official Wistar gatherings usually came each Saturday evening following the regular Friday meeting of the APS, but similar gatherings were so common that Philadelphians came to lump them all under the category of Wistar parties.[2]

Although toward the end of his first year in Princeton, Henry had scouted the APS library and Philadelphia's various scientific, technological, and literary institutions, he first fully experienced the camaraderie of Franklin's Philadelphia during his next visit, in early November 1833. His initiation into Philadelphia's inner scientific circle came through a chance invitation to an unofficial Wistar party. One evening, thwarted in his intention to call on Robert Hare when the eminent chemist turned out not to be at home, he was walking along the street when he happened to encounter John Vaughan. An English expatriate whom Henry had previously met, Vaughan reigned at age 77 as the patriarch of Philadelphia's scientific community, mooring not only the APS but also the Wistar gatherings; he was on his way to a related party being given that evening by a prominent citizen and invited Henry along. "I was very politely received," Joseph told Harriet about the party, "and introduced to most of the company which consisted of the principal & most celebrated *savants* of the city." After remarking that he met not only the elusive Hare but also Alexander Dallas Bache, he added that the gathering "consisted entirely of men and was one of the kind called 'Whistar Parties.'" Before the group dispersed at eleven o'clock, Henry had accepted invitations "to dine with Prof. Bache & to take tea with Dr Hare." Perhaps the party and these social engagements with Philadelphia's scientific gentry contributed to Henry's decision, a month before his next visit to Philadelphia, to have an Albany tailor make him a new dress coat. "I am not altogether pleased with the appearance of my costume," Joseph had complained to Harriet, "and must indeavour for the sake of appearances to improve it."[3]

On the day after the party, Henry spent the early afternoon at the University of Pennsylvania with Bache, professor of natural philosophy and chemistry. The two then retreated to the professor's home, where Henry met Bache's wife, Nancy, "an interesting little Lady who assists him in all his magnetic observations." "I suspect they have no children," Joseph correctly surmised, "since the lady has so much time for philosophical pursuits." In the late afternoon, Henry returned to the university and spent the next five hours with Hare. They began with a detailed inspection of Hare's chemical apparatus, laboratory, and lecture hall, which the professor had patterned after the Royal Institution's well-known hall. "He says that his lecture room is the best furnished of any in the world," Joseph reported to Harriet, "and that his experiments are made on a larger scale than they are in any other place." "I remarked to him that I feared he was giving me too much of his time and that I was encroaching on some of his engagements; his answer was that he had never exhibited his apparatus to one who more readily understood them & that he ment to make the most of me. To this compliment I could only say that I was very much interested in the subject & highly gratified by what I saw."[4]

In later months and years, Henry's relationships with Vaughan, Bache, and Hare would deepen, especially with the former two. Perhaps in part because of his own modest background, he particularly found impressive Vaughan's and Bache's close ties to Benjamin Franklin. In Vaughan's case, Henry felt that it was noteworthy that the "old gentleman" had been "an inmate of Dr Franklins family when in France and was presented by the Dr to the Royal Family as one of the Dr's family." When Vaughan died in late 1841, Henry represented Princeton at the funeral and, afterward, in a note of sympathy to Vaughan's nephew, revealed the depth of his feelings toward the man: "Mr. Vaughan always received me with the kindness of a Father. . . . I loved the old gentleman while he lived and will cherish his memory now that he is dead." As for Bache, when writing a letter of introduction for him to Scotland's James Forbes in 1836, Henry emphasized that Bache was a direct descendent of Franklin, though he mistakenly called him Franklin's grandson rather than great-grandson. Henry wrote of his, by then, close friend, who was his junior by nine years: "No person of his years is more highly esteemed in this country than Professor Bache. He is a Grandson of Dr Franklin, a graduate of the military school at West Point; for several years has been Professor of Chemistry and Natural Philosophy at the University of Pennsylvania and in this situation has exerted an important influence on the scientific character not only of Philadelphia but also of this country." Henry might have added that Alexander Dallas Bache had graduated first in his West Point class and was descended from leading Philadelphia families on not only his father's but also his mother's side; whereas the Bache name traced through his father to Franklin's daughter's husband, his mother's name of Dallas was distinguished in its own right, especially for political might and civic service.[5]

A year after Vaughan had escorted Henry to the party where he met so many of Philadelphia's scientific luminaries, Vaughan had joined with Bache and four other associates in nominating Henry for APS membership. Actually, the initiative for the nomination had come in back-to-back appeals from Benjamin Silliman and Charles Hodge. In a letter to Vaughan urging Henry's election, Silliman explained: "Mr Henry is one of the most promising of our natural philosophers & is already distinguished for some very fine advances in Science." After extolling Henry's experimental investigations, Silliman added: "His private character is also excellent & I am sure the society does not possess a member of his age of greater merit nor any member of more promise." In a letter drafted a few days after Silliman's, Princeton theologian Hodge, a Philadelphia native who had married a great-granddaughter of Benjamin Franklin, was more brazen about the APS offering membership to Henry. Writing to his brother, a prominent Philadelphia obstetrician and APS regular, Hodge simply asserted: "I want Professor Henry elected a member of your philosophical society. He is one of the first men in the country, so say Profs. Silliman & Renwick & so thinks every body here." Hodge ended this directive on a more personal, but no less emphatic, note: "I think him myself a genius, who would do honour to the Royal Society of London, or any other body of the kind."[6]

Henry, while aware of the shortcomings of the timeworn APS, was enthusiastic about the organization's future. He felt that younger members such as Bache were reinvigorating the society, moving it from an era of fallowness to one of renewed productivity. He also trusted that these younger leaders would not be "quite as exclusive as their predecessors," thereby overcoming the membership's traditionally elite and local character. Thus, when Vaughan informed Henry of his election in early January 1835, Henry considered it a true honor and assured Vaughan that he would seek to prove himself worthy of the distinction. Writing more candidly to Bache, Henry acknowledged feeling some personal pride in the honor, but he insisted that he valued his election primarily because of the greater opportunity for joint research with Bache and other members. Torrey, who was elected simultaneously with Henry and two other colleagues, reacted more cynically, joking with Henry that the notoriously cliquish society finally merited the all-encompassing adjective *American*. In particular, upon learning of his own election, Torrey sarcastically asked his friend: "Have you been dubbed with a diploma from the—I must *now* say—AMERICAN philosophical Society? They have had the penetration to discover my modest merits, & as you so greatly resemble me (I mean in *modesty*) I suppose you have likewise had this blushing honor put upon you!" For the earnest Princeton natural philosopher, APS membership was not a joking matter. To his own brother, he could reveal the depth of the pride only hinted at to Bache: "I felt it something of an era in my life to join a society formed by the immortal Franklin and to subscribe my name in the same list with those of Jefferson, Rittenhouse and many others of the same class." The current president of the APS, Henry marveled, literally "sits in the same chair which Franklin once occupied." What Joseph felt to be the profundity of the honor led him even to warn James to conceal his election from his former Albany colleagues for fear of rousing envy.[7]

Regarding the more practical advantages of APS membership, Joseph further explained to James that he now enjoyed access to the society's outstanding collection of scientific books and journals. The sizable gaps in Princeton's collection, particularly the lack of current scientific journals, made this access essential. To ease Henry's visits to the APS library, Vaughan, the society's official librarian and treasurer, provided a locked drawer on a table for the itinerant professor's personal use. During a week-long visit in spring 1835, Henry typically would have breakfast with Vaughan before beginning by 6 A.M. to transcribe notes from pertinent books and articles. With a full slate of evening engagements and meetings, another benefit of APS affiliation, he seldom got to bed much before midnight. Membership also served Henry by providing him with a public forum for presenting papers and abstracts as well as a formal mechanism for staking priority in new discoveries. In particular, by periodically submitting his research notes to an APS committee for private review and authentication of dates, Henry hoped to ensure his priority in potentially innovative but still incomplete investigations. Previously outpaced in major breakthroughs by Moll, Faraday, and Forbes, he had grown protective of his investigations. Thus, he had the committee review his laboratory notebooks four times in the years around 1840 as he plugged away at the research that would lead to the later articles in his series, "Contributions to Electricity and Magnetism." During an APS meeting in 1842, for example, after giving an oral report of what would become the fifth contribution in his electrical series, he left his notebook with the committee. "The book contains," Henry advised Bache, the chair of the committee, "a record of all the experiments placed under their proper dates. I wish these inspected by the committee as my vouchers for the time these were submitted to the society."[8]

As a final and more mixed benefit, APS membership furnished Henry a creditable outlet for publishing his findings: the APS *Transactions.* The benefit was mixed because, while Henry could count on publishing his major work under the imprimatur of the esteemed APS, the printed volumes of the *Transactions* were slow to appear and had only limited circulation. Indeed, in 1835, Henry deemed it necessary to personally send copies of the first two entries in his series—the description of his battery and the extended analysis of self-induction—to twenty-two of the nation's leading scientific researchers, fifteen old acquaintances, and three Princeton trustees. (The absence from the list of prominent European natural philosophers probably reflected not only Henry's insecurity about intruding on these luminaries in an impersonal manner but also the cost and inconvenience of sending packages overseas.) He told Silliman in 1839 that he particularly regretted not issuing the self-induction article in the Yale professor's *American Journal* where it would have become "much more widely known." As he further explained to Silliman, he had elected to publish under the auspices of the APS to help Bache, Hare, and other principals "resucitate the Phil Society"—an effort that he reassessed in 1839 and found, without intending to sound egotistical, to have been reasonably successful. A measure of the revival's success had come in 1838 when the APS added a more informal journal, the *Proceedings.* Henry would personally find the *Proceedings,* issued every quarter, to be, as he told Vaughan, "of great importance in the way of securing priority." Although creation of the new journal helped Henry personally, it still did not eliminate the problem of disseminating APS reports in a timely manner. As late as 1846, Silliman's son—Benjamin, Jr., who now helped edit the *American Journal*—complained to Henry that issues of the *Proceedings* were not arriving regularly.[9]

During the same 1833 trip in which Vaughan introduced him to Hare and Bache, Henry became associated with another of Philadelphia's technical societies boasting a national reputation. Bache helped draw Henry into the Franklin Institute, a decade-old organization that embodied its namesake's commitment to bringing scientific knowledge to bear on "the mechanic arts." Wanting to go beyond the institute's original reformist goal of merely educating artisans and mechanics, Bache was striving to pragmatically link scientific and technological practices and so help to foster American inventions and patents. He soon enlisted Henry in an institute project to appraise a purportedly improved mariner's compass. Jacob Green, Henry's longstanding Philadelphia friend, also reflected the institute's scientific ambitions when he exhibited before the group one of Henry's electromagnets. In a more personal gesture, when Henry was looking for ways to highlight his earlier discovery of self-induction following Faraday's unexpected announcement, Bache used the institute's *Journal* to publish speedily the abstract of Henry's newest experiments and the accompanying comment on his priority. By 1844 Henry assumed a leadership role in the institute when he chaired a special committee to investigate the explosion of a twelve-inch gun, the "Peacemaker," aboard the Navy's new steam frigate, the *Princeton*—an explosion during a public demonstration that killed the secretary of state, secretary of the navy, and four other men. "I was never more impressed," Henry commented to Bache during the thick of the Philadelphia-based investigation, "with the important exercise of mind which the members of the Institute receive in the discharge of their duty as members of committees of enquiries of this kind."[10]

In 1836, when a circle of colleagues affiliated with the Franklin Institute invited Henry to join them for experiments in an open field on the outskirts of Philadelphia, Henry found himself participating not merely in a group that had adopted Benjamin Franklin's name or even his guiding philosophy. He was also directly taking part in an extension of Franklin's most famous nonpolitical experiment: flying a kite to investigate atmospheric electricity. Although "much fatigued" by the first day's trials and although the sky had been "perfectly clear," Henry succeeded with the Franklin Kite Club in drawing sparks from tandem kites at the end of about one mile of wire. Seemingly smitten with a fascination for Franklin's electrical studies, Henry was also beginning about this time to use his Princeton facilities to delve deeply into an invention related to the kite experiment—Franklin's celebrated invention of the lightning rod. Even the roof and wire assembly that Henry rigged in his Princeton home to analyze lightning in the comfort of his study mimicked a similar apparatus that Franklin had contrived to detect electrical storms by activating a bell in his own study. In two anonymous letters in 1838 to the *Newark Daily Advertiser* criticizing an alleged new design for a lightning rod, Henry apparently went as far as to speak for Franklin, though playfully, when he signed the letters with only the letter "F." And whereas in his lectures and experiments he had always been sympathetic toward Franklin's single fluid theory of static electricity, by the early 1840s he declared himself a definite adherent of an amended version of the theory—a theory out of favor with Faraday and other European researchers. Also by the early 1840s, he was actually using in his laboratory "the old Dr Franklin battery"—a historic assembly of twenty-four connected Leyden jars that he probably acquired through Bache, Vaughan, or the APS. In light of these involvements and views, it is not surprising that Henry's contemporaries began to identify him with, as one acquaintance expressed it to him, "your great prototype, Franklin."[11]

Into the 1840s Americans' increasing identification of Henry with Franklin went hand in hand with their rising image of Henry as the nation's most ingenious electrical investigator. Just as the rising scientific image boosted the Henry-Franklin association so too did the increasing invocation of the association boost the image. Nowhere was this mutual reinforcement clearer than in the keynote address that Robert M. Patterson delivered on the highly public occasion of the APS's four-day, centennial celebration in 1843. Appearing before a large, national audience that happened to include leaders of the Presbyterian Church (who were in Philadelphia for their own meeting), Patterson spoke as not only an APS vice president but also the director of the United States Mint—a prestigious Philadelphia post held earlier by Patterson's father and originally by David Rittenhouse. Toward the end of his address, he proclaimed the continuing glory of the APS by singling out Henry's electrical achievements and comparing them favorably to Franklin's. Although embarrassed by this public adulation, Joseph confided to Harriet: "Dr. Hodge and Dr. Maclean appeared to enjoy the compliment to me as a professor of Princeton very much since it was given before the clergy from every part of the United States." Following inclusion of Patterson's speech in a special centennial volume of the APS *Proceedings,* a Princeton paper reprinted his flattering comparison of Henry with Franklin; ever modest in public, however, Henry intervened with the editor of the *Newark Daily Advertiser* not to republish the piece. The modesty would be less apparent in the autobiographical sketch that first appeared anonymously in 1871 in the *Princeton Review.* There Henry used Franklin's acclaimed electrical work as a criterion for a flattering evaluation of his own research program. He branded his own early, electrical investigations "the first regular series on Natural Philosophy which had been prosecuted in this country since the days of Franklin." Evidently, Henry had taken to heart the many reminders of Franklin's legacy that he had encountered in Philadelphia—Vaughan's and Bache's personal backgrounds, the APS, the University of Pennsylvania, the Franklin Institute, the Franklin Kite Club, and Patterson's centennial speech.[12]

In 1836 there was nothing unusual about fifteen scientific devotees meeting in a meadow and lining up with hands joined to form a human gauge to estimate the magnitude of a spark drawn from a kite far overhead. Group scientific endeavors were endemic to the early nineteenth century—another remnant, in part, of the Enlightenment enthusiasm for nature study. Experiments and measurements provided the participants with opportunities as much for genial social interchange as philosophical inquiry. From the traipsing about the countryside of a band intent on enumerating geomagnetic dips to the gathering of a throng resolute on electrifying the corpse of an executed criminal, Henry seemed to relish these cooperative ventures. He also was a spirited booster of the era's many formal, social groupings, from the Albany Institute to the American Philosophical Society. But even with all these opportunities for interplay, he and his closest colleagues still perceived a need for another, more intimate group. Thus was born "the club." "It is to consist," Henry informed Torrey, "of a few Individuals associated for mutual support and instruction." The core group included not only Henry, Torrey, and Bache but also brother-in-law Stephen Alexander and three colleagues working in Philadelphia during the mid-1830s: mathematician and astronomer Edward H. Courtenay; meteorologist James P. Espy; and geologist Henry D. Rogers. Except for Espy, who was at least a decade older than any other member, all shared birth dates between 1796 and 1808. Henry explained to Torrey that the new group would meet three or four times annually at sites of mutual convenience. "It has no name," Henry further advised his friend about the group, "and must be kept secret so as to excite no jealousy &c."[13]

For a few years, at least, "the club" did provide its members mutual support and instruction. Although the meetings were sporadic, the members managed to help each other in matters ranging from exchanging technical information to securing jobs. For example, Rogers drew on a recent London visit to call Henry's attention to Wheatstone's study of electricity's velocity, and, in turn, Henry drew on his background in meteorology to feed Espy data on dew points for his developing theory of storms. Similarly, Henry took advantage of his contacts with the New Jersey governor to help land Rogers an assignment as head of the state's first geological survey. Despite repeated appeals from Torrey, however, Henry was unable to arrange a job at Princeton for Torrey's promising young protégé, botanist Asa Gray. Notwithstanding the beneficial role of "the club," by 1840 it was fading because of members' relocations and changing interests. Bache commiserated with Henry over the circle's demise: "Here a few years ago we had Courtenay & his astronomy, Espy & his meteorology, Rogers & his geology, you and I electrically and magnetically engaged, meeting to rub each other up. Now where is the club?" From the original group he singled out Henry as the sole member to retain his earlier commitment to scientific exchange: "You the only one of the set precisely as you were then, saving *years* & experience, gray hairs & babies."[14]

Although strongly drawn to fraternal groupings dedicated to scientific inquiry, Henry also joined at least one all-male association that dispensed entirely with scholarly talk and concentrated only on fellowship. Within months after Bache nostalgically lamented the collapse of "the club," he was helping to induct Henry into the Vaughan Club. Founded in the late 1830s, this Philadelphia group had a simple purpose: to honor its leading member and namesake, John Vaughan, through the annual sharing of exotic wines.

Henry was an avid crusader for temperance, but since at least 1833 he had distanced himself from what he considered the zealots; he limited his mandate to the protection of untainted youth from alcoholic beverages rather than the reform of confirmed drinkers or the elimination of moderate drinking. Whatever his rationale—or rationalization—after being received into the club on a Wednesday evening in mid-July 1840, he stayed until nearly midnight with Vaughan, Bache, and a dozen other revelers who were toasting, singing, orating, and sampling an amazing collection of wines.

The evening's festivities started when a member "marshaled" the new bottles that each participant had just contributed to the club's already impressive stock. Henry presented a sherry, an "Amontillado of 1785." Other notable additions included, according to the evening's assigned chronicler, a Madeira that had belonged to Chief Justice John Marshall and a "Butler Wine, considered the ne plus ultra by many" that had cost "155 dollars the demijohn." Later in the evening, not long after a member had sung the "Vaughan Ode," the most exceptional wine made its appearance: "the Old ancient brown Sherry—of 1680—the wine which was recorded in virginia at the time of the Revolution as 'the oldest in the country.'" As the evening wound on, according to the chronicler, someone "proposed the health of Professor Henry & accompanied the toast with some felicitous & complimentary observations to which Professor Henry responded." Then came the requisite salute to Franklin, "the illustrious and the good," with Bache wittily retorting for his great-grandfather. Finally, with "the bottles being tested, and the judgment of the Members being more than ripened by the excellence of the entertainment . . . the Club adjourned after a most agreeable Session, and one of unmixed harmony & good fellowship." For better or worse, Henry would be unable to avail himself of the Vaughan Club for long; the gatherings would cease with the death of John Vaughan at the end of 1841. Henry would, however, continue to avail himself of the group's main staple. Five months after his evening of revelry, as Christmas and New Year's Day approached, he confided to James that the family had stocked "a lot of good things" for the holidays, "including several bottles of the pure juce of grape"; he coyly added, "but as we are temperance people this must not be said aloud."[15]

In late 1833, when the Albany *Argus* ran a story on a professor named Henry becoming president of a southern college, some of Joseph's old friends presumed that he was the new appointee. In fact, the news story involved another educator named Henry, and Joseph quickly instructed his brother to correct his friends' misconception. "Such an appointment," he insisted, "would not be congenial to my tastes, talents or acquirements. I made a resolve more than fifteen years ago to devote my life to the pursuit of science and no situation however lucrative or honorable could induce me to depart from my resolution."[16] During the coming few years, although no offers of administrative jobs would test Henry's resolve to devote his life to science, a bid would come that would sorely try his commitment to practicing science at the College of New Jersey. In mid-1835, a distinguished southern university would try to lure him away from Princeton with a professorship that was unquestionably more lucrative and perhaps more honorable. The offer would come through a contact in Philadelphia, the city that was providing increased professional exposure for Princeton's rising star.[17]

In Philadelphia, Henry had impressed a fellow visitor, a prominent mathematician from the University of Virginia. He was so impressed, that he later persuaded the board of his university to offer Henry a professorship in natural philosophy. This was not just any professorship at the university founded by Thomas Jefferson, but a position that respected natural philosopher Robert Patterson had recently relinquished to become director of the United States Mint. The Virginians offered Henry an appealing package: an attenuated teaching load that would allow ample time for research, a well-stocked cabinet of philosophical instruments, a "handsome house" for his family, and an annual wage of about $4,000. This was one of the nation's highest academic salaries, well above Henry's current earnings of $1,200. Though $1,200 was the top salary being paid a Princeton professor, Henry apparently considered the sum inadequate; in his unsigned 1871 autobiographical sketch, he characterized his income at the time of the Virginia offer as "scarcely sufficient to support his family, and to meet other demands upon him." Nevertheless, after much soul searching, encouragement from colleagues such as Hodge, and negotiation with Princeton's trustees, Henry politely declined the lucrative Virginia appointment. He explained to his suitors that, previously, he had made a personal commitment to Princeton's officers to stick with the school as long as he was receiving adequate support and fair treatment—which he was. Ultimately, however, his decision to remain was contingent on Princeton's trustees authorizing additional perquisites. Joseph described to James his informal negotiation with the trustees, in which he told them, "if they would immediately give me a new house and raise my salary[,] I would stay." In salary Henry was requesting an additional $300 annually, an increase of 25 percent.[18]

Henry found the dickering with Princeton's trustees to be nerve wracking, as he delayed his response to Virginia while trying to get assurance that his conditions would be met. "Had I been longer at Princeton and did I not conceive that I was under some obligations to the Institution," he confided to James, "I would not have hesitated to accept the call from Virginia. I do not much like the idea of living in a slave state or going much farther to the south. Still these considerations would not alone have weighed against the salary and other advantages of the place." Perhaps the large company of Princeton students from southern states, along with his own vacillation if not antipathy toward African Americans, inured Henry to Virginia's slave heritage. Decades later, in the autobiographical sketch, he confirmed that, in rejecting the offer, Virginia's status as a slave state had not seriously influenced his decision. In particular, writing after the Civil War, he recalled discounting a prescient warning made during the negotiation; a Princeton trustee had predicted, "if you live to the ordinary age of men, you will see a rupture between the North and South, and in that case your position in Virginia may not be agreeable." Lingering questions about the Virginia job became moot when Princeton's trustees acceded to Henry's wishes for a new house and an increased salary. Although he soon profited from the $300 boost to his annual income, he and his family would have to wait until September 1838 for the new house. A "very pleasant commodious house" is how Henry would describe the family's new dwelling. Spacious and handsome, the main section of the rectangular, brick house featured a pillared front porch bracketed on either side by two large windows, an array of five large windows across the face of the second story, and four substantial chimneys. There was even room for Henry's own study, which sported elegant woodwork and tall windows. Workers built the house in accord with Henry's master plan for the campus, along the quad's edge next to the cleared site of the family's original dwelling.[19]

With Henry's star in the ascendency, only a few months elapsed before another school, also in Virginia, was courting him. In a curious twist, the College of William and Mary, the nation's second oldest college, tried to persuade Henry to fill a professorship of natural philosophy that had just become vacant following the occupant's move to the post that Henry himself had refused at the University of Virginia. According to the official who was pleading William and Mary's case, the school could not quite match the salary offered by the University of Virginia but it could provide Henry with greater independence and leisure to pursue his personal interests. Moreover, concerning "pride of feeling & patriotic associations," the official invoked Jefferson's name to suggest that the College of William and Mary simply outshone its upstart, cross-state rival: "If the University be the offspring of Mr Jefferson's old age, this is his alma mater. Here he laid the broad foundation of his fame, as did most of the Illustrious men of our Commonwealth." These pleadings, while eloquent, did not sway Henry.[20]

What did again sway him were feelers that his Philadelphia friends extended during the next summer concerning a vacancy at the University of Pennsylvania. Bache not only helped alert Henry to the vacancy but, what is more important, created it by accepting the presidency of Girard College, a heavily endowed school being planned for poor, white, male orphans in Philadelphia. While taking part in experiments with the Franklin Kite Club in July 1836, Henry learned that, if he so desired, he would be elected professor of natural philosophy and chemistry at the University of Pennsylvania. He wrote to Harriet warning that he might be facing the same type of personal "perplexity" that he experienced with the University of Virginia offer; at the moment, however, he was averse to moving to Philadelphia though it "would doubtless be very desirable to us all." Back in Princeton, Torrey, who was a steadfast Presbyterian, argued against the move because his friend would be forsaking a Christian enclave for a presumably irreligious throng. "I don't like the idea of his removing to Philadelphia," Torrey groused to Asa Gray, "for I think that the men of science there will have an unfavourable influence upon his spiritual welfare."[21]

A month later, after receiving a formal invitation from University of Pennsylvania trustee Robert Patterson—and after "much deliberation"—Henry declined the appointment. He told Patterson that, while tempted, he decided against the move for reasons ranging from his personal "dislike to change" to "additional inducements which my friends in Princeton now hold out for my remaining." Foremost, the inducements likely included an extended trip to Europe, a venture very much in vogue among aspiring American educators and scholars. Ostensibly, the trip would enable Henry to buy much needed philosophical instruments for the college; but it would also allow him to establish personal relationships with leading European scientists and to observe firsthand their teaching and research methods. The idea of the trip had been percolating for some months, apparently arising as a possibility through the Virginia negotiation. Indeed, Henry later recalled that the negotiation precipitated not only the salary increment and new house but also the European trip. Thus, before Princeton vice president Maclean could have received intimations of the Pennsylvania offer, he wrote to Henry acknowledging that, "if you are willing, arrangements must be made for you to proceed next spring to Europe, to spend there as much time as you desire, & procure whatever apparatus we need." The idea of the trip, which the board of trustees soon formally approved, probably received an additional boost when Bache announced that, as part of his preparation for leading Girard College, he would be leaving in the autumn to survey educational programs in Europe. By remaining as a professor in the village of Princeton, Henry had assured his passage to London, Paris, and Edinburgh.[22]

Henry would set sail from New York en route to London in late February 1837—a year notable for America's first great depression and Victoria's succession to the British throne.[23] But before he could depart for a trip that would extend from February through October, he would spend months making preparations. In the preceding December, as he celebrated what he thought was his thirty-seventh birthday (he was actually thirty-nine), he began to secure various letters of introduction to ensure smooth entry into European scientific and social circles. Indeed, he gathered letters from the country's leading figures not only in science but also in government. Fitting in a quick trip to Washington—his first trip to the nation's capital—and using contacts that he had made in Albany, he visited New York politician and President-elect Martin Van Buren to obtain letters addressed to the U.S. diplomatic ministers in London and Paris. He also used the occasion to set up an "interview" with John Quincy Adams, the nation's sixth president and now a Massachusetts representative. When he was about to sail in February, however, he discovered that his letters from, by then, recently inaugurated President Van Buren might not suffice for gaining entry into France. In response, he simply jotted off a note to Washington requesting a formal passport from another old Albany acquaintance—the nation's attorney general and acting secretary of war.

In his usual, conscientious manner, Henry not only arranged for Stephen Alexander and Albert Dod to cover his summer classes on astronomy and architecture but also compressed his own, year-long natural philosophy course into three months. The contracted schedule necessitated evening lectures and this led in mid-winter to the occasional problem of the hall being, as one senior remarked, "too dark to take notes."[24] He also arranged for his family to "board in the country" near the Alexander's old hometown of Schenectady. Although he needed immediate cash and had to "call in all the funds I can claim as my own"—$900, secured by a note to the Alexander estate—he was confident that the trip "will not in all probability cost me much more than my ordinary living." Henry anticipated receiving not only "a commission from the college to purchase apparatus" but also supplemental funds from benefactors. Torrey, who lived in New York and taught at the College of Physicians and Surgeons when not teaching during summers in Princeton, booked a stateroom for Henry on the New York packet *Wellington,* a "noble vessel" about to depart from New York on its inaugural transatlantic passage. He also reserved a stateroom for a young man who would be Henry's traveling companion, twenty-year-old Linn DeWitt, who had been under Henry's charge in Princeton since the death of his father, the surveyor general of New York.

Already anxious and tired as he spent a few days in New York awaiting the departure of the *Wellington,* Henry probably found no solace in the title of an article that a colleague entrusted to him at the last minute for distribution in Great Britain: "On the Gales and Hurricanes of the Western Atlantic." A special $40 premium that he had paid to ensure that his life insurance policy remained in effect while abroad perhaps brought some peace of mind. Unable to sleep on the Sunday evening before embarking, he stayed up until 11:30 P.M. writing letters to his sister, brother, and wife. Apparently, Nancy had been fretting about the dangers associated with her brother's pending voyage. Earlier in the month, Joseph had asked James to tell their sister "that she must not take trouble on trust; it is bad enough when it comes." Now, the night before the voyage, Joseph personally repeated the message and went on to console Nancy with Christian assurances of God's providence:

> It is not proper to make ourselves unhappy in anticipating troubles. The All Wise disposer of events will order all things aright; and whatever happens will be for the best. A voyage to England is now considered a matter of little consequence and the danger in such a vessel as the one in which I sail, is not more than that of a trip to Albany in a steam-boat. Is not the same Providence our protector on sea as on land? We are not sure of our lives from moment to moment, and while we should be prepared at any time to die we should also recollect our lives are in His keeping who holds the winds in His hands and who controls the rage of the elements.

He closed with a benediction: "I will join you in beseeching the God of all Mercies—if it be consistent with his will—that I may be safely restored to you and my family and that you may all be under His guidance and protection during my absence."

While the patriarchal implication that he served as the family's earthly protector was familiar, the open expression of Christian sentiment was unusual for Henry, even in a private letter to his own sister. Although committed to essential Presbyterian teachings—and their Calvinistic underpinnings—he was reluctant to broach matters of personal faith. It would typically take an event of the magnitude of a death or birth to elicit his religious feelings. For example, an affinity for the related doctrines of predestination and foreordination—that an all-powerful God predetermines persons' eternal destinies in death and personal fortunes during life—showed a few years later in his responses to the passing of a friend's daughter and the birth of his own niece. In the first instance, he assured the bereaved father that not only does such a death serve humans by making certain "that our attachment to life may not be too strong" but also that the child is "beyond our sympathies" because "its destinies are in the hands of a just God." In the latter case, on the occasion of his niece's birth, he matter-of-factly wondered aloud to his brother whether "weal or woe" would be "the destiny of the little woman." Henry developed the theme of avoiding too strong an attachment to life when he later wrote to Harriet about the impending death of Stephen Alexander's wife, stricken in the prime of motherhood with a fatal disease. "Truely this world is not our abiding place," he counseled, "and it behoves us so to live as if we were merely the transient inmates of a way house awaiting for the arrival of the vehicle which is to bear us hence." Although our earthly existence holds little significance compared to our afterlife and though we have no way of knowing if we have been elected for salvation, we must, nevertheless, "strive to do our duty and leave the result without too much solicitude to the disposition of a higher power." A few days before his sister-in-law finally died, Henry enjoined her "to throw herself entirely on the merits of her saviour"—that is, as a sinful human, simply to trust in the redemptive power of Christ's crucifixion, realizing that nothing she could do would earn or ensure salvation anyway. Such submission was necessary, he explained, because even "saints are impure in the sight of the Righteous Judge" and "can only be saved through the merits of a Saviour." Similarly, in the consolation he offered when Bache's brother died at age thirty-five in an accident, Henry revealed a belief in the innate corruption and sinfulness of humans. Calling attention to "the downward tendancies of our nature and the thousands of our fellow men who sink into their graves a disgrace to themselves and the race to which they belong," Henry comforted Bache by explaining that his brother's death was merciful in that he died as a dutiful family and professional man with an "unsulied name." Henry was normally so tight-lipped about his religious feelings that he would not join Princeton's First Presbyterian Church until

1844. And although this formal step would require a public declaration of faith, he apparently would instruct the pastor henceforth never to ask him to participate in prayer meetings, supposedly because of his "reluctance to public speaking." He seemed more comfortable in a dispassionate role such as trustee of the church, a part that he accepted two years later, or as teacher of the doctrines of natural theology, a role that he annually reprised in his college lectures.[25]

On the eve of setting sail and leaving his loved ones, however, Henry had become caught up in the poignancy of the moment and allowed his religious feelings to surface. Thus, later that night in his letter to Harriet—an impassioned tribute to her for behaving bravely at his departure (not like "a weeping paralized female")—he would also close with another Christian benediction. "And now my Dearest on earth," he intoned, "I consign you to the protection of Him who never forsakes those who put their trust in him aright." The next day, he imposed on the pilot of the steamboat towing the *Wellington* out of the bay to post a final note to Harriet: "All the canvass is set and every thing appears to give us a prospect of a pleasant voyage."

London, Paris, and Edinburgh 14

The voyage turned out to be a routine, late winter crossing, but the North Atlantic was still rough enough to keep Henry mildly seasick for all nineteen days. Once settled in London, however, he found almost everything to his satisfaction. "My reception thus far in England," he wrote to Harriet at the end of his first week ashore, "has surpassed my most sanguine expectations. Every person I have met has treated me with the greatest kindness and attention." He explained that he and DeWitt were sharing a suite of two bedrooms and a parlor "in a very pleasant and fashionable part of the city"; he hastened to emphasize, though, that he had chosen this site not because of its poshness but merely because it happened to be near the scientific riches of the Royal Institution. This hint of a puritanical aversion to self-indulgence became more pronounced as he further explained to Harriet that, because she was not along to share in the experience, he felt obliged to deny himself the pleasure of fully enjoying "the wonders of London, this modern Babel." "I deeply regret that circumstances did not permit you to accompany me. I could then enjoy my visit unalloyed. I now feel I must devote myself almost exclusively to science and give only a hasty glance at other matters of interest."[1]

The hasty glances briefly multiplied when Henry, ten days after arriving in London, finally linked up with Bache and his wife, who were also visiting the city during their European tour. Although Henry had dutifully spent his first days getting acquainted with the scientific resources and notables of the Royal Institution, the Royal Society, King's College, and the Royal Military Academy at nearby Woolwich, he quickly fell in step with the Baches as they went sightseeing. He later explained to Harriet that he had merely taken advantage of an Easter lull in scientific activity. That is, he had used the lull "to kill off a few of the lions of the city," by which he meant touring the Tower of London, Westminster Abbey, and other main attractions. He also went to the theater, his adolescent passion, and saw a performance of *Macbeth;* he jotted in his travel diary that, while "pleased" with the production, he failed to "experience as much pleasure as formerly in the amusement of the stage." Two months after this Easter burst of urban entertainments, as he prepared to leave London for Paris, he would reveal a twinge of guilt as he again voiced to Harriet his precept of self-denial. "Were you with me My Dear Dear Harriet I would enjoy this visit very much, but I now feel that any thing that I do in the way of mere pleasure is not proper and that I must confine my attention to the main object of my visit."

The main object of his visit—science—did claim most of his time. To his gratification, he discovered that many members of the physical science community in and near London were familiar with his name. Indeed, during his first days in the city, he came across a copy of his multicoil electromagnet on display in a public gallery and talked to a prominent instrument maker who had filled many orders for "Henry" magnets. Lying on the instrument maker's counter was even a recent, electrical journal with an article whose title referred to Henry's work on self-induction. Similarly, when he later visited the office of the *Philosophical Magazine*—the journal on which he had cut his teeth when it still went by the name *Annals of Philosophy*—the main editor volunteered that, minutes before, he had run across Henry's name in a German journal. Henry found to his surprise that he did not usually need to present his formal letters of introduction when first calling on his London colleagues because, as he was informed during one visit, "my name was a sufficient introduction." (Outside of scientific circles, the letters remained important. His note from President Van Buren to the U.S. minister in London proved useful in assuring "very attentive" treatment from the minister.) Not only his reputation but also interventions by two hospitable London contacts eased Henry's entry into local scientific households and halls; John Vaughan's elderly brother and middle-aged nephew, both well-connected Londoners, took Henry under their wings just as their relative had in Philadelphia.

During his two months in and near London, Henry enjoyed mingling with an imposing group of scientific trendsetters whose writings and accomplishments he already knew well. His daily schedule overflowed with activities involving prominent researchers and scientific authors such as John Daniell, Peter Barlow, William Sturgeon, Charles Wheatstone, Charles Babbage, William Ritchie, Michael Faraday, Mary Somerville, Baden Powell, Samuel Christie, and Edward Sabine. He even crossed paths with out-of-town visitors, including Edinburgh natural philosopher James Forbes, the twenty-eight-year-old who had been Henry's primary European contact for the past three years, the two men having exchanged meteorological and geomagnetic data and apparatus.[2] Reflecting his continued interest in the practical arts as well as basic science, Henry's schedule also included tours of state-of-the-art factories and industrial installations ranging from a ship-yard to a brewery.

Henry involved himself particularly with Wheatstone, with whom he seemed to have a natural rapport, and Faraday, with whom he shared so many technical interests. Five years younger than Henry, Wheatstone occupied the Chair of Experimental Physics at King's College; Henry first met him, however, at a music store that he ran with his brother. Wheatstone soon familiarized his new American friend with his latest work on subjects such as musical sound, the electric spark, the velocity of electricity in conductors, the mechanical simulation of the human voice, and the telegraph. In his travel diary, Henry matter-of-factly described Wheatstone's telegraph; the American apparently was unperturbed that his British colleague was about to patent a different version of an instrument that, back in Albany and Princeton, he had helped pioneer. Unlike his own electromagnetic telegraph, Wheatstone's was a galvanic device that used separate circuits to deflect various magnetic needles (in effect, galvanometers) that indicated letters of the alphabet. Henry also seemed unperturbed that Wheatstone had fabricated a variant of a "relay," again using a magnetic needle, for restoring the strength of telegraphic signals traveling great distances; Henry later recalled telling Wheatstone that in Princeton he also had devised a relay, but one using an electromagnet to transmit large mechanical effects over long distances. Whereas Henry did not take unusual notice in his diary of Wheatstone's telegraph or relay, he did take detailed notes on a related electrical law about which Wheatstone was knowledgeable—Ohm's law. Earlier, in Princeton, Henry had unsuccessfully sought information about this German law.[3] Wheatstone, ahead of most of his English associates, had grasped the significance of the law and could lend Henry a recent French explication. Henry would soon be integrating into his own studies Ohm's versatile, algebraic relationship, extensions of which Henry had come close to revealing in his own investigations of "intensity" and "quantity" electromagnets. A month into his London visit, already indebted to Wheatstone, Henry sent home a glowing appraisal of his new friend: "Mr Wheatstone of whom I have before spoken as having devoted much time to me is the most 'talented' person as far as my observation extends in London." In what could pass as a self-criticism, however, Henry judged that, while Wheatstone "has more new projects than will suffice for the labour of a long life," he unfortunately "does not finish one thing before he begins another."

No one could claim that Faraday left projects unfinished. With major achievements in chemistry and natural philosophy, he held sway at the Royal Institution as, in Bache's words, "the genius loci"—the prevailing spirit of the place. In contrast to Henry and Wheatstone, who apparently achieved an immediate rapport, Henry and Faraday seemed to approach each other in a more measured and reserved manner. As a courtesy Faraday gave the American visitor full access to the facilities and offerings of the Royal Institution. Henry took advantage by attending a morning series of technical lectures that Faraday was offering on metals. He also attended two lectures in a popular series that Faraday was giving Saturday afternoons on the ancient Greek elements of earth, water, air, and fire. Henry was struck by the turnout for the Saturday program, commenting in a letter home that the street approaching the Royal Institution was "crowded with the carriages of the gentry." Inside the institution, he further remarked in his journal, "The room was crowded by ladies many of whom were very busily engaged in taking notes." Using his journal to register a quibble about a particular, questionable fact in the afternoon lecture, Henry went on to offer begrudging praise for Faraday's expository skills: "Mr F is deservedly a very popular lecture[r] but does not surprise or strike one with the depth of his remarkes or the power of a profound mind but more by his vivacity of manner & his hapy illustrations as well as inimitable tact of experimenting." Later, back in the sanctuary of his own Princeton lecture hall, Henry would depict the Royal Institution as "a society established by our countryman Count Rumford for the amusement of the nobility & gentry of London."[4]

A month after arriving in London and about ten days since beginning to attend the lectures, Henry received an invitation from Faraday to have breakfast at the Royal Institution in his and his wife's living quarters. During the get-together Faraday took time to show his guest a few of the institution's historical artifacts before he apparently had to leave to attend to other obligations. Henry used the visit to inventory in his diary a sampling of the Institution's electromagnetic instruments, and he also spent much of the morning chatting with Sarah Faraday and her two visiting brothers. She showed Henry her husband's extensive collection of portraits and letters from, as Henry expressed it, "the most eminent men of science in Europe." Moreover, she not only gave him a duplicate portrait of one of his favorite "men of science" but also agreed to accompany him the next day on a visit to the person pictured in the portrait: Mary Somerville. Already favorably disposed to Somerville because of her coverage of his electromagnet in her popular textbook, Henry had come to appreciate her more after meeting her a week earlier and discovering that she was a native of Scotland. "I was not aware of this circumstance," he recorded, "and claimed kindred with her as having descended myself from the land of lakes." In coming years a report that Henry drafted would cite Somerville as an example to support the contention that young women in the United States could profitably be taught mathematics without compromising "all that is amiable and interesting in the female character." Despite her "profound attainments" in mathematical science, Somerville not only remains "well-skilled in music and in painting" but also "has lost nothing of that sensibility and sweetness of character which ever made her the admiration of her acquaintances, and now render her the loved as well as repected of her family."[5]

A side result of calling on Somerville with Sarah Faraday was that Henry gained insight into a reclusive side of Michael Faraday's personality. During the afternoon's conversations Sarah Faraday revealed that her forty-five-year-old mate was "very fond of novels; not desposed to go into company but associates principally with a few persons; denies himself to every person three days in each week; [and] never dines out except when commanded by the Duke of Sussex at the Anniversary of the Royal Institution." Aware of these quirks, Henry was probably not shocked when his British colleague suffered a breakdown in the early 1840s. One Princeton student recorded Henry commenting that Faraday had stopped doing research "owing to failing (mental!) health."[6]

A few days after breakfasting at the Royal Institution, Henry involved himself in an experiment with Wheatstone that would eventually draw Faraday out of his sanctuary. In particular, Henry and Wheatstone devised an experiment to produce a result that Faraday and others had unsuccessfully sought: generating a spark from a thermoelectric current, a current notorious for its weak "intensity." Researchers had hoped to detect the spark to provide further confirmation that the well-known phenomenon of thermoelectricity—a current induced by a flow of heat—had a common "identity" with electricity from more familiar sources. Henry and Wheatstone envisioned generating the elusive spark using a thermoelectric "pile"—a zigzag sequence of two, alternating metals that produced a current when the junctures along the top of the sequence were heated and those along the bottom cooled.[7] In his diary Henry recorded that, having failed in Princeton to evoke a spark using a smaller pile, he "immediately resolved" to repeat the experiment on realizing that, through Wheatstone, he had access to much larger piles. Soon after this decision, however, Wheatstone learned and passed on the news that Italian investigator Vincenzio Antinori had recently achieved the desired result; Antinori was the same investigator who, with Leopoldo Nobili, had extended Faraday's work on mutual induction and, ahead of Henry, had used a magnet to produce a spark. In spite of this news, Wheatstone and Henry went ahead with their experiment. As with Henry's geomagnetic, corpse, and kite experiments in the United States, the London thermoelectric experiment was a group endeavor—providing opportunities as much for social interaction as scientific inquiry. And the group would grow.

The experimental plan involved hooking one of King's College's large piles to a local reproduction of one of Henry's own special coils of copper ribbon that, in contrast to traditional coils of wire, produced heightened self-inductive effects. The ribbon coil would accentuate the subtle thermoelectric current and, feasibly, produce detectable sparks. The coil actually came from the laboratory of Wheatstone's distinguished colleague at King's College, John Daniell—a chemist who, according to Henry, had a pleasant disposition and "constantly laughs." But before Henry could use the coil, he had to overcome a potentially embarrassing technical glitch. Daniell had forewarned Henry that, though he had built the device in accord with a description of Henry's prototype in Silliman's *Journal,* he had been unable to get it to work properly. Confident that he could readily correct the malfunction, Henry volunteered to repair the coil as Daniell and Wheatstone waited, presumably busying themselves with other tasks at the college. Unfortunately, after two hours of tinkering, Henry realized, much to his "mortification," that the coil showed no improvement. "This placed me in rather an aquard position before strangers, since the failure would almost lead them to doubt the veracity of my printed statments." Doubling his resolve, he completely disassembled the coil and rebuilt it, all the while doubting the wisdom of venturing "to experiment in a strange Laboratory and amoung those to whom I had always looked up as masters in Science." To his relief, he ended with a coil that worked "sufficiently well to call fourth the admiration of all present." Wheatstone and Henry, with the assistance of Daniell, could go on with their attempt to evoke a thermoelectric spark.

In his diary Henry remarked that his success with the coil had buoyed him so much that he felt confident "in taking the lead" in the main experiment. He arranged the apparatus and, with Wheatstone and Daniels applying ice and a hot poker to opposite sides of the pile, he interrupted the wire connected to the coil in hopes of drawing a spark. After a string of failures, and Henry's further adjustments to the apparatus, the trio finally witnessed the spark.

Bache had arrived in time for the trials but had gone to a nearby room when the endeavor seemed to stall. Henry commented that a jubilant outburst drew Bache back and that he "afterwards made much sport" over the trio's exuberance: "Prof Daniell flourishing the poker[,] Wheatstone with the ice[,] and I jumping as he said in extacy." Although Antinori had already achieved a similar result, Henry took solace not only that his group was first in England to produce a spark from thermoelectricity but also that the group had obtained the spark using the self-inductive principle (along with the ribbon coil) that he had discovered in the United States. He was "somewhat vexed," however, that merely a lack of proper apparatus in Princeton had prevented him from carrying out the thermoelectric experiment two years earlier. The obstacle had been, he claimed in an aside in his journal, an instrument maker's failure to provide him as promised with a large thermoelectric pile. Meanwhile, the resourceful London trio arranged to show their successful experiment to Faraday. His inclusion in the group would give Henry an opportunity not only to evaluate firsthand the famous experimenter's investigative style but also, privately, to savor victory over him in a minor, technical dispute.

On an evening in late April, two days after the successful thermoelectric test, Faraday joined Henry, Wheatstone, and Daniell in the latter's laboratory at King's College for a reprise of the spark experiment. Faraday apparently brought along a coil of his own—a large coil of wire wrapped around an iron core, in contrast to Henry's coreless coil of copper ribbon. Using a stronger pile than before, the experimenters tried to elicit sparks with Henry's coil and Faraday's more conventional device. "The spark was readily produced by the ribbon coil but *not by the wire one of Dr Faraday,*" Henry wrote in his diary, emphasizing the inadequacy of Faraday's coil. Curiously, however, when the experimenters checked the two coils by connecting them directly to a galvanic battery to produce customary, self-inductive sparks, the reverse occurred: Faraday's wire coil, with its iron core, produced a more vivid spark. Henry offered an explanation of the reversal in the coils' capabilities. He suggested that, although the galvanic current could magnetize the iron core of Faraday's coil and thus boost its ability to produce a spark, the thermoelectric current could not. But Faraday disagreed about the thermoelectric current's deficiency and went on to substantiate his position with a seemingly decisive, supplemental experiment. Henry acquiesced, but not before saying that his own experimental findings from Princeton seemed at odds with Faraday's supplemental experiment. A few days later, however, Henry was with Faraday at the Royal Institution when Faraday repeated a variation of the supplemental experiment, this time without success. At that moment Henry revived his original explanation and suggested that Faraday's seemingly decisive, supplemental experiment involved a spurious effect produced by other factors. This time Faraday agreed. Indeed, Henry made a point of noting, "so far was he convinced of the justness of my chritisism on his experiments" that he expunged a related paragraph from an article that Wheatstone was ready to publish in the *Philosophical Magazine* on the Italian and English spark experiments.

To be sure, Henry's disagreement with Faraday over a mere supplemental experiment had minimal scientific significance. Indeed, by the time Henry drafted his anonymous, encyclopedia article in 1847, he had tacitly conceded that a large thermoelectric pile could not only produce a spark in a ribbon coil but also, as Faraday had contended, magnetize soft iron. Even the main experiment—the detection of the spark—had only secondary significance in light of Antinori's prior announcement. The negligible import of the disagreement perhaps accounts for Faraday leaving no known commentary on the incident. The minor import of the spark experiment also perhaps accounts for Wheatstone's cursory treatment of the experiment in his article in the *Philosophical Magazine,* in which he also covered Antinori's detection of the thermoelectric spark; rather than singling out the specific contributions of Henry or other London contributors, he simply commented that Daniell, Bache, and Henry helped in the experiment. For Henry, however, his involvement with this experiment took on deeper significance.

Henry apparently perceived that he had not merely held his own but actually prevailed in a consequential scientific trial—a trial that allowed an untested American to match wits with international "masters of Science," including his erstwhile rival, Faraday. In other words he felt that he had proven himself and, in the process, bolstered the standing of American science in the eyes of his august London colleagues. Thus, when he later returned to the United States, he probably had a hand in a report carried by a New Jersey newspaper that embellished his role in the thermoelectric experiments while discounting Faraday's: "Professor Henry, of Princeton it is stated, attracted much attention in England by performing some electromagnetic experiments in public, which Mr. Faraday had attempted in vain."[8]

Also after returning home, Henry soon would be regaling his students with anecdotes about, as Theodore Cuyler recalled, the ways in which Faraday and other European scientists "had shown him the highest honor."

Cuyler, who entered Princeton in 1838, remembered that Henry was so wont to introduce into his lectures the phrase "when I was in Edinburgh" or "when I was in London," that this type of phrase "became a byword with his pupils." After moving on to Washington and the Smithsonian Institution, Henry continued to enthrall aspiring young scientists with stories of his London triumphs. Physicist Alfred Mayer—an intimate protégé, who was a teenager when Henry first befriended him in the early 1850s—recalled that "Henry loved to dwell on the hours that he and Bache had spent in Faraday's society." "I shall never forget Henry's account of his visit to King's College, London," Mayer elaborated, "where Faraday, Wheatstone, Daniell and he had met to try and evolve the electric spark from the thermopile. Each in turn attempted it and failed. Then came Henry's turn. He succeeded: calling in the aid of his discovery of the effect of a long interpolar wire wrapped around a piece of soft iron. Faraday became as wild as a boy, and jumping up, shouted: 'Hurrah for the Yankee experiment.'" Besides portraying Henry as outshining his inept London colleagues and winning their praise, this version of the King's College episode misassigns Faraday's particular coil to Henry and Henry's celebratory leap to Faraday.

Even Bache retailed an embellished account of the episode. Seven years after the event, in a speech that touched on Europeans' lack of appreciation for American science, he transformed Henry's disagreement with Faraday into a titanic clash that pitted an underrated American practitioner against his celebrated European rival. "In one case at least," Bache told his audience, "I have the best reason for knowing that an American philosopher was able to engage in a keen encounter on his own subject with the most admired of British philosophers and to come off the victor. I witnessed the encounter with mixed emotions; the philosopher was not only a countryman, but a cherished friend. More talent than that evening was collected in the laboratory of King's College does not often come together: there were Daniell, and Wheatstone . . . and first among equals Faraday. They came to see the spark from the Thermo-electric battery which was then a new result. Faraday and Henry differed about a result in electro-magnetism, as such men differ, mildly but firmly." Bache then summarized the disagreement, saying that Faraday seemed to be carrying the argument—until the two disputants met again at the Royal Institution, when "Faraday found and admitted that he was wrong." The American had prevailed in the international contest.

Meanwhile, back in London during late April 1837, Henry continued his diary entry about the thermoelectric experiments. Having observed Faraday's investigative style at King's College and the Royal Institution, he sketched an appraisal: "I was happy on these two occasions to have an opportunity of seeing Mr F. experiment in private and was much pleased with his method. It is however altogether of the tentitive or emperical kind. He attemps many experiments and collects facts with great rapidity. These he afterwards arranges but does not deduce many experiments appriori from known principles." This assessment was a mild rebuke. It reflected Henry's personal preference for tentative hypotheses in devising experiments and disdain for a reputed, Baconian, inductive gathering and ordering of raw facts. Continuing with his appraisal, Henry suggested that another of Faraday's distinguishing attributes was his agility in concocting useful apparatus from common materials. Finally, on a loftier note, Henry observed: "Mr F appears to be deeply imbued with the true spirit of philosophy. I remarked that our only object in investigation should be the determination of truth; his answer was yes but this is the most difficult to learn."

In late May, after ten hours on a steamboat from London, Henry found himself among "the inhabitants almost of an other planet." He had arrived in France, where, to his surprise, women shared in manual labor, people dined out-of-doors, and mourners decorated graves with flowers. Reaching Paris, he found everything to have "a strange aspect." Wide boulevards contrasted with dingy, cramped side streets, and serene parks and plazas offset rows of shops selling every imaginable article. One shop even specialized in "female articles of mourning." He declared to Harriet that, while the Tuileries Gardens in the heart of the city were exquisite, they had a startling feature; in particular, "every part of the garden is ornamented with marble statues in many instances perfectly naked and which are thus at first rather revolting to an american taste but which are beautiful specimens of modern art." Even the rooms that he and DeWitt had rented, located in the Latin Quarter near the city's main scientific institutions, had unusual furnishings, including "an article of furniture very convenient but not known in america called a *Table de nuit*."[9]

After London's drab soot-covered buildings, the bright palaces and other resplendent Parisian structures rekindled Henry's architectural sensibilities. Seemingly having repressed the aversion to self-indulgence that he had expressed to Harriet in London, he unrepentantly spent his first days in Paris visiting "the lions" of the city. These landmarks ranged from the recently completed Arc de Triomphe to the Palace of Versailles on the outskirts of the city. A friend from Albany who was studying in Paris served as guide. Although the immense Palace of Versailles awakened Henry to the full meaning of "Regal Splendor," the royal family itself left him unimpressed. Happening to glimpse a contingent of the family in Paris—and being struck by each member's plain features—he told Harriet: "The great in circumstances are not always the favorites of Nature either in mind or body." The royal family aside, Henry was enjoying himself. "I am disposed to be much better pleased with Paris than London as a place," he wrote back to Bache, particularly pointing out Paris's lower prices. He was even enjoying a ritual that had boggled him when first encountered in London: "management of the tea-pot." Later writing to Harriet while "under the reviving influence of a few cups from the alembic which distills the exhilerating but not intoxicating liquid," he lightheartedly but meticulously detailed the "practical philosophy" of the teapot.

As for partaking of natural philosophy in Paris, Henry gradually realized that he faced two constraints: the French language and Parisian workloads. Although he labored diligently to supplement his French reading skills with oral skills—and though he quickly gained in fluency with the help of a tutor—he recognized that he would never duplicate the close relationships that he had enjoyed in London. Compounding the language barrier were the busy schedules maintained by Paris's leading, more established researchers. "The French Savans of the older class," he complained to Bache, "are so much occupied in the business of teaching, also in politics and legislation (they are all members of the Chamber of Deputies) they have no leisure for attention to strangers." Further complicating his reception as a stranger was his oversight in not equipping himself with proper letters of introduction, documents that he had found largely unnecessary in London. In spite of these constraints he either attended lectures of or arranged interviews with leading researchers. He also visited a meeting in Paris of the Academy of Sciences (Académie des sciences), which boasted a core of salaried members as a branch of the national Institute of France (Institut de France). To Harriet he confessed that it felt disconcerting to be "in the very presence of those whose names are associated with my earliest recollections of Science and who have always before appeared almost to belong to another age." These luminaries included Antoine-César Becquerel, Claude Pouillet, Dominique Arago, and Joseph Gay-Lussac. Through a lecture by chemist Gay-Lussac, Henry fulfilled a request made by Daniell and Faraday to gather information on a rumored, new process to solidify carbonic acid—that is, to create dry ice, the solid form of carbon dioxide. Gay-Lussac not only detailed the process but identified the Parisian inventor, whom Henry later visited in anticipation of reporting to Daniell and Faraday.[10]

Toward the end of his second month in Paris, Henry "began to make up for lost time" by starting, at last, to gather fuller scientific information and to establish deeper relationships. The turnabout came when he began to associate with younger, less ensconced researchers in Paris. Among this set was Macedonio Melloni, an exiled Italian specialist in radiant heat who was Henry's age. Perhaps Melloni and Henry related well because both were foreigners coping with an often inaccessible Parisian scientific establishment. Whatever the reason, Henry took special pleasure in learning about Melloni's version of the thermomultiplier, a device that measured radiant heat from weak or distant sources by illumination of a thermoelectric pile connected to a sensitive galvanometer. Henry had used a Melloni pile in the initial thermoelectric experiment at King's College; later, in 1845 with the assistance of Stephen Alexander, he would adapt Melloni's thermomultiplier to show that an individual sunspot had a cooler temperature than the surrounding solar disk—an innovative application of the thermomultiplier that helped launch Angelo Secchi's more thorough, better-known solar studies in Rome.[11] Possibly through Melloni, Henry also met another prominent physicist about his own age, Swiss researcher A. A. De La Rive, who was passing through Paris.

As his stay in Paris drew to a close, Henry became heartened by his access to younger researchers such as Melloni. The arrival of the Baches in Paris perhaps also lifted his spirits, giving him a more positive perspective on his Parisian scientific initiatives. Reviewing his experiences in London and Paris, he would boast to Harriet that, by drawing on the notes in his diary, he "could furnish more material for sketches in the line of English and French science than anyone who has of late years visited Europe since I have been usually fortunate in gaining admission to intimacy with most of the men actually engaged in science in London and Paris." In hindsight, however, he would recognize that the next phase of his trip—an extended tour of mainly Edinburgh, Glasgow, and Liverpool—was comparatively more rewarding. "I was much delighted with this tour," he would acknowledge after experiencing Scotland, "and now regret that I stopped so long in Paris."[12]

Henry's stay in Paris did enable him to achieve a basic goal of his European trip: the acquisition of scientific instruments for teaching and research. Although he purchased some items in London, he acquired the bulk of his 151-item inventory in the shops of Paris's acclaimed instrument makers. But he first had to reconcile himself to a shortfall in funds. Princeton officials had led him to believe that he would sail to Europe with an instrument budget of $5,000, but at the last minute they provided only $500 with the assurance of sending an additional $500. The shortage reflected, Henry thought, the shortsightedness of the Princeton leadership. It also reflected unlucky timing. The Princeton officials chose not to forward the additional $500 when American financial institutions started to reel from the great depression of 1837—a debacle that triggered for Henry a lifelong concern with issues of finance and political economy. In Paris, Henry resigned himself to supplementing the college's original $500 with his own money. Buying items ranging in price from a 10-cent lens to a 150-dollar electrostatic friction machine, he compiled a bill that exceeded his college allotment by $910. Whereas he would recoup $100 of the excess expenditure soon after returning to Princeton, he would wait until 1844 before gaining assurance that the college would reimburse him for the remainder plus interest; he achieved this concession from the still strapped college only after having received another job offer from the University of Pennsylvania. But Princeton's board of trustees, continuing to lack surplus funds during subsequent years of financial strain, did not actually reimburse him until 1865; by then, long established as the head of the Smithsonian, Henry himself had been a Princeton trustee for a year. Initially compounding the delay in recouping his expenses was a delay in receiving the shipments of not only the instruments but also books that he had purchased. A main, prepaid order of instruments from Paris was a year late in delivery, impeding his research program on static electricity. Also, this and another order arrived incomplete. "So much for the honesty of the French," Henry griped to Bache. By the early 1840s, however, with an instrument budget created by a five-dollar fee levied on the members of each senior class, Henry was again placing orders with the French instrument makers, but now insisting that payment follow delivery.[13]

Henry planned to conclude his European travels by visiting colleagues in Scotland and other parts of Great Britain. Specifically, he intended to remain through September for the Liverpool meeting of the British Association for the Advancement of Science (BAAS). After leaving Paris, however, he first made a brief tour of Belgium and Holland. He was traveling by himself, having parted company with DeWitt earlier in Paris.[14] Reaching Antwerp near the border of the two North Sea nations in late July, a passport problem coupled with persistent dysentery (that he attributed to the water in Paris) persuaded him to curtail his sojourn and take the next available steamboat back to London. But earlier he had fit in some sightseeing, chiefly in and around Brussels. A sidetrip to Waterloo, the site of Napoleon's decisive defeat 22 years earlier, proved particularly poignant. Shown the battlefield by a villager who had witnessed the carnage, Henry was moved to recall a melancholy verse penned about the site by Lord Byron, one of his favorite poets, and to reflect on the irrationality of war. Noting that tens of thousands of soldiers had died, he asked in a letter to Harriet: "And for what? To gratify the pride, the lust for power of a few by nature no better than the meanest that fell. I actually grew sick as I stood on the mound which covers the remains of two thousand of the slain and in imagination faintly realized the horror of the battle." Eleven years earlier, a mass grave of British soldiers at the ruins of Fort Erie had similarly moved Henry, but for a different reason; he had been stirred by how quickly time was erasing all evidence of the valorous dead combatants. More recently, when first visiting London, his experience at Fort Erie had come full circle. He had breakfasted with British Major Edward Sabine and learned that the eminent expert on terrestrial magnetism (whose needles Henry had used) had served at Fort Erie in 1814—against the Americans. Pathos yielded to nationalistic feelings about war, however, when Henry interjected in his diary that Sabine had built a breastwork "taken by the americans at the gallant action on the part of our troops."[15]

Back in London after two and a half months on the Continent, Joseph wrote Harriet a lengthy letter tinged with not only nationalistic pride but also chauvinistic prejudice. He decried the treatment of lower-class women in Europe even as he excused the treatment of African Americans in the United States. Convinced that the United States outshone Europe regarding "the condition of females in the lower classes," Henry detailed the plights of the European women. He explained that, whereas French women labored in fields and toiled in artisans' shops, some Belgium women even earned their livings by collecting dung along the roadsides. With London images fresher in his mind, he described the hardships of that city's women in language that approached the vivid characterizations of Charles Dickens—an aspiring young novelist who was then writing *Oliver Twist* and who soon would become one of Henry's favorite authors.[16] "During the coldest and wetest weather," Joseph recounted in one example, "you will find a female stationed at a cross walk with a birch broom in one hand and the other supporting an infant[,] constantly sweeping the stones[,] asking a penny of every passenger. The greater number of these pass her by unheaded, occassionally one gives her a copper."

From the plight of these lower-class women, Henry's train of thought carried him to the few "English Ladies" that he had met. Mention of these "very intelligent" women led him, in turn, to comment on the frequency with which they pressed him about African Americans. "They cannot understand our prejudices relative to them," Henry protested about these British abolitionists, "and cannot see why they should not have all the privaledges of a white man." Henry apparently became so exasperated by a particular woman's lack of appreciation for the white American's perspective, that he had to pose what he considered an impertinent, rhetorical question: "I was obliged on one occasion to ask a lady if she could ever think of marr[y]ing a black man. She said she certainly would not fancy the colour but that there should be no legal objection to prevent those who pleased from marr[y]ing together." In a similar vein Joseph had mentioned to Harriet, with some disbelief, that an American friend then in London had with him a black servant who presumably could marry a white woman. The friend was Henry James, a recent student at the Princeton Theological Seminary whom Henry earlier had tutored in Albany. On this stopover in London, Henry was boarding with James, whose family wealth enabled him not only to travel with a servant but also to have donated a sizable sum, about $1,000, to Henry's decimated instrument fund. "He brought with him," Henry pointed out, "a black man from Albany as a waiter who is a very good fellow in the serving line. He attracts much attention and is quite [a] Lion among the lower classes[,] eats with the family with whom we lodge[,] and would find no difficulty were he not married in getting a white wife." Such possibilities probably troubled Henry, who during this period was a financial contributor to an enterprise encouraging and facilitating the return of free black Americans to Africa. Most of Princeton's academic and theological leaders would have joined Henry in believing that this strategy of the American Colonization Society was progressive and benevolent. While not condoning slavery, the program's backers justified the relocation of blacks to Liberia by maintaining that it offered them better living circumstances.[17]

Henry was especially agitated by a new book, of which he had just read a review, whose English author savaged the United States for its deplorable treatment of slaves. "She gives some stories relative to the horors of slavery which are really revolting but which cannot be true," Joseph reassured Harriet. He acknowledged that some abuses probably take place, but these occur only along the nation's frontiers "where civilized and savage man come in approximation" and normal social controls breakdown. Along these isolated frontiers, "an English tourest may find ample materials to make a book of american atrocities which with a litle colouring and well chosen examples may render us black in the eyes of the world as crime can make us." Although Henry personally resisted being rendered "black" in any sense of the word, he ended this disposition on European women and African Americans contritely. He admitted that the United States not only possessed flaws but also has had as a nation little impact on Europe.

> There are many things connected with America which are much to be regretted and which every reflecting traveler must feel. When at a distance[,] we see with less predijudice the good and bad qualities of our home[,] and no person can have a proper ide[a] of America who has not viewed it from a distance and in connection or in comparison with other countries. We have too exalted an opinion of our influence and the share of attention we occupy in the minds of Europe. We are too far removed to exert any direct influence and the great mass of people in France and England are too ignorant of our affairs to be much influenced by them. The French have but litle respect for the Americans. They hate the English but do not despise them; they have no more love for the Americans and far less respect.

On an even more bitter and probably personal note, he added: "The Americans in Paris receive no attention, they are never admitted into fashionable society."

Henry lingered in London only a little over a week before heading north to Edinburgh, but he put the time to good use. Soon restored to health by the ministrations of Henry James, he joined Faraday on a three-hour excursion along a new, imposing railway line, accompanied by the engineer in charge of construction. He also caught up on Wheatstone's latest achievement with his galvanic telegraph, which now transmitted messages 13 miles, and made a careful drawing for his own later use of Daniell's imaginative design for a battery of more constant output.

One-and-a-half days on a steamboat brought Henry to Scotland, the country of his parents' births. Landing near Edinburgh on the east coast, he maintained his quick pace of scientific mingling and gleaning with a prompt visit to the Royal Society of Edinburgh. John Robison Jr., the secretary of the society, was among those local scientific leaders who, in contrast to their counterparts in Paris, received Henry "with much kindness and attention."[18] Faraday's brother-in-law, whom Henry had met at breakfast with the Faradays in London, also escorted Henry around Edinburgh, introducing him to the city's scientific gentry. Unfortunately, Forbes, the natural philosopher with whom Henry had been in regular communication, was traveling on the Continent. Whereas thermoelectricity had figured prominently in Henry's London interactions, optics preoccupied him during much of his stopover in Scotland. The topic came up in discussions with Thomas Henderson, the Astronomer Royal for Scotland, and William Nicol, an authority on polarization.[19] Henry also encountered optical issues, but of a more practical bent, when a colleague invited him on a steamboat excursion arranged by Scottish officials to allow a visiting French expert to examine local lighthouses. The expert, whom Henry had met in Paris, was the younger brother of Augustin Fresnel, a pioneer of the wave theory of light who also had devised an ingenious lens that transformed lighthouse design.

At one lighthouse visited by the party, Henry seemed enchanted to discover a poem entered in the register earlier in the century by Sir Walter Scott. Previously, Henry also inadvertently encountered Scottish history when he happened into the Sunday service of St. Giles' Cathedral; this church was John Knox's cradle, during the 1560s, of Presbyterian religious and political authority in Scotland. The veteran American tourist also engaged in some outright sightseeing— now that, since experiencing Paris, he apparently had moderated his misgivings about indulging in the pleasures of travel. The landmarks that he visited ranged from the monument honoring poet Robert Burns to Edinburgh's main royal palace. "Of course I did not fail to visit the Palace," Joseph nonchalantly informed Harriet, "but as my tour is one of science and not of general and usual sightseeing, I did not pay my respects to this piece of antiquity until the day before my departure."

The highlight of his scientific tour in and around Edinburgh had came when he took a sidetrip to the estate of Sir David Brewster. A distinguished, fifty-five-year-old researcher in optics, Brewster lived south of the city near Melrose Abbey, a medieval ruin immortalized by Sir Walter Scott, who also had resided nearby until his death in 1832. Brewster received Henry cordially, assuring him that a letter of introduction was unnecessary because he knew his name, having just summarized his experiments for an encyclopedia article on magnetism. The two natural philosophers seemed to have much in common. For example, both doubted the efficacy of strict Baconian scientific method and both had been "mortified" by other investigators anticipating their experimental findings. After tea on the first day, Brewster entertained Henry by reading aloud letters penned by Isaac Newton, part of a major collection of largely unexamined documents that Brewster was using to prepare a biography of the revered natural philosopher. By this late stage of his British travels, Henry was no stranger to artifacts associated with Newton. Earlier in London and Edinburgh, in what seemed obligatory rituals performed for guests, Henry's hosts had paraded before him relics ranging from Sir Isaac's telescope to a panel from his bookcase.[20] But Brewster's collection of documents provided a more intimate glimpse of the author of the *Principia.* The letters, particularly a love letter composed by Newton, led Joseph to comment to Harriet: "These were quite amusing and conclusively proved that profound and abstract as were the speculations of this Prince of Philosophers he still belonged to the *genus homo.*"

Henry also enjoyed the many optical experiments and effects that Brewster showed him. "The information given me by Sir David," he wrote to Harriet, "was precisely the kind for which I came to this country: the methods of original experimentation and those things which cannot be learned from books." Fixated on the optical experiments, Henry apparently took little note of his Scottish colleague's associated theoretical conviction that light involved the emission of particles rather than, as most natural philosophers were coming to believe, waves in an ethereal medium.[21] On their second and final day together, so intent were Brewster and Henry on viewing experiments that, though horses were saddled after breakfast for an intended ride to Sir Walter Scott's homesite, they left the horses stand until late afternoon; the two men had become distracted by a progression of experiments made possible by the sun's unexpected emergence in a previously cloudy sky. Finally, realizing the lateness of the hour, Henry rode alone to the homesite at a "gallop," took "a hasty glance," and returned "just in time for the evening coach back to Edinburgh."

Next, traveling west from Edinburgh to Glasgow with a leisurely canal ride, Henry continued his scientific tour during the two weeks bracketing late August and the opening days of September. Glasgow's prominent chemists, active in both research and manufacturing, preoccupied much of his time as he again received "kind attention." He even joined in a "short excursion . . . to the Highlands" with visiting German chemist Justus von Liebig. With the Liverpool BAAS meeting looming, however, he soon left Glasgow and began his trek back south toward England. He chose his route to fulfill a personal quest: to visit his Scottish relatives. Having sought family names and locations from brother James before sailing for Europe, Henry located kin on his mother's side in Ayrshire, a west coast county. He also found a possible distant relative on his father's side who maintained the likely original spelling of the family name, Hendrie.[22] After spending a few days socializing, particularly in the coastal town of Girvan—where his mother had been born in 1760—he finally boarded a steamboat for Liverpool and the last phase of his European venture. Joseph had been telling Harriet for more than a month that, if it had not been for the pending BAAS meeting, he would have ended his travels and sailed directly back to the United States. Although experiencing "a loneliness and longing for home," he realized that prudence and expediency required him to remain for the meeting—he was already in Great Britain and probably would never have a similar opportunity. Resigned to complete the trip, he wrote in response to a letter from Harriet: "You say you will never part with me again. I will never require you to do so and should I return safely I will have done with wandering and be content to settle down for life in any situation which Providence may designate."

15 Measuring Up

In September 1837 the city of Liverpool, near England's northwest coast, was host to the seventh meeting of the British Association for the Advancement of Science. Growing in reputation, the association was attracting to its gatherings interested laypersons and scientific specialists from not only Great Britain but also continental Europe and North America. Helping Henry negotiate the resulting "immense concourse of people" were Daniell and Faraday. "Faraday was of great importance to me," Henry later told Bache, "since he made it a point to come to me even through a crowd and introduce me generally to all arround. These introductions I found had more weight than even Mr F. appeared to attach to them."[1]

Henry's formal participation in the weeklong meeting was threefold: joining in a discussion about cast-iron magnets held in the mathematical and physical science section; making a brief presentation on U.S. canal and railway systems to the mechanical section; and reading a paper on static-electric lateral discharge to the mathematical and physical science section. He had drafted the paper in early July at Melloni's request but never presented it, as intended, to the Paris Academy, partly because of translation problems. For the Liverpool meeting, however, he resurrected the paper and found it to be "well received." Indeed, a few decades later, he would look back with pride to a discussion that, he claimed, the paper soon provoked within the British government about the correct design of lightning rods for vessels in the British Navy. He was referring to a dispute in which William Sturgeon, an opponent of the accepted naval design, sought to reinforce his case by invoking Henry's paper on lateral discharge. Contrary to Henry's later implication, however, his paper neither instigated the flap nor determined its outcome—an outcome sealed when Faraday and Wheatstone supported the accepted design for naval lightning rods. At best, Henry's paper played an ancillary role in the dispute.[2]

During the BAAS meeting Henry generally enjoyed "the same kind reception" that he had received throughout Great Britain. At the conclusion of the proceedings, he particularly appreciated the honor of being included among four foreign participants singled out for special recognition. The others were: Liebig, the productive German chemist with whom he had spent time in Glasgow; De La Rive, the Swiss natural philosopher whom he had met in Paris; and Gerrit Moll, the Dutch physicist who had anticipated him in announcing an enhanced electromagnet.

One incident, however, blemished Henry's week of otherwise affable interactions. In his talk to the mechanical section, he had presented a succinct overview, illustrated with a map, of American "internal improvements" involving railways, canals, and navigable rivers. As he was leaving the podium to appreciative applause, a member of the audience happened to ask if he knew how fast American steamboats could travel. After pointing out that he spoke not as a "practical engineer" but merely as someone who had "lived on the banks of the Hudson," Henry replied that he personally had ridden a steamboat that made the 150-mile passage between New York and Albany, with a boost from a favorable tide, in nine hours. To Henry's chagrin, as he later told Bache, one of the section's vice presidents then "jumpt up" and pronounced the Hudson figures to be suspect. The disputant was Dionysius Lardner, a professor at London's University College and prominent science writer with a particular interest in steam propulsion. Lardner declared that he and the audience "did not want popular information on the subject" and, moreover, that he "did not believe" that any steamboat had ever achieved the speed claimed by Henry. At this point the session's moderator intervened, chastising Lardner for his disrespectful treatment of "a gentleman and a stranger." Although Lardner's face became flushed and "the whole room was in a state of commotion," Henry remained calm and responded with a high-minded disclaimer. "I then stated," he told Bache, "that I regretted I could give no more definite information, that I wished my communications to receive no favor on account of my being a Foreigner, that truth and Science should know no country—&c &c." These noble sentiments drew not only applause but also a robust defense of Henry's scientific credentials from one of his Edinburgh acquaintances. Henry's figures turned out to be accurate and he emerged from the clash vindicated. "The conduct of Dr. Lardner was reprobated by the majority of the persons present," Henry observed, "and during my stay in Liverpool not a day passed afterwards without some persons introducing themselves to me and mentioning the affair."

Although resolved in Henry's favor, the incident stung him and led him to draw some pessimistic lessons—a reaction heightened perhaps by the self-doubts and insecurities that he retained from childhood. His first lesson derived from the contrast between his embarrassing treatment as an American visitor in the mechanical section, where few participants knew him, and his positive reception in the mathematical and physical science section, where many knew him well. As he explained to Bache a few weeks after the meeting, "it is rather a hazardous affair for an American to make a communication unless he be well known and his communication addressed to the Section in which his Friends are found." Henry was suggesting that the British attitude toward Americans, unless ameliorated by a personal relationship, smacked of disrespect if not condescension. This suggestion led him later to draw a related lesson: that the British tended to patronize their American colleagues. Recounting the steamboat incident in follow-up letter to Bache a year afterward, Henry discerned "a great prejudice . . . against Americans." Although he granted that the prejudice was perhaps partially justified because of the low caliber of the typical American scientific visitor to Great Britain, he resented the attitude. "They treat us with great kindness and I have no doubt but many of the persons with whom you and myself associated had a real respect for us and were prompted by proper feelings to show us the attention we received, but still there was in my case the appearance of a little hesitation in allowing the same merit in public as in private." He attributed the patronistic perspective to a combination of "the influence of the aristocratical and political Institutions of the country" and "the general low opinion" of the British toward "American science and literature." In fact, Henry had experienced other instances of what could be considered British condescension toward American science. Three months earlier, for example, a London editor subverted his effort to counteract a negative review, tinged with anti-American sentiments, of a report that Bache had prepared for the Franklin Institute on steam boiler explosions.[3]

Henry himself, however, had not arrived in Europe free of prejudices toward the comparative practice of science in the United States and Europe. Indeed, he arrived with a defensive outlook, a predisposition revealed, for instance, in an 1835 letter of recommendation for an American colleague. There he implied that the best, freethinking American researchers rightfully guard against following doctrinaire Europeans and, instead, engage in independent thought grounded on firsthand investigations. Commending his colleague, he wrote: "He is one of those who, not content with retailing the untested opinions and discoveries of European Philosophers, endeavor to enlarge the boundaries of useful knowledge by experiments and observations of their own." Part of this defensiveness traced to Henry's staunch allegiance to American republicanism—a national affinity, born in the American Revolution, for freedom of expression and self-governance but aversion to repressive monarchies. During his European travels, not only had he disparaged the French royal family as offering little "either in mind or body," but he sarcastically labeled the King of England "his most greacious and *illiterate* majesty." (He was speaking two months before the death of England's King William IV, whom Victoria would succeed.) Once back home, he cuttingly attributed his inability to cross the border between Belgium and Holland to "the stupid old *Dutch* King."[4] Indeed, Henry's and Bache's later, jingoistic slants on the King's College thermoelectric experiment—slants in which an underrated American, Henry, unexpectedly bests his acclaimed British colleagues, Daniell and Faraday—reflected a combination of American defensiveness toward Europeans and perceptions of European condescension toward Americans.

Henry's clash with Lardner also contributed to a souring of his attitude toward national scientific meetings and organizations in the BAAS mold. Although he had gone into the Liverpool meeting optimistically anticipating that it would be the crowning event of his trip, he left slightly disenchanted. Certainly, his general experience was positive. But he came to realize that, contrary to his expectation, the BAAS meeting failed to provide a meaningful formal forum for reporting substantive scientific information. Instead, the meeting served primarily as a "scientific Jubilee." That is, it provided "a kind of festival for men of science for mutual stimulous." Part of the reason for the dearth of substantive scientific exchange was, Henry felt, the association's inclusion of lay dabblers alongside expert practitioners. As he later expressed this point, the quality of presentations was uneven because the British had committed themselves to a meeting "where every person has the liberty of making a communication and where the merest *Sciolist* and the profound *savant* are on the same apparent level."

The association's inclusiveness, however, had an advantage: it led to a large membership and correspondingly substantial revenues for the BAAS. In Henry's opinion the association was appropriately directing these revenues to meritorious research projects. He elaborated a year later, telling Bache that only those members with solid scientific reputations controlled the research funds. "The great body of the members have no voice in the management of the Institution, and in this respect the society is quite as aristocratical as the government of the nation." Elaborating still further, he deviated from his normal republicanism by endorsing this "aristocratical" denial of equal participation of common members in society governance. Without the proven practitioners assuming authority, he cautioned, "the *third* and *fourth* rate men would soon controll the affair and render the whole abortive and ridiculous." Although voicing these sentiments of exclusivity, Henry viewed himself as committed to a republican scientific order in which all practitioners worked as equals—all practitioners, that is, who qualified as genuine contributors, not mere dabblers. Thus, a few years later he told a London colleague that science's "tendencies are republican in the proper sense of the term"; the proper sense was that "those who honestly labour to advance its cause should be recognised as Brothers in what ever part of the wor[l]d they may be found." In Henry's republic of science, power should lie in the hands not of all laborers but of the honest laborers.[5]

At the close of the Liverpool meeting in mid-September 1837, Henry returned briefly to London. En route to the city, he paused in Manchester and Cambridge, where John Stevens Henslow, a botanist later famous as Charles Darwin's mentor, showed the American visitor around the venerable colleges of Cambridge University. Whereas Henry admired the Gothic architecture, he was "amused" by the college's "antiquated" ceremonies and procedures. Henry had felt obligated to revisit London because he had agreed to help Bache with a transatlantic project on terrestrial magnetism. In particular, he had consented to take readings in London and then, later, Princeton with two magnetic needles the friends had christened "Rusty" and "Bright"; though still active in geomagnetic research, Henry was increasingly deferring to Bache's initiatives. In London, Henry also needed to purchase a few, final pieces of apparatus.[6]

Finally, on October 2, after six and a half months in Europe, Henry sailed for home aboard the packet *Toronto*. He was "wretchedly ill" from seasickness during the first week—and he experienced the "considerable danger" of a thunderstorm so violent that the tip of every spar and mast on the ship glowed with "an electrical brush" (to watch, he had inched his way onto the deck while clutching the rigging)—but he ultimately judged the unusually quick passage of twenty-six days to be "pleasant." On the day before the sailing, he also judged the European trip to be a success. "I leave England well satisfied with my reception and also with the profits I have gained by my tour," he told a French friend. "The acquaintances I have formed in Europe will I hope be of much service and pleasure to me during life and should I be able to do anything more in the way of science I will have an opportunity through my acquaintances in this country and on the continent of making it immediately known." He later wrote to Faraday, "I left your country with warm feelings highly gratified with the kindness I had received, and with the unreserve and liberality with which I was instructed in various branches of science." According to Henry James, Faraday reciprocated the warm feelings. James reported to Henry that Faraday "spoke in the highest terms of you in every way—and reverted so frequently when I saw him to the subject, and always in a manner so cordial and evidently affectionate, as to shew that he remembered your visit with very great pleasure."[7]

His quick return across the Atlantic notwithstanding, complications at the end of the voyage meant that Henry was unavoidably late in getting from New York Bay to his family. A warm reunion, however, compensated him for this last-minute delay. Six-year-old William simply doted on his long-absent father. "The boy waited to see me until he could sit up no longer," Henry commented soon after rejoining the family; he further remarked, William "was with me before I was out of bed and has been with me ever since." Nine months earlier, on the eve of his voyage, Henry had given one of his most valued possessions to William—his copy of Gregory's *Popular Lectures,* a chance acquisition of his late teenage years. Inscribing the copy to "my son," he had written on the inside cover that the book had "under Providence exerted a remarkable influence on my life." In particular, the book had led him "to resolve at the time of reading it that I would immediately commence to devote my life to the acquisition of knowledge." Now, more than two decades later, his dedication—and a packet ship—had carried him fully into the mainstream of international science.

The impressions that Henry brought home from the BAAS meeting would have repercussions, both professional and personal. His warning to Bache about third- and fourth-rate members of the BAAS taking control of the research money had implications for the United States and scientific societies in general. Because he believed that scientific pretenders still predominated in the United States, Henry opposed the creation of a national scientific association with open meetings in the British style. Nevertheless, as he reminded Bache a year after the BAAS gathering, such initiatives were afoot. At the Liverpool meeting itself, a Harvard medical professor had broached with Henry the possibility of starting a parallel American association; now, a year later, the professor was attempting to garner support for the plan through the American Philosophical Society and key individuals such as Henry. "Much has been said since my return of the propriety of a meeting of the kind among us," Henry advised Bache, "but I am convinced a promiscuous assembly of those who call themselves men of science in the country would only end in our disgrace."

Henry had been discouraged by the nation's rampant "charlatanism." "I often thought of the remark you were in the habit of making," he told Bache, who was still in Europe, "that we must put down quackery or quackery will put down science." He recounted examples of the "disgrace being inflicted on the country" by scientific pretenders. The abuses ranged from extravagant claims for electromagnetic motors to unfounded assertions on behalf of Samuel Morse that he was responsible for "the entire origin" of his lauded, new telegraph. In another case, one that particularly threatened "the welfare of American science," Henry had felt obliged to write an anonymous article in the *Princeton Review* rebutting a scientifically groundless theory of terrestrial magnetism and navigation that a physician had persuaded Congress to endorse. He even had to complain to the normally steadfast Silliman for including in the *American Journal* "a mass of trash relative to electricity and electro magnetism which would disgrace the annals of electricity." Although disheartened by these instances of national inadequacy or outright quackery, Henry was not without a plan and prognosis for the future. "I am now more than ever of your opinion," he assured Bache, "that the real working men in the way of science in this country should make common cause and endeavour by every proper means unitedly to raise our scientific character, to make science more respected at home, to increase the facilities of scientific investigations and the inducements to scientific labours. There is a disposition on the part of our government to advance the cause if this were properly directed. At present however Charlatanism is much more likely to meet with attention and reward than true unpretending merit." In this dubious climate Henry was wary of having a British-style association in the United States. Others, particularly the influential Philadelphians, shared his doubt and the idea faded.[8]

Similarly, in the early 1840s, Henry and his allies would be cool to the formation in Washington of the National Institute for the Promotion of Science, an organization partially controlled by politicians and government bureaucrats. "I do not like the plan of uniting science and party politics," Henry advised a Harvard colleague, "and I cannot acknowledge the right of the Political Gentlem[en] at Washington to call a meeting of the cultivators of science in our country." Without the sustained support of these select "cultivators," the National Institute also faded, losing visibility in the late 1840s. Indeed, when Bache addressed the institute's varied membership during the 1844 meeting (in the same speech recounting how an American, Henry, had prevailed over Faraday in the London thermoelectric experiment), he urged only a nominal role in the United States for not only the institute but also scientific societies in general. The emphasis should not be on holding national meetings but on providing practitioners with the resources to do research and journals to keep abreast of international developments.[9]

Henry's aversion to "promiscuous" scientific assemblies, whether involving charlatans or politicians, reflected a professional ethic emerging in his personal circle of internationally attuned researchers. In an 1841 nod to this nascent professionalism, he would advise the editor of a journal covering phrenology and mesmerism that the topic of animal electricity can nowadays be studied legitimately solely by experts with the requisite credentials. "Philosophical discoveries in the present advanced state of science," he cautioned, "can only be made by those who have prepared themselves by long study for the purpose, and have served, as it were, an apprenticeship to the business of experimenting." Although professionalization among scientific aspirants in the United States would occur subtly and slowly through the nineteenth-century, Bache and Henry would, by 1843, be sensing an ongoing heightening of expectations regarding the skills and experiences of scientific practitioners. Again criticizing Silliman and the *American Journal*—whose quality and subscriptions had slipped—Henry would now fault his longtime Yale champion for not keeping abreast of "the change in the character of American science which has been going on for several years unperceived and unsuspected by the good people at New Haven." Bache and Henry envisioned a simple solution for Silliman: placing experts such as themselves fully in charge of the journal's content in their respective areas of specialization. (In fact, Silliman would increasingly depend on his son, Yale chemist Benjamin Silliman Jr., and his son-in-law, Yale geologist James Dana, to revitalize the journal.) Another sign of the emerging professional ethic would be Henry's intensifying conviction in the mid-1840s that American colleges, when hiring physical science professors from pools of qualified candidates, give preference to those with established records of "original research." The proven researcher, Henry maintained, not only brings enthusiasm and integrity to the classroom but also expertise and ambition to the laboratory, thus advancing simultaneously the nation's educational and scientific capacities.[10]

The link between science and integrity also played a part in a more personal consequence of the BAAS meeting. Henry's clash with Lardner engendered in Henry a lingering bitterness toward the widely read English science writer and prolific editor of a popular series of books. Probably nothing would have come of the bitterness, except that Lardner relocated to the United States in late 1840 and spent the next few years giving public lectures, especially in Philadelphia and New York. Moreover, he came to

Henry's home turf for a reason that aggravated Henry's animosity: to sidestep a scandal that erupted in England over an affair with a married woman. Henry sardonically advised Torrey that it was time for him to leave New York now that "the Gay Lothario" was lecturing in the city; in a similar vein, he added that "Philadelphians have some reason to crow" because they had denied Lardner support and encouragement. In fact, despite the scandal, Lardner was flourishing in the United States as a public lecturer. And when the popular Englishman realized that at least some Americans resented his treatment of Henry at Liverpool, he offered a circumspect public accounting of the incident, attributing his conduct to a misunderstanding. One of Henry's Princeton contacts was in the audience for this 1842 Philadelphia lecture and, in a summary for Henry, described Lardner's tone as divided between defensive and apologetic.[11]

Henry purported to be indifferent to Lardner's public accounting. In an 1842 letter to John Stevens Henslow, whom he had met at the end of the Liverpool meeting, he said he took "no notice" of Lardner's Philadelphia lecture. Moreover, he professed to hold no personal resentment against Lardner but merely to pity him for the misfortunes he created for himself through the illicit affair with the married woman. The misfortunes of "this paramour" included, according to Henry, being snubbed by the American scientific elite and being reduced to lecturing in second-rate theaters that alternately booked jugglers and public dancers. Henry insisted to Henslow that, though Lardner possessed talent, he lacked "an essential element of scientific character[:] a sacred regard to truth." He also contended that nobody who is "guilty of the crime of which he is charged" could be trusted as a scientific commentator.[12]

Despite his denials to Henslow, Henry did take notice of Lardner's public accounting. Stephen Dodd, an 1846 Princeton graduate, recalled Henry detailing for his class how he had been "rebuffed by Dr. Lardner, who refused to receive his statements as the truth, but who afterwards was constrained to make most ample apology." And, despite pitying Lardner for being reduced to lecturing in public theaters, Henry had earlier strained to construct a self-satisfying justification when he himself spent three weeks in 1839 in New York City presenting six lectures to a nightly audience of up to one thousand. Indeed, his conscious motive in making the appearances had been the prospect of earning the sizable sum of $600, money needed for settling into the family's new Princeton house. Consistent with his later attitude toward Lardner lecturing, however, Henry had entered into his own New York series warily. At the time, he had rationalized that the lectures were not public but restricted to members and "female relatives" of the sponsoring Mercantile Library Association. He also had stressed: "To be considered a lecturer is not in the line of my ambition." And while on stage expounding on electromagnetism, he had taken the opportunity to denounce the associated "quackery in the city of New York," thus producing "something of a sensation." But even in 1842, a few months after Lardner's public accounting, Henry accepted another invitation and an honorarium of $50 to present an evening lecture in Philadelphia to the Mercantile Library Company. True to form, however, he emphasized to the organizers that he typically "declined all invitations to lecture to a popular audience" and was accepting only because of the group's persistence.[13]

Although Henry's self-righteousness toward Lardner might appear to have been born merely of personal insult suffered at Liverpool rather than impartial indignation, both motivations seemingly informed his attitude. He apparently had convinced himself that Lardner's flouting of both a "sacred" commitment to truth and a moral rule of conduct precluded him from producing trustworthy scientific analysis. During this era at Princeton, Charles Hodge was analogously preaching that the correctness of a person's understanding of moral and religious subjects depended on the rectitude of the person's character. "Remember," Hodge was instructing his theological students, "that it is only in God's light that you can see light. That holiness is essential to correct knowledge of divine things, and the great security from error." Henry elaborated this theme in his Princeton geology course in the 1840s when he lectured on the connection between geological truths and Christian revelation. He stressed to his students that, "although bad men have endeavored to pervert the truths of this science, these truths themselves have been discovered by pious and good men." "We must make a distinction," he continued, "between an amateur and a real geologist. I have no hesitation in saying that geology rightly understood will be found to agree with the Bible rightly interpreted." Similarly, in an oblique reference likely to Lardner, Henry told a colleague at the Royal Society: "I have long been of the opinion that as a general rule no person whose moral feelings are not properly exercised can secure a high and lasting reputation in the persuit of science. His blind love of self unless restrained by a proper cultivation of the heart will sooner or later induce him to do something inconsistant with that perfect justice, in reference to the claims of others, which is strictly demanded of those who would aspire to an elevated station in the Republic of Science."

In his more explicit references to Lardner, Henry was similarly insisting that the popular lecturer's moral depravity precluded him from producing trustworthy scientific analysis. For Henry this declaration was not toothless rhetoric but a call for action against Lardner. In fact, he exhorted a colleague at Harvard, who was involved in an unrelated dispute over a chemistry textbook, to resist buttressing his case by invoking Lardner's familiar name. "I hope you will not bring forward the authority of Lardner to establish any of your positions and that for the sake of the cause of good morals in our country he will receve the countenance of no Person of your standing amoung us. It is true that we are all liable to be misled by our passions and therefore it may be said we should be charitable in regard to those who may have given way to temptations but it should be recollected that just in proportion as we are in danger of vices the greater is the necessity of visiting every case with the more severe condemnation." In Henry's mind this moralistic judgmental condemnation was a dispassionate response unrelated to the personal indignity suffered at Liverpool. As he would pronounce to his students while lecturing on geology and revelation, "Man is a fallen being."[14]

Approaching age 40, Henry had successfully crossed from the local, regional, and national settings of Albany, Princeton, and Philadelphia to the world venues of London, Paris, and Edinburgh. Just how well he would fare on the international scientific stage became clearer two years later, when Faraday joined Wheatstone in nominating him for the Royal Society's prestigious Copley Medal. The two Londoners sought to honor their close American associate for his past achievements in electromagnetism and, particularly, his recent article on the inductive effects of static electricity—the substantial, third article in his series that he completed after returning home. Unfortunately, the full Council of the Royal Society did not concur. Although accustomed to choosing two recipients for gold medals each year, the council selected only one of the five nominees in 1839, a prominent botanist with the British Museum. Whereas the sensed prejudice against American science might have contributed to this decision, Wheatstone later revealed to Alfred Mayer the official, more tangible reason for Henry being passed over. Although Faraday and Wheatstone had initially written a letter "urging that the Copley medal, that laurel wreath of science, should be bestowed on Henry," the leadership had elected to postpone the award, contingent on additional contributions from the American. "On further consultation with members of the Council," Mayer recalled Wheatstone explaining, "it was decided to defer the honor till it would come with greater éclat, when Henry had continued further his researches in electricity."[15]

Although the other unsuccessful aspirants for 1839 would eventually gain renomination and garner Copley Medals, Henry would never again even be in the running. His third article on static electricity would turn out to be his last major, original publication in electromagnetism. He did come to add two more publications to his series, but the fourth would be primarily a painstaking elaboration of the previous article, and the fifth would remain merely a preliminary abstract of his multifaceted ongoing investigations. To be sure, in his research on electricity and magnetism, Henry achieved and sustained the high respect of especially those foreign colleagues with whom he had close personal relationships, including preeminent practitioners such as Faraday, Wheatstone, and Daniell. However, in his published research record, he never progressed to the point of winning top recognition among more impersonal, elite segments of the broader, international scientific community—among multidisciplinary, disinterested juries expecting "greater éclat" such as the Royal Society Council. Thus, although he cultivated his overseas contacts through the remainder of his career, he failed to attain even an honorary, foreign membership in either the Royal Society or the Paris Academy, the foremost scientific societies of the two countries with which he maintained strongest personal and professional ties. The Royal Society rejected his nomination in 1842 and 1846—the last time discounting the backing of Daniell, who, unpropitiously, died before the election. In turn, between 1852 and 1870, the Paris Academy refused him six times.

After Henry's death, in an obituary for the London *Academy*, physicist Charles Tomlinson, the Dante lecturer at University College in London, perhaps captured the attitude of the English inner circle toward Henry. Tomlinson opened the obituary by assigning the American to what he deemed an appropriate international rank. He declared that Henry "furnishes one of those examples, not unfrequent in a free country, of a man rising by his own efforts from a comparatively obscure position to one of eminence, at least among his fellow-citizens. Not that he was unknown to science in Europe, for his name was seen from time to time in company with those who were discussing the scientific topics of the day; but, in common with many other members of the great army of science, he was a useful subaltern, well fitted to obey orders, but not capable of originating the plan of a campaign." Tagging Henry as a subordinate or secondary practitioner, Tomlinson found annoying what he took to be American attempts to attribute European breakthroughs to him:

Some of his countrymen claim for him discoveries which are usually assigned to European philosophers; but no one, we think, can admit the justice of such a claim, any more than the character of an original mind for Prof. Henry in that most difficult of all intellectual employments—original research, where a man, though traveling in the dark, must have the sagacity to know the true path and to keep in it without going astray or stumbling. When one of those master-minds has done this, and by his superior sagacity has thrown light upon an unknown and hitherto unsuspected path into the wide domain of nature, it is no disrespect to say that men of inferior capacity may easily enter thereon, and assist the original discoverer in matters of detail. This is precisely what Prof. Henry did.

As an example, Tomlinson cited Faraday's original discovery of the creation of secondary currents through galvanic induction (a current in one wire inducing an oppositely flowing current in another wire) and Henry's follow-up detection of additional, higher order currents (using multiple, adjacent coils). Ironically, Tomlinson was disparaging a finding that Henry had published forty years earlier in the paper that had triggered the Copley nomination—the packed, third paper in the APS series. The finding was an easy target for Tomlinson in that Henry had acknowledged openly that he was extending Faraday's "admirable discovery." In pressing his critique, Tomlinson chose his language carefully, seemingly to avoid the charge of intellectual elitism or British chauvinism. "This result, and similar results in various branches of physical science, are meritorious," Tomlinson concluded about Henry's research, "but do not display any great originality of mind, or raise their author above the respectable and useful workers who may be reckoned by hundreds, while the Franklins and the Faradays can only be estimated as single units."[16]

In 1861, writing to a junior American colleague, Henry himself had contrasted his own and Faraday's research regimens. By pointing out the dissimilarities, Henry highlighted a personal trait that contributed to the deficiency of his overseas recognition. He wrote: "The maxim given by Faraday to a young friend was '*finish* and *publish*'. I have, myself, lost much credit for various researches, by not adopting this rule." "On the other hand," he added, in what was to some extent a rationalization, "it is much more desirable that a work should be done well, than that it should be done quickly." Henry must have found some consolation when, in 1849, and now Secretary of the Smithsonian Institution, Faraday's Royal Institution designated him officially an Honorary Member—a distinction limited to thirty non-British scientists.[17]

Proper Compensation 16

In late February 1846, as Henry drafted a long overdue letter to Charles Wheatstone in London, he had no inkling that before the year's end he would become the first Secretary of the new Smithsonian Institution. Inundated as ever at the College of New Jersey in Princeton, he did not even know that members of the U.S. Senate and House were finally moving to cap years of deliberation on the best use for Englishman James Smithson's bequest to create "an Establishment for the increase and diffusion of knowledge among men." Of course, Henry knew about Smithson's half-million-dollar bequest. In fact, a week before Henry met Wheatstone in March 1837, he and Bache had dined with Richard Rush, the distinguished Princeton-educated diplomat who was in London to facilitate the conveyance of the bequest to the United States. All the same, as Henry later recalled about the London dinner and the possibility of personal involvement in the Smithsonian enterprise, "nothing was farther from my thoughts."[1]

What Henry did convey to Wheatstone was an underlying discontent with his existing level of reward and recognition—a level that Henry deemed meager, particularly in light of his standing as arguably the most productive scientific researcher in the United States. To be sure, in reestablishing contact with his London colleague, Henry reported being "at present very pleasantly situated at Princeton." However, throughout the letter, he either hinted at grievances or openly expressed resentments. He couched these complaints in terms of a lack of support for basic science in the wider American society, but he also related this deficiency to his own immediate circumstances.

On a national level the lack of deserved reward and recognition supposedly showed in the scarcity of Americans engaged in meaningful, original research. "The truth is," Henry complained to Wheatstone, "we are overrun in this country with charlatanism. Our newspapers are filled with the puffs of quackery and every man who can burn phosphorous in oxygen and exhibit a few experiments to a class of Young Ladies is called a man of Science." Rekindling a gripe from earlier years, he declared that a main factor inhibiting true practitioners from obtaining deserved remuneration and appreciation was the nation's inattentiveness to international copyright laws. This lack of concern encouraged publishers to pirate foreign textbooks rather than finance original works by American authors. "You can readily see how this effects the science of the country," he explained, "take for example my own case—it is true I have done but little[,] but that little I think you will say is more than any one else has done in this country in the way of original research[,] yet I have scarcely any popular reputation. . . . the want of an international copy right prevents my furnishing for my class and the classes of other colleges a text book in which I might set fourth my own claims and thus receive a proper compensation in money and fame for my labours."

Compounding the financial component of the copyright problem was, as Henry mentioned to Wheatstone earlier in the letter, the "inadequacy" of his Princeton salary. This issue loomed large now that he faced the prospect of paying for his four children's educations. And exacerbating the problem of public recognition was the tendency of unmindful persons to usurp credit for his hard-gained research findings. "I early resolved if possible to earn for myself by patient industry an honourable reputation and to do full justice in all cases to the claims of others," he explained in an earlier draft of the letter, "and while I have [en]deavoured to act in accordance with the last resolution I have not always had strict justice done me." Henry offered Wheatstone two recent, personal examples that highlighted the joint problem of inadequate reward and recognition. One involved the roundabout path that led the president of Harvard to single out Henry for a job bid. The other pertained to a perceived professional slight from telegraph magnate Samuel F. B. Morse.

Henry's immediate reason for writing to Wheatstone was to thank him for indirectly triggering an invitation to assume Harvard's Rumford Chair. This professorship was endowed by the same American expatriate, Benjamin Thompson (better known as Count Rumford), who had founded the Royal Institution of Great Britain. Apparently, Harvard's new president, Edward Everett, while earlier in London, had learned through Wheatstone of Henry and of the esteem in which the English held his scientific investigations. This awareness eventually contributed to Harvard's decision, six weeks before Henry wrote to Wheatstone, to extend the job feeler.[2] "A little commendation from the other side of the water often does wonders for the reputation of an american character," Henry sardonically mused. That an otherwise well-informed American needed to go to Great Britain to gain appreciation of a fellow American's scientific investigations reinforced Henry's point that, lamentably, he enjoyed little popular reputation in the United States.

Although Henry spared Wheatstone the details, the Harvard offer threw into sharp relief another problem that Henry had mentioned in the letter: his inadequate Princeton salary. Harvard's generous offer, which at least matched his current pay while requiring much less teaching, gave him pause about remaining in Princeton. For the past few years he had been complaining that normal living expenses were forcing him to overspend his Princeton salary of $1,500 by between $200 and $300 annually. Indeed, one of Henry's students recalled: "The salaries of the Professors at that time were so low, that it was said Mrs. Henry made his suits." During the winter of 1843-44, Henry had become so frustrated with his Princeton compensation—a salary, unchanged since the mid-1830s, and free use of a house—that he was ready to accept a position at the University of Pennsylvania (the vacancy created when Bache left to head the U.S. Coast Survey). He made his acceptance contingent on receiving a salary of $2,700. Although the negotiations eventually died when Pennsylvania could not quite match this amount, the dickering did provide an opportunity for Henry to extract new financial concessions from Princeton: a $3,000 life insurance policy and a $1,500 promissory note to cover the debt, with interest, that he had incurred in Europe buying instruments.[3] Now, two years later as he wrote to Wheatstone, he related that he was declining the job offer from Harvard but monitoring another potential position. He confided that, although he would be reluctant to leave Princeton after being "so kindly treated by the trustees," there existed "one situation . . . I would attempt to get." In particular, he put Wheatstone on notice that he might call on him for a letter of recommendation if Robert Hare retired from his professorship at the University of Pennsylvania. "In point of salary," Henry commented, "this is the best in the country and does not occupy but half the time of the professor."

Henry was intent not only on improving his and his family's material lot but also on improving his public reputation. As he drafted the closing paragraphs of his letter to Wheatstone, he broached the issue of reputation when he called attention to a book that he was forwarding to his London colleague. The book was *The American Electro Magnetic Telegraph: With the Reports of Congress, and a Description of All Telegraphs Known, Employing Electricity or Galvanism*. Its author was Alfred Vail, an 1836 graduate of the University of the City of New York who, as partner of the older Samuel Morse, had helped refine and finance the Morse telegraph. In the book Vail depended heavily on extracts from other publications to describe the Morse telegraph—that is, "the American Electro Magnetic Telegraph"—and to trace the general history of the device. While chronicling the main European and American developments, Vail not only exaggerated Morse's contribution but also omitted Henry's role—except for inserting a fleeting reference to Moll's and Henry's strong electromagnets.[4] In writing to Wheatstone, Henry did not explicitly state his grievance regarding Vail's book, but merely declared: "I am displeased with the production and intend to inform Mr Morse if he suffers any more such publications to be made by his assistants he will array against him the science of this country and of the world." Henry began to unveil the general contours of his complaint, however, as he went on to characterize Morse as an ingenious inventor who, nevertheless, completely lacked the ability to discover original scientific principles. "I have given him from time to time information on the subject of electricity," Henry added, as he further uncloaked his complaint, "but I think in the future I shall be more cautious of my communications." Although he granted that Morse might not have had any hand in framing Vail's fragmentary narrative, Henry had alerted his London colleague to his concern with garnering credit for his contributions. He also had alerted Wheatstone to a perceived slight that marked the beginning of an irreparable fracture in a previously cordial relationship with Morse.

A five-mile reel of copper wire signified the cooperative relationship that had existed between Henry and Morse from at least 1838. Originally, a year earlier, Morse had combined this exceptionally long reel with a similar one to display the capability of his recently improved electromagnetic telegraph to transmit messages through ten miles of spooled wire. In 1838 another of Morse's partners, Leonard Gale, lent the large reel to Henry, with Morse's approval, for use in the induction experiments that would lead to the third in Henry's APS "Contributions to Electricity and Magnetism." Gale—a scientific friend of Henry's who had enabled Morse to improve his telegraph by alerting him to Henry's Albany findings on electromagnets—could lend the reel because Morse was busy in Europe trying to secure patents for his telegraph, having already staked a tentative claim for a U.S. patent and tantalized Congress with his clever invention. When Morse returned to New York City in 1839, he wrote to Henry that he would like to consult him "on points of importance bearing upon my Telegraph." "I should come as a learner," he entreated, "and could bring no 'contributions' to your stock of experiments of any value, nor any means of furthering your experiments except, it may be the loan of an additional *5 miles of wire,* which you may think desirable to have." Meanwhile, he asked if Henry noticed any technical hitches in his particular telegraphic design—one that used an electromagnet to record, through mechanical linkages, a series of dots and dashes that stood for letters of the alphabet. Mentioning that French experts could find no scientific problems, Morse added that he valued Henry's judgment as highly as the Europeans' opinions. "I think that you have pursued an original course of experiments, and have discovered facts more immediately bearing upon my invention than any published abroad."[5]

In agreeing to Morse's Princeton visit, Henry remarked that he was unaware of serious problems with the invention. "I am acquainted with no fact which would lead me to suppose that the project of the electro magnetic telegraph is impractical," he assured Morse. "On the contrary I believe that science is now ripe for the application and that there are no difficulties in the way but such as ingenuity and enterprise may obviate." Henry's only caution to Morse was that he might need to modify the apparatus to ensure transmissions over great distances. The meeting in Princeton apparently went smoothly. Henry wrote to Bache: "I had a visit from Morse and on the whole was much pleased with him. I gave him my opinion freely in reference to his telegraph."

Through the early 1840s Henry and Morse settled into a pattern of congenial interchange. The son of a Massachusetts minister, an 1810 graduate of Yale, and later a painter with London training, Morse had turned to electromagnetic telegraphy only in the mid-1830s as his artistic career waned. Although six years older than Henry and college educated, Morse acted deferentially to the established Princeton professor. The deference probably reflected Morse's spotty background in basic science. He had penetrated the science of electricity and magnetism mainly through interactions with Silliman at Yale and Dana in New York City, including being in Dana's audience for his 1827 lecture series unveiling a Sturgeon-type electromagnet (the same series that Henry may have attended as a beginning professor at the Albany Academy).[6] Morse's deference also probably reflected a desire to cultivate Henry. Morse could not only gain critical technical information but also secure the Princeton professor's endorsement, which would be useful in enlisting project backers. In turn, while affable toward Morse, Henry maintained a detached demeanor in his dealings with the ex-painter, one of many hopeful but not always credible inventors who were beleaguering him during these years.

Henry's ambivalence came to the fore in 1842, when Morse requested his influential friend to write a letter to help persuade Congress to finance a fifty-mile telegraph test line. Henry consented. But in strongly recommending that Congress support the pilot project, Henry interjected that Morse's invention, while deserving, held only a secondary position in the hierarchy of creative endeavors. That is—in a familiar but begrudging refrain that shifted credit for the invention to, by tacit implication, his own basic research—he relegated Morse's telegraph to the subordinate status of a mere practical application dependent on more critical, fundamental scientific principles. He began by offering his assurance that "the science is now fully ripe" for applying to telegraphy the recently discovered principles of electromagnetism. Moreover, he emphasized, "I have not the least dout, if proper means be afforded, of the perfect success of the invention." He warned, however, that "little credit can be claimed" for the invention of the telegraph, "since it is one which would naturally arise in the mind of almost any person familiar with the phenomena of electricity." Nevertheless, although innovators in England and German had devised their own serviceable telegraphic systems, Morse's version still stood out as the best and, thus, deserved support. Henry suggested that Morse, by advancing his project "at the proper moment when the developments of science can furnish the means of certain success" and by translating the project "into practical operation," had established "the grounds of a just claim to scientific reputation as well as to public patronage."

Morse doubtless concluded that Henry's endorsement of the project more than compensated for the professor's aspersion about the telegraph's derivative nature. The inventor quickly sent his thanks, commenting: "You will not consider it flattery when I say that I value your opinion on this subject as highly as that of any scientific man living." There followed, in the early 1840s, visits of Henry to New York City for inspecting the telegraph, which he found impressive, and of Morse back to Princeton. Henry provided advice on technical issues—he would later contend that it was decisive advice—and Morse reciprocated by offering use of his apparatus for Henry's experiments. Throughout these years of collegiality, Morse remained deferential and Henry detached. For example, Morse responded with forbearance in 1843 after Henry belatedly apologized for not even acknowledging two personal invitations to join other scientific leaders in a major test of the telegraph. "I had already made to myself your apology," Morse reassured Henry, "for I know that gentlemen in your station have not their time always at their disposal."

As long as Morse and his coworkers treated Henry with proper respect, it seems that Henry was happy to help them with what he viewed as their deft application of his own scientific principles. By autumn of 1845, however, his attitude began to sour. He had read Vail's *American Electro Magnetic Telegraph,* in which, from Henry's perspective, the author had done him a disservice. He had neglected to cover Henry's contributions to telegraphy in general and to electromagnets in particular; he also then had the gall to reprint the recommendation to Congress that Henry had written for Morse. By late February, a week before complaining about the book to Wheatstone, Henry railed against its omissions while lecturing to his natural philosophy class. Student John Buhler jotted in his diary: "HENRY sticks it into MORSE. Says M's Assistant *Vail* has lately published a book—purporting to be a history of *The Telegraph* & hasn't mentioned him atal in it—although it was through communications & instructions freely made by him—that M's telegraphic scheme came to a consummation." Henry also expressed his disgruntlement to his and Morse's mutual friends, portraying himself as the object of "manifest injustice." Though Morse was well out of earshot in Europe, Henry's rant soon reached Vail. While visiting Philadelphia to open a telegraph line to New York City—Morse's Magnetic Telegraph Company was beginning to flourish—Vail was shocked to discover that Henry was complaining that *The American Electro Magnetic Telegraph* had "done him great injustice." He was unable to pinpoint the nature of the injustice until he encountered a Princeton student in Philadelphia who contended that Henry had invented the telegraph; later, Vail verified that Henry had been routinely "intimating to his class" that he was the true inventor. Having little background in basic science and caught up in the practical success of Morse's invention, Vail remembered being perplexed: "These reports from Prof Henry shook my previous good opinion of his integrity—and [I] could not but think either that I was misinformed or that he was envious of Morses reputation and celebrity in the invention of a machine the elementary principles of which were the every day study of Prof. Henry."[7]

When Morse returned to the United States—and when "insinuations of the injustice" continued—Vail enlisted his senior partner's help in drafting a letter to Henry seeking resolution of the misunderstanding. Using Morse's forthright and conciliatory prose, Vail told Henry he had heard the rumored complaint and now suspected, after consulting with Leonard Gale, that the complaint had something to do with an inadvertent neglect of Henry's electromagnetic researches. Vail explained that in preparing the book, despite a sincere effort, he had been frustrated not only in finding a published account of Henry's contributions but also in arranging for a mutual acquaintance, one of Henry's student assistants, to obtain an account directly from Henry. He also absolved Morse of responsibility for any oversight and urged Henry to point out omissions so that they could be corrected in an intended future edition. "I am not only willing but most desirous to have anything on the subject you may be willing to communicate; for I consider you as standing at the head here of this branch of Science and I wish that ample justice may be done to your labours not only on your own acc[t] but on that of the country which also derives honor from them."

Although Vail (through the words of ghostwriter Morse) had acknowledged his oversight and offered to rectify the omissions, "I did not think fit," Henry recalled, "to make any reply." Not only did he decline to respond, but—according to Morse and Vail—he continued to revile them. The ongoing backbiting dismayed and dumbfounded the two entrepreneurs.

A decade later Henry revealed his reason for not answering Vail's conciliatory letter. He had felt certain not merely that Vail had done him injustice but also that Morse had sanctioned Vail's blatant omissions. That is, he had believed—and still believed—that Morse, through Vail, had knowingly slighted his contributions to telegraphy. Apparently, Vail's affirmation in his letter of a continuing unfamiliarity with the details of Henry's contributions had made the professor even more implacable in his conviction that the two telegraphers had failed to properly recognize his contributions. Not even a disarmingly frank and contrite follow-up letter from Morse himself could appease Henry. In this letter—which Morse wrote in October 1846, three months after Vail's message, and after Henry's complaints to mutual associates persisted—Morse expressed "astonishment" that Henry was linking him to "the alleged injustice." He stressed to his old colleague that any slip on Vail's part had been unintentional and that he himself had played no substantive role in the publication of the book; nevertheless, he stood ready to make amends. He explained:

> I certainly am unconscious of ever doing you the slightest injustice, in thought, word, or deed. On the contrary, I am sure of entertaining only the most exalted opinion of your genius, and your labors, and on all occasions whenever, your name was mentioned, or when I spoke of you, I have always thus expressed my feelings of respect, and *affection,* and, therefore, in view of this matter, I confess to feelings of deep regret and mortification. I can truly say that if I had heard that my own brother was thus entertaining, such unfounded suspicions of me I could not feel more surprize or more unfeigned regret.

Private letters between Morse and Vail support Morse's contention that his involvement with the book was inconsequential. While it is true that he cautioned Vail against disclosing technical details that might thwart his petitions for European patents—especially details regarding the telegraph's essential "receiving magnet"—he never even intimated that the prohibition should apply to Henry's basic, fifteen-year-old findings on electromagnets. Indeed, neither Morse nor Vail ever mentioned Henry's name in their correspondence on the pending book. Moreover, Vail resisted Morse's call to withhold select facts from his book, whose subtitle promised a "description of all telegraphs known."[8]

Henry at least deigned to send a note to Morse acknowledging his letter, but he still refused to believe that Vail and Morse were guileless. When Morse finally met with Henry in the early winter to try again to end the discord, Henry supposedly exploded at the mention of Vail's name. Morse remembered Henry contemptuously exclaiming: "Mr. Vail? I will have nothing to do with Mr. Vail. What right has Mr. Vail to write the history of electromagnetism? He knows nothing about it." Henry later commented on this outburst of "indignation." A decade after the incident, he still believed that his ire had been justifiable, particularly in light of his conviction that Morse had been responsible for the book's omissions. In contrast, when Morse later reflected on the episode, he marveled that Vail's minor, unintentional oversight had triggered Henry's outrage—against not only Vail but himself. Vail admittedly published his book "with but a slight notice of Prof. Henry, but certainly not with a *slighting* one." Morse also marveled that Vail's and his candid supplications had failed to assuage this previously cordial colleague. "A grievance so inveterate, that has withstood so many efforts to allay and to remove, must surely be of a most serious character," Morse sarcastically remarked. For Henry the grievance was serious. Bedeviled throughout his career by questions of due credit and priority in discovery, he prickled at the thought of this latest perceived slight. In response he dismissed Morse as a ruthless inventor who had opportunistically gleaned the telegraph's crucial ideas from the existing scientific record. This was not an impersonal record but a record that Henry himself had spent years painstakingly helping to create and about which he had grown extremely proprietary—in part, no doubt, because of his longstanding personal insecurities.

Implacable, Henry held firm in his perception of Vail and Morse's public slight and undervaluation of his achievements. Harboring such adamant feelings, he likely took solace in knowing that Morse, Vail, and their backers were not the only telegraph developers in the nation who might provide appropriate recognition and reward. Indeed, about the time he first became aware of Vail's omissions, a rival camp of telegraph developers had offered to pay for his expertise in telegraphy. Whatever his motivations, he had accepted the offer. His cooperation with the competing telegraphers was the first in a series of similar (but henceforth nonremunerative) involvements that Morse would come to consider hostile collaborations. As a consequence, over the next few years, Morse would begin reciprocating Henry's distrust and animosity. Both parties, in other words, would eventually be alleging personal grievances.

American inventor Royal E. House had designed the competing telegraph—a "printing telegraph"—and a railway firm was interested in its potential. In two actions that occurred about the same time and were probably more than coincidence, Henry not only evaluated the new device favorably for the firm but also began to disparage Vail's book. In the fall of 1845 Henry judged House's electromagnetic telegraph to be workable and, moreover, clear of patent infringements. His payment for this consultation must have seemed a windfall, a welcome supplement to his perennially inadequate Princeton salary. Unfortunately for his finances, the payment would remain his only monetary gain from the increasingly profitable enterprise of commercial telegraphy. He used this sole payment in a personally meaningful manner, however, passing it on to Harriet as a gift. "Permit me to present you with the enclosed note, my fee from the Railway company for inspecting the magnetic telegraph," he wrote to Harriet on Christmas eve, 1845, "and to request that you will expend it in any way you may think fit with the single restriction that a good black silk dress for yourself be one of the articles purchased with it." Five years later, when Morse and Vail were still trying to fathom their falling out with Henry, Vail reminded Morse of an old rumor concerning the professor's acceptance of the fee. Henry purportedly had expressed a self-serving—if not vengeful—reason for agreeing to provide the railway company with a favorable appraisal of the House telegraph: "That he rec^{d} nothing from Morse for advocating the interests of the Electro Magnetic Telegraph, whereas the House people had voluntarily paid him a handsom sum!"[9]

Again, it was probably no coincidence that during the months following Vail's publication but preceding his conciliatory letter, Henry adopted a self-aggrandizing tone in drafting his anonymous article on electricity and magnetism for the supplement to the *Encyclopædia Americana*. Because the pending volume would reach a similar popular audience as Vail's book, Henry himself could counter the book and set straight his own research record.

And he apparently had no compunction about recommending the article to even close scientific colleagues without revealing his authorship.[10]

Whatever Henry's precise rationale in certifying the House telegraph and touting his own researches in the encyclopedia article, by late 1845 and early 1846 he had grown discontent with his existing level of reward and recognition. As he implied to Wheatstone, he was intent on improving his material lot and his public reputation. The possibilities of more lucrative professorships at Harvard and Pennsylvania tantalized him, whereas perceptions of Morse and Vail's public slight vexed him. Also, he could not help but notice that these telegraph entrepreneurs were poised to make fortunes off a device that he had helped pioneer. Even his closest friend, Bache—who was younger—was now enjoying high salary and prestige as head of a major scientific bureau in Washington, the Coast Survey, charged with charting the nation's coastline. Meanwhile, back in Princeton, Henry—approaching age forty-nine—had stalled not only in publishing but even in pursuing his electromagnetic researches, including his favorite studies of the induction of currents using static electricity. Recently, Faraday had again scooped him, this time in isolating magnetic effects on polarized light.

For a knot of personal reasons, then, the Princeton professor was singularly susceptible to offers that might boost his career. True to character, however, although the Smithsonian call came at a personally opportune moment, Henry would justify his acceptance—justify it not only to others but also to himself—in terms of duty to the nation and science rather than of self-fulfillment. He hinted at this sublimation of his personal feelings by impersonal considerations when, after summarizing for Wheatstone his private complaints, he extrapolated to the national arena: "I live in hope however that we will be able to bring about a better state of things in this country. Bache as you know is now at the head of the Coast Survey and indirectly has much influence at Washington. I have also a number of personal friends in the dominent party and some in the cabinet and with our united influence we hope to effect some changes for the better." Ten months later, primarily through Bache's behind-the-scenes maneuvering, Henry would be positioned better than any other American to sway the course of the nation's science.

17 On to the Smithsonian

In early September 1846, as the new academic year was getting underway at Princeton, Henry composed a lengthy letter to Bache. Hastily drafting the long letter during a single afternoon, Henry arranged for someone—probably a favorite student whom he had summoned to his study—to take dictation. Transcribing Henry's straightforward prose, the student would soon have realized that his professor was advancing a plan for the proposed, new Smithsonian Institution.[1]

In dictating his plan Henry was merely responding to a request from Bache to comment on a portentous bill that both houses of Congress had recently passed—a bill to create the institution. This was overdue legislation, coming ten years after the announcement of Smithson's bequest; eight years after the receipt of the half million dollars; and two years after the beginning of heightened congressional debate on the disposition of the original sum plus a large accumulation of interest. The debate had become spirited, enmeshing politicians ranging from former president John Quincy Adams to the up-and-coming Stephen A. Douglas. The discussion had grown inventive as well, generating possible missions for the institution ranging from an agricultural teaching and research center to a national library. Although the bill's authors eventually sidestepped most organizational details to secure a consensus in the House and Senate, they did create an organization that, while formally and functionally linked to the government, retained a degree of autonomy. Administering the organization would be a Secretary—a presiding official similar to the secretaries heading departments of the president's cabinet—who would answer to a quasi-governmental board of regents. The regents, in turn, would answer to a strongly governmental "Establishment."

The fifteen-person board, headed by a chancellor elected from within the group's ranks, was responsible for fleshing out the institution's programs and overseeing resulting operations. Bache was one of the six private citizens on the board who joined with three senators, three representatives, the mayor of Washington, the chief justice of the Supreme Court, and the vice president, George M. Dallas, who was elected chancellor. Dallas also happened to be the uncle of the politically well-connected Bache, who carried the Dallas family name as his own middle name, and even answered to that name rather than Alexander among his kin. Vice President Dallas also served on the Smithsonian establishment—a higher-level assembly that further included President Polk, his cabinet, the chief justice, the patent commissioner, and the mayor. Whereas the board, with its heavy representation of legislators, would attentively pilot Smithsonian operations, the establishment, composed mainly of officers from the government's executive branch, would remain relatively inactive.[2]

When Henry prepared his suggestions for Bache, he recognized the possibility that his friend might share them with the board. He was unaware, however, that his own name was already among the names being touted in Washington for the secretaryship.[3] Before receiving Bache's request, he had been inattentive to the debate over the institution. He even sent a follow-up letter to Bache fretting that his suggestions were probably "too crude to be of much use to you." Later, he recalled: "At the time of writing these suggestions, I had not the least idea of being a candidate for the situation of Secretary; indeed I had paid but little attention to the law, and supposed that the Secretary was merely a clerk, and not what in fact he really is, the Chancellor of the Institution." Bache, however, already had him in mind for just that post. At the first meeting of the board of regents, immediately after receiving Henry's tentative but clearheaded suggestions, Bache used the suggestions to secure his friend's standing as "the most prominent candidate" for the important office.

Although Henry readied his recommendations without much forethought, he did candidly offer not only his general vision for the organization but also his specific preferences for programs. In fact, the letter contained the basic principle that would inform Henry's vision of the Smithsonian for the next three decades. He began by restating Smithson's instruction that the new institution should foster the increase *and* diffusion of knowledge. With this twofold charge in mind—and with the presumption that *knowledge* meant *scientific knowledge*—Henry went on to urge his guiding principle: that the goal of increasing knowledge should take precedence in the new Smithsonian Institution over the goal of diffusing knowledge. "The increase of knowledge is much more difficult," Henry insisted, "and in reference to the bearing of this institution on the character of our country and the welfare of mankind much more important than the diffusion of knowledge. There are at this time thousands of institutions actively engaged in the diffusion of knowledge in our country, but not a single one which gives direct support to its increase." In contrast, he continued, England and France provide secure institutional homes and solid incentives for scientific researchers, including protection of their ideas under international copyright laws. Furthermore, England and France were not the only nations outpacing the United States. "There is no civilized country in the world in which less encouragement is given than in our own to original investigation," Henry declared, "and consequently no country of the same means has done and is doing so little in this line."

When Henry insisted that the Smithsonian accent the increase of knowledge, he had in mind not only the relative neglect of such an emphasis but also its inherent utility. He believed unconditionally that original scientific research spawns tangible, practical applications. He had come to this view through early exposure to European books such as Gregory's *Popular Lectures* and Parkes's *Chemical Catechism* and to scientifically sympathetic mentors such as T. Romeyn Beck at the Albany Academy and the Albany Institute. As a professor at the academy and later in Princeton, Henry routinely taught that natural philosophy generates significant practical benefits in everyday life—including such technological strides as the steam engine. His Philadelphia colleagues, inheritors of Benjamin Franklin's Enlightenment trust in human progress through scientific advancement, reinforced Henry's utilitarian bent. At the Franklin Institute, for instance, Bache had been seeking to bring scientific principles to bear on technological practice. As Henry now readied his suggestions for the Smithsonian, however, he felt certain—as he had for the past twenty years—that the message of science's usefulness was being ignored. In the United States, he contended, both the scientific charlatan and the nonscientific inventor dramatically overshadow the true scientist.

The strength of Henry's conviction reflected not only two decades of personal frustration with the exigencies of basic research in the United States but also his current exasperation with Morse and Vail. He still had their supposed transgressions on his mind, having recently received (and snubbed) Vail's apologetic letter. His grudge against the two telegraph entrepreneurs undoubtedly gave immediacy to his call for emphasizing basic research. Thus, after underscoring the nation's neglect of research, he shifted to the converse of the problem—the nation's disproportionate and misguided attention to practical applications. Henry likely had in mind popular accounts of the telegraph—particularly Vail's recent book—that glorified the technical invention while disregarding the underlying science. "Indeed," he complained to Bache, "original discoveries are far less esteemed among us than their applications to practical purposes, although it must be apparent on the slightest reflection that the discovery of a new truth is much more difficult and important than any one of its applications taken singly." Smarting under the perceived slight from Morse and Vail, Henry wanted the Smithsonian to redress the national misplacement of values by bolstering support for basic research.

In the remainder of the letter, Henry listed concrete suggestions for fostering not only the increase but also the diffusion of knowledge. Although incisive, the suggestions were neither the first to be proposed for the institution nor in all cases the ones that would be implemented during Henry's tenure as Secretary. The congressional bill itself, while vague on all organizational issues except the mandate for the establishment, the board of regents, and a Secretary, had at least set basic parameters for the institution's mission. One surviving remnant of an earlier, otherwise discarded version of the bill required the new Institution to assume responsibility for the natural history specimens and other artifacts from the U.S. Exploring Expedition. Under the command of Lieutenant Charles Wilkes, the expedition had collected the items from 1838 to 1842 along the west coast, on Pacific islands, and in Antarctica. By the time the final bill passed, the legislators had extended this charge to include government collections of art and sundry curiosities. The legislators, nevertheless, left unspecified how and when the Smithsonian should begin administering these diverse collections. Along with this museum directive, the final bill also contained a library behest. The framers of the bill generally concurred that the Institution should function as a national, public repository of books, including all books carrying a United States copyright. Whereas they set a high dollar figure on the maximum annual expenditure of Smithsonian funds for library purposes, they undercut this initiative by neglecting to require that any money at all actually be spent for books and related expenses. Ironically, the legislators provided firmer direction on the museum, library, and related missions simply by specifying the functions of rooms in the projected new Smithsonian Building. Allowing a lofty sum for its construction, the legislators directed that the structure accommodate not only the natural history and art collections but also a library, chemical laboratory, and lecture rooms.[4]

It was in response to these varied and, at times, murky legislative mandates that Henry drafted his tentative plan. Drawing on firsthand experiences with the leading scientific organizations of Europe—including the Paris Academy of Sciences, with its elite nucleus of salaried members—Henry believed that the new institution could best achieve the goal of increasing knowledge by assembling and supporting a small company of the nation's leading scientific researchers. He explained that this select group should represent scientific rather than nonscientific fields because of Smithson's own scientific involvement. That is, having been an active scientific investigator, Smithson presumably "intended by the expression 'for the increase of knowledge' an organization which should promote original scientific researches." As for the secondary goal of diffusing knowledge, Henry suggested that the institution concentrate mainly on publishing and distributing widely a series of journals that tracked progress in not only scientific but also other fields. Covering the physical sciences to the fine arts, the initial issues would provide up-to-date, systematic overviews of the fields. Later issues would build on these foundations, providing regular reviews of subsequent worldwide developments.

Henry's scheme for the journals had a double root. First, it arose from his longstanding personal commitment to keeping abreast of the latest scientific publications. Over the years he had filled a stack of notebooks with records of his readings. Second, the scheme reflected his recent, deep involvement in writing updates of subjects for the supplementary volume to the *Encyclopædia Americana.* At the time of his letter to Bache, he was engrossed in the encyclopedia project, wrapping up various articles that he had undertaken following his main entry on magnetism.[5] Two days after mailing the letter to Bache, he sent a follow-up note highlighting his plan for the journals, characterizing it as perhaps his most important suggestion for the pending institution. Emphasizing how the multidisciplinary journals could bring the institution renown at home and abroad, he commented specifically on the effect of the projected journal devoted to physical science. Arguing that Silliman's venerable but recently declining *Journal* "is no longer[,] if it ever was[,] the exponent of American Science," he envisioned a bright future for a new Smithsonian periodical dedicated to tracking developments in physical science. "It might be inferior in original articles to some of the foreign journals," he told Bache, "but it would be far superior to any of them in the posting up of the descoveries of the world."

Although convinced of the efficacy of the review journals in spreading knowledge, Henry expressed reservations about the main mechanisms of diffusion contained in Congress's Smithsonian bill. In particular, he objected to the bill's requirements that the institution serve as a major library, a national museum, and a provider of public lectures. Maintaining that Smithson had intended the institution to benefit the entire nation if not the world, he worried that the proposed library, museum, and lectures would minister to only the local residents of Washington. He granted, however, that scaled-down versions of these activities could further the institution's scientific mission. Specifically, he acknowledged the usefulness of a low-budget "working library," a natural history museum that is "gradually formed," and short courses of lectures given annually by each Smithsonian researcher "to keep up popular interest in the institution." Indeed, during earlier years in New York City and Philadelphia, he had delighted in Peale's popular museums that adeptly juxtaposed collections of natural history artifacts and fine art paintings with displays of scientific instruments and demonstrations of experiments. And in London, he had seen Faraday juggling scientific research and popular lectures at the Royal Institution (an establishment founded, in a curious reversal of Englishman Smithson's later transatlantic role, by American Benjamin Thompson). Finally, he told Bache that only a small portion of the Smithsonian money should go to a new building. He preferred that the Institution simply operate out of existing public facilities. Reflecting a Calvinist preference for good works over material ostentation, this lifelong Presbyterian declared: "It should be remembered that the name of Smithson is not to be transmitted to posterity by a monument of brick and mortar, but by the effects of his institution on his fellow men."

When Bache and Henry exchanged letters about the Smithsonian bill in early September 1846, each man was writing to not only a professional colleague but also an intimate friend. But they were unlikely friends, judging solely by their backgrounds. Although they were both the oldest sons in their families, they differed in age and came from radically different backgrounds. Bache, who had enjoyed the advantages of a stable, patrician upbringing, was nine years younger than Henry, who had endured the disadvantages of an erratic, plebeian home life. Drawn together by their scientific interests and concerns while hobnobbing in Philadelphia and traveling in Europe, the friends happened to confide the depth of their mutual affection in the late winter of 1844. This interchange occurred not long after Henry had helped secure Bache's appointment as superintendent of the Coast Survey and, in turn, Bache had counseled Henry on possibly filling the Pennsylvania professorship left vacant because of the appointment. In the wake of these mutual favors, Henry assured Bache, the past leader of fledgling Girard College and now influential head of a two-hundred-person federal agency: "I do not esteem you because I hope ever to advance my personal interests through your friendship. I had the same regard for *Professor* Bache as for the President of Girard College or the Superintendent of the Coast Survey. I desire to remain in terms of close intimacy and friendship with you for other motives than those which may be founded on the mere reciprocity of kind offices. My warm attachment to you is not only a source of great pleasure to me but of great importance to my moral character, I desire to cherish the feeling on its own account." Bache responded even more expansively: "It is commonly said that friendships formed later in life are not of that warm character which boyhood forms & this is doubtless generally true, but the exception is a strong one in our case for I do not feel towards my own brothers more love than towards you my very dear friend."[6]

Nevertheless, their unfeigned friendship did not preclude advancing each other's "personal interests." In the half year preceding passage of the Smithsonian bill, Henry seized another opportunity to help Bache. He boosted his friend's public standing by writing a popular overview of the Coast Survey in which he accentuated the organization's national utility in mapping coastal areas. Because Henry urged ample federal financial support of the survey, Bache could later distribute reprints of the article to members of Congress to promote appropriations.[7]

Henry would have insisted, of course, that Bache's personal interests in the Coast Survey and other scientific enterprises coincided with American scientific interests. Bache would have reciprocated this sentiment regarding Henry's personal agenda. After all, from the time of their European travels, if not their earlier Philadelphia soirees, they shared similar aspirations for American science. In his 1844 address to the National Institute for the Promotion of Science, Bache had even foreshadowed Henry's plan for the Smithsonian Institution. Specifically, in this appeal for strengthening American science, Bache had called for providing practitioners with, first, the resources to do research and, second, journals to keep abreast of international developments. Two years later, when Bache positioned Henry for the Smithsonian secretaryship, he certainly felt that he was acting in the best interests of American science, not merely the personal interests of his friend. In fact, in this case, he was not only protecting the interests of American science and his comrade but also guarding his and Henry's friendship. After all, the secretarial nomination carried with it the potential to reunite them in Washington. Following Bache's move from Philadelphia to the capital, Henry repeatedly stressed during his visits to Philadelphia that he missed his old companion and scientific cohort. As a temporary remedy, only a few weeks before passage of the Smithsonian bill, he had enjoyed an extended, summer reunion with Bache and his wife in Washington, ostensibly to counsel Bache on a sensitive personnel matter. Soon after, with the secretarial appointment looming, the opportunity arose for Bache and Henry to serve the cause of American science while simultaneously and more permanently serving the cause of shared Sunday dinners and parlor chats. Until his death in 1867, Bache would remain Henry's only true confidant in Washington. Also, the two would continue to use their combined authority to leverage the course of American science, eventually enlisting a few other well-placed colleagues in their campaign, forming an informal coalition known as the "Lazzaroni."[8]

At the first meeting of the board of regents, in early September, Bache floated Henry's name for Secretary. He later told Henry that the regents probably would have elected him on the spot if they could have been certain of his acceptance. But acceptance was a moot issue since Henry knew nothing of the possibility. He first learned that he had become a candidate—the leading candidate—during a Philadelphia rendezvous with Bache during the next few weeks. Apparently, Bache used the meeting to sketch a strategy for further action. He would assume responsibility for advancing Henry's candidacy while, at the same time, Henry would refrain from involving himself personally or from commenting publicly on the matter. Although Henry accordingly revealed his candidacy to only Harriet and intimate friends—assuring them that he had not yet decided on accepting an offer—word still got out. Indeed, the *New York Herald* reported his candidacy. Colleagues and acquaintances were soon asking about his Smithsonian intentions. Faithful to Bache's instructions, however, he remained noncommittal or silent about his prospects and designs.[9]

Meanwhile, Bache intensified his behind-the-scenes campaign. He straightaway solicited letters in support of Henry's candidacy from European scientific leaders—a group that included his and Henry's close colleagues James Forbes, Michael Faraday, and David Brewster. In his request to Forbes, Bache sketched his vision for Smithson's bequest—a vision that harmonized with Henry's, of course, rather than that of the bill's framers: "As one of the Board I am most anxious that this fund should furnish means of scientific research in our country & that the institution should thus supply a want which all of us feel to exist unsupplied by our Colleges & Universities." He informed Forbes that, to set "the right tone" for the new Smithsonian Institution—"our true national institute," as opposed to Washington's presumptuous National Institute for the Promotion of Science—the regents needed to elect Henry as the first Secretary. Faraday responded quickly and affirmatively to a similar appeal. "As far as I have any ability to judge," he wrote, "I think that Henry is just the man to be placed in such a position as that you describe." Brewster followed suit, adding punch to his testimonial by trailing his signature with a long string of academic and professional titles. (Whereas Faraday insisted that his letter remain private, Brewster made no such stipulation; after Henry's election, a Washington newspaper would reprint Brewster's laudatory recommendation.)[10]

Apparently, Bache nudged the board toward delegating fuller authority to the office of Secretary and toward seeking a Secretary having a scientific rather than secular background. He accomplished this as a member of a committee that the regents had established at their first meeting to work out organizational details for the new institution. By the time of the regents' next meeting, in late November and early December, however, much political and personal maneuvering had occurred, leading to at least twenty-seven nominations for the secretaryship. But the committee's sharpened job criteria narrowed the viable candidates to a handful of prominent *scientific* figures. When the regents finally voted on December 3, they divided their ballots among only three contenders. Of the twelve votes cast, one went to Charles Pickering, who had been the lead zoologist on Wilkes's U.S. Exploring Expedition. Four went to Francis Markoe, the politically well-connected but, in certain eyes, controversial corresponding secretary of Washington's National Institute for the Promotion of Science. (Some of Markoe's detractors felt that he had mismanaged the National Institute and considered him a threat to the Smithsonian's scientific welfare.) The remaining seven votes went to Henry, giving him a minimal majority. The board, in a show of collegial solidarity, reacted by "unanimously" approving Henry's election. "I have the honor to inform you that you were this day elected Secretary of the Smithsonian Institution," the board's clerk immediately announced to the Princeton professor.[11]

The day after his election, probably just hours before learning through the clerk's letter that a vote had even taken place, Henry was growing anxious about the regent's deliberations. "I leave my cause entirely in your hands," he wrote to Bache, fretting that affairs were not unfolding according to design. "Do *not* be troubled on my account," he continued, remarking that he had ambivalent feelings anyway on becoming Smithsonian Secretary. On the one hand, if the regents elected him, he would accept the position and "enter on the duties with enthusiasm and with confidence of making the name of smithson familiar to every part of the civilized world." On the other hand, if they did not choose him, he would return "with redoubled ardour" to his scientific writing, motivated in part by a desire to show the public that his Smithsonian backers had been justified in supporting him. Although he conceded that he would accept the secretaryship if offered, he reminded Bache of a few stipulations. During the institution's formative period, he would prefer the appointment to be temporary, allowing him the option of returning to Princeton if the position proved unsatisfactory. Otherwise, if the regents insisted on a permanent appointment, he wanted recompense at a high enough level to protect his family's welfare—a salary of at least $3,500 and a house. This level of recompense was reasonable, he told Bache, because not only had Princeton agreed recently to boost his $1,500 salary but Harvard had renewed its efforts to lure him into the Rumford Chair by offering up to $2,500.[12]

Having mixed feelings about leaving Princeton, Henry had been insisting to family, friends, and colleagues—and would continue to insist through the remainder of his career—that he had exerted no personal effort to land the secretaryship. For example, he had made no direct overtures to even a close Albany colleague who was on the board of regents, Gideon Hawley—the colleague who, two decades earlier as an Albany Academy trustee, presented the welcoming remarks at the ceremony inducting him as a new professor. At the end of his anxious note to Bache, however, he did indirectly press his own case. In particular, he provided fresh evidence of his scientific prowess for Bache possibly to relay to the regents. Describing an original investigation of what seemed to be heat waves and their wavelike interference with each other, he remarked: "Since little things are sometimes important you may if you think fit mention my recent experiment." He also called Bache's attention to the just published encyclopedia supplement containing his anonymous but self-promoting article on electricity and magnetism. Two weeks earlier, he had similarly prompted Bache. After expressing pleasure with Faraday's complimentary letter of recommendation ("sufficient of itself to repay most of the anxiety I have felt on the subject"), he had suggested that Bache solicit a similar endorsement from Harvard's president.[13]

As the board's deliberations continued, Henry began to harbor a penchant for the Washington position. Thus, in a mid-October letter to a colleague at the Albany Academy—in a passage that he eventually deleted, apparently in accord with Bache's injunction of silence—he envisioned himself a self-assured, effectual Secretary: "I am not naturally very confident in reliance on my own talents but the success which has attended my efforts thus far in life under the guidance of a merciful providence has given me considerable confidence in my ability to suceed in what ever I strongly direct my mind to and inreference to this institution if it be put on the proper footing and I can have the right kind of men as coadjutors I have little fear of the result." Later, with the decision imminent, he was even more blunt to his brother, asserting that he deserved the job. After telling James that he would take the position only if the regents met his salary request, he declared: "If the institution is to be of a scientific nature and scientific reputation formed on scientific descoveries is to be the ground of choise then I am entitled to the situation."[14]

Largely because of Bache's gambits, the new Smithsonian Institution was to have a significant scientific component to its mission, and its first Secretary was to be an accomplished scientific researcher. Also thanks largely to Bache's brokering, the Secretary-elect had not only a confidential understanding that he could return to his Princeton professorship but also a robust $3,500 salary and $500 allowance for rent (until his permanent living quarters were ready in the new Smithsonian Building). Bache was overjoyed with the regent's offer. Appealing to his friend's sense of duty, he urged immediate and unconditional acceptance. "Science triumphs in you my dear friend & come you *must*. Redeem Washington. Save this great National Institution from the hands of charlatans. Glorious result. . . . Your position will[,] rely upon it[,] be most favorable for carrying out your great designs in regard to American Science." Bache closed this entreaty with even greater ardor: "Come you *must* for your country's sake. What if toils increase & vexations come. Is a man bound to do nothing for his country, his age. You have a name which must go down to History the great founder of a great Institution. The first Secretary of *the* American Institute."[15]

There was one catch—one condition of Henry's appointment. To placate the bloc of regents still committed to creating a library, Henry would need to nominate an assistant secretary who would serve as librarian. Moreover, he would need to nominate the regents' choice for the post: Charles C. Jewett, an up-and-coming librarian and professor at Brown University. "You will conciliate them by nominating him," Bache matter-of-factly informed his friend. He explained that not only was Jewett a good choice but also the appointment would enable Henry to delegate the library responsibilities. In the meantime, Bache encouraged Henry to accept the regents' terms and travel to Washington to settle details. "I have assumed to speak for you," he reminded his friend, "so I pray you say or do nothing until you see me. You will not of course dishonour my drafts." He elaborated, saying that he understood the board's temperament and could coach Henry on approaching the group. "Your confidence in me & the carte blanche you gave me," he assured Henry, "have not I am sure been used in any way that you would not like."[16]

Henry replied immediately, exclaiming to his confidant and emissary, "you have worked wonders." He pledged that he would accept the offer unconditionally, including Jewett's commission as librarian, and would travel to Washington shortly. Henry's reply and another note from Bache crossed in the mail. In the new note Bache called attention to an article in a leading Washington newspaper trumpeting Henry's appointment. With the article in mind, Bache summarized public sentiment: "The cry is now huzza for Henry. I say huzza for Science which means the same."[17]

The article itself, in the *National Intelligencer,* began by reviewing the regents' latest initiatives. These included persuading President Polk to appropriate "the whole of the large reservation of land called the *Mall* to the purposes of the Institution." The Mall, though an integral element in Pierre Charles L'Enfant's original design for the city of Washington, remained an undeveloped, boggy, rectangular acreage sprawling westward from the Capitol toward, on its northern edge, the White House. The second half of the newspaper report concentrated on the regents' selection of Henry as Secretary and sought to show why he was the best choice for perhaps the most crucial appointment to date in the nation's cultural and intellectual history. The article contained an exceptionally flattering—and what would became a moderately widely reprinted—biographical sketch of Henry written by "a scientific friend." The friend—whom Bache identified as Sears Walker, a former Philadelphia colleague and now a Washington astronomer—opened his paean by ranking "the philosopher of Princeton" as second only to Benjamin Franklin in the nation's scientific pantheon. Becoming more specific, he accorded Henry the particular public recognition that he had been futilely seeking from Morse and Vail. "If we ask who gave to the electro-magnet of soft iron now used for the telegraph its present form, and discovered the laws by which its effective power could be made active," Walker wrote, "the answer is JOSEPH HENRY." Shifting from what he took to be Henry's Albany findings to his Princeton researches, Walker next bestowed on Henry the honor of discovering electromagnetic induction—or, at least, Walker's blurry impression of mutual induction and self-induction. "One of the most important discoveries of recent date, that of the identity of the laws which regulate electric and magnetic and electromagnetic induction, was among the early fruits of his researches at Princeton. If Franklin discovered the identity between lightning and electricity, Henry has gone further and reduced electric and magnetic action to the same laws."

Never mentioning Morse or Faraday, Walker went on to herald Henry's international reputation, which, he surmised, rightly led the regents to select the Princeton natural philosopher. "It is the man that gives dignity to the office," he pointed out, "and not the office to the man."[18]

Two weeks after this two-part article appeared, the *New York Observer* ran a lengthy paraphrase of the section summarizing the regent's latest actions. Instead of including Walker's biographical sketch of Henry, however, the editors attached a letter from their own special correspondent. Written in Washington back on December 3, the day of Henry's election, the letter was from the brother of the editors of the *Observer,* a Presbyterian-affiliated weekly. Although he identified himself in print only as "M," the correspondent was Samuel Morse. Writing two weeks after his conciliatory message to Henry regarding Vail's telegraph book, Morse apparently intended the public letter to back what he had privately told Henry: that he had the utmost respect for Henry and intended to make amends for any perceived (but unintended) slight. "As there is a well founded anxiety in regard to the character which the new National Institution, the Smithsonian Institute, is to assume," Morse began in his public letter to his brothers, "I am sure it will gratify you, as well as the friends of science throughout the country, to learn that this day the Trustees of the Institute have elected unanimously Professor Joseph Henry, of Nassau Hall, as the Secretary of the Institute." Morse continued: "By this choice, to the most responsible, and I may say the highest scientific post in the country, the Trustees have but given utterance to the universal voice of the scientific world. The Trustees deserve the thanks of the community for their impartiality, and the judiciousness of their selection. I fear not the arousing of any jealousy in his contemporaries, when I assert that no man in the country has all the qualifications for this high trust, in a greater degree than Professor Henry." After explaining that Henry had not sought the secretaryship but that the regents had solicited him on the basis of "merit" and "native nobility of mind," Morse ended by expressing his confidence that Henry would accept the post.[19]

Whether Henry ever saw the *Observer* article—and realized the identity of "M"—he did meet with Morse soon after, while both men were in Washington. This was the emotionally charged meeting in which Henry recoiled at the mere mention of Vail's name, but the old colleagues seemingly reached an understanding and parted on civil terms. Henry received the impression from Morse that any future edition of Vail's *American Electro Magnetic Telegraph* would include coverage of Henry's contributions to telegraphy. A few months later, however, the civility vanished when Vail issued a new run of his text—in the original, unaltered form. Vail and Morse considered the printing merely an expeditious continuation of the initial 1845 publication, undertaken for recouping remaining costs. Noticing a new date on the later printing, Henry judged it a new edition—an 1847 edition lacking the promised revision. He was irate at the perceived betrayal. The fracture had become irreparable between the natural philosopher and the two inventors—a natural philosopher who was now the highly visible Secretary of the Smithsonian Institution. Vail soon felt the Secretary's sting. For the May ceremony to lay the cornerstone of the Smithsonian Building, Secretary Henry persuaded the organizing committee to drop Vail's book from the original list of historical items to be included in the stone's hollowed center.[20]

As news of Henry's appointment spread, congratulatory letters poured in to Princeton. Typically in their messages, Henry's colleagues, former students, and friends followed the newspapers' leads and characterized the secretaryship as the nation's preeminent scientific or scholarly post. In responding to these flattering appraisals, Henry downplayed the prestige of his new station and repudiated any implication of personal or even professional fulfillment. Instead, insisting that he had made no effort to obtain the post, he voiced an altruistic rationale for conceding to others' calls to serve as Secretary. He had accepted, he maintained, out of duty to the nation and science. This rationale, with its connotation of self-sacrifice and moral obligation, contrasted sharply with the concerns over salary and recognition that Henry had repeatedly raised in his more intimate correspondence with Bache and brother James. He had told James that he would not entertain the suggestion of moving to Washington unless the salary were generous; he also had said that, if award of the secretaryship were dependent on scientific reputation, then he was undoubtedly "entitled to the situation." The rationale also contrasted with Henry's earlier hints to Wheatstone about his discontent with existing levels of reward and recognition—a discontent manifested in his flirtations with jobs at Pennsylvania and Harvard and in his grudge against Morse and Vail. Nevertheless, in responding to the congratulatory letters, he consistently cited duty, rather than self-fulfillment, as his motive for accepting the post. In particular, he claimed to be selflessly responding to scientific colleagues' urgings to save the Smithsonian bequest from being squandered. And he vowed to withdraw from the secretaryship, as a matter of honor, if he failed in his mission. Resorting even to a slavery metaphor, he told Asa Gray, "I am sold for the present to Washington."[21]

Clearly, Henry believed the rationale, believed that he was selflessly fulfilling a moral obligation. For decades he had been placing duty and altruism at the core of his personal value system. This stance was an expression of not merely his innate temperament but also probably the psychological shadings traceable to the childhood experience of his father's alcoholism. Henry emerged from his youth with a deep need for approval and affirmation. But he masked public expressions of this egocentric side of his personality. Instead he imposed on himself a regimen of service to others. He consistently invoked what he perceived to be his moral "duty," rather than personal desire, as his motive for following a particular course of action.

His commitment to duty and altruism also echoed contemporary social and religious mores. From Albany through Princeton he had been moving in circles that mixed a patriotic concern for the welfare of one's fellow citizens with a Calvinistic commitment to humanitarian outreach and social reform. These republican and Protestant values translated into Henry's personal but self-denying crusade to improve the nation's science and thus boost the commonweal. He was bent on promoting, as Bache had expressed it, his "great designs in regard to American Science." A few months into his Smithsonian tenure, he reminded even Harriet that in accepting the secretaryship he was discharging a duty that entailed subordinating his own "ease." And he was experiencing feelings more of concern over meeting expectations than of pride in attaining the berth. "I can truly say that my appointment has not been to me a source of self congratulation," he affirmed to Harriet, "for though on some occasions and at some moments I may have felt a little proud of my advancement yet you can bear me witness that the prevailing feeling has been one of deep solicitude as to the responsibility I have assumed."[22]

Henry's seemingly inconsistent stances—grumbling in private about salary and recognition while sublimating in public his pride in achievement—partly reflect an expected contrast in his public and private voices. Throughout his career and in step with contemporary values, he had routinely couched his more public statements in elevated, even moralistic language. Conversely, he restricted his more self-revealing comments to intimate colleagues and family members. Altruism and self-concern were each part of Henry's nature. In accepting the secretaryship, then, he displayed no personal inconsistency in simultaneously wanting to improve the nation's science *and* his own lot and reputation. Nevertheless, in the weeks following his election, he gradually embellished his public rationale for accepting the post.

Specifically, Henry supplemented his original claim—that he merely was responding to his colleagues' calls for saving the Smithsonian bequest—with the additional claim that he just happened to be the *only* scientific practitioner available for the job. That is, he was intimating that, because of a lack of other viable candidates, neither he nor the regents had a choice in his appointment. Responding at the close of December to another letter of congratulation, he elaborated this point as he offered a self-abnegating reconstruction of the events leading to his selection: "The office to which I have been elected is one of great responsibility and which has not been saught by me. Indeed it was decided that I was the only available scientific candidate and my name brought forward before the Board of Regents before I was made acquainted with the thoughts and intensions of those who acted in the matter. I resolved from the first not to move in the affair and in the event of my election to be guided by what I should consider my duty to the country and the world after a careful study of all the circumstances. . . . I was fully convinced that unless I accepted the office the noble bequest of Smithson would be in danger of being squandered on chimerical and unworthy objects."[23] In fact, although Henry did not initiate his own candidacy, he did press it later, when he prompted Bache to pass information to the regents. Also, although Henry might have considered them suspect, other scientific leaders had been in the running for the secretaryship, most prominently Markoe and Pickering.

In continuing his response to the late-December letter, Henry offered an even more self-effacing explanation for accepting the post: divine providence. The decision was out of his hands not only because the regents had no other choice but also because the deity guided the choice. While this steadfast but reserved Presbyterian had long harbored an affinity for Calvinistic foreordination—which in Henry's belief allowed for free will and individual moral responsibility—the immediate stimulus for invoking providence perhaps came in an earlier congratulatory letter from a former Princeton colleague, now a practicing Presbyterian minister. "I see the hand of Providence leading you to a post," the minister had suggested, "where you will have great opportunities of showing that Christianity is worthy of aid from philosophy. My prayer is, that you may eminently subserve the cause of sound and evangelical religion."[24] Whether or not Henry viewed the secretarial post as a pulpit to preach the gospel of Christian natural theology or merely to proselytize government officials to support science, he did view the regents' ready acceptance of his provisos for taking the job as a divine sign. That is, "as all my requirements as to salary and nominal connection for the present with Princeton college were complied with[,] I looked upon the whole as a Providential indication that it was my duty to accept the appointment." Further affirmation came in mid-December, when, on traveling to Washington to confer with the regents, he succeeded beyond his "most sanguine expectations" in persuading them to reconsider some of their plans for the Smithsonian in favor of his alternatives.

A trust in divine providence and Calvinistic foreordination would continue to figure prominently in Henry's attitude toward the secretaryship. A few weeks into his Smithsonian tenure, he would tell Harriet: "I have resolved to first study carefully what is my duty and then to do it fearlessly[,] relying on a conscience void of offense for justification of my acts[,] leaving the result to the direction of a kind providence." He added that, by making his sincerest effort and relinquishing the outcome to providence, he could rest assured that even "failure is for the best in the long run or it would not be permitted to take place." So pronounced was his reliance on what he perceived to be the cues of divine providence that, a few months later, he considered surrendering the secretaryship. In particular, he again felt forced to abandon himself to the presumed plan of his deity—whatever the still-cloaked plan might turn out to be—after Robert Patterson and other Philadelphia friends offered him an attractive, alternative position at the University of Pennsylvania. The chair long held by chemist Robert Hare, who had finally retired, was the sole position Henry had fancied before receiving the unexpected Smithsonian bid. A lucrative, prestigious, and not too demanding professorship that allowed time for research, it had the added advantage of being in Philadelphia. Joseph and Harriet strongly preferred this scientifically and socially congenial city over Washington. Believing that he risked losing his honor and reputation if he forsook his Smithsonian appointment merely for personal gratification, he felt that he could justify backing out of the appointment only if a providential imperative compelled him to take the Pennsylvania post. "If it is the design of Providence that I should be a candidate," he mused to Harriet, "some way will be opened for me to retire with honor from the Smithsonian[;] if not[,] I must then do the best I can with the position in which I have been placed[,] no doubt for wise purposes[,] since the position is one not of my own seaking."

When authorities representing the institution and the university seemed to sanction a compromise in which Henry could retain minimal, essential supervisory responsibilities at the Smithsonian, Henry almost persuaded himself to accept Hare's professorship. However, when Smithsonian chancellor Dallas opposed the compromise (an opposition seconded by Bache), Henry dutifully reconciled himself to the secretaryship. Subsequently, however, he let it be known to associates that, having preferred the Philadelphia over the Washington assignment, he had made a large personal sacrifice for, as he variously expressed it, "the cause of American science" and "the good of the Smithsonian." The sacrifice, he rationalized, gave him license to insist on the implementation of his scientific plan for the Smithsonian. It also gave him grounds for asserting that, though he might not succeed in carrying out his plan, he would "at least deserve some credit for the attempt." Key regents were aware of Henry's seeming sacrifice and acknowledged that the institution should view itself "as deeply his debtor."[25]

Henry's success with the regents on first meeting them in early December represented more than providential intercession. It also represented Bache's canny coaching. Thanks largely to family connections, Bache understood Washington politics. He enjoyed a particularly close relationship with his uncle, George Dallas, the Democratic vice president and Smithsonian chancellor. He also numbered among his intimate contacts brother-in-law Robert J. Walker, who was secretary of the treasury in the Polk administration and who would be responsible for dispersing Smithsonian funds. Although usually mum on party preference, Henry had earlier revealed to Bache that he also leaned toward Polk's Democratic ticket. And Henry maintained his own moderately strong political connections, a carryover from his years of mingling in Albany with influential New York politicos. Most prominent among Henry's contacts was Polk's secretary of war, William L. Marcy, a former three-term governor of New York. With both Bache and Henry attuned to Washington politics and its pitfalls, the two confidants even took the precaution of having Henry move out of Bache's home, where he first stayed on arrival, and relocate to a boarding house. This avoided the appearance of being, as Joseph confided to Harriet, "too much with Bache in the way of planning the Smithsonian." In ensuing years, as the friend's alliance became more apparent in Washington circles, pundits would charge that Bache controlled not merely the Coast Survey but also the Smithsonian.[26]

Henry was encouraged by his reception. The Washington-area regents with whom he conferred were sympathetic to his plan of emphasizing scientific over library and museum initiatives. He soon realized, however, that they felt only partially able to accommodate the plan because of the strictures in the original Smithsonian bill. Most distressing to Henry was their obligation to spend as much as all of the bequest's accumulated interest—about a quarter of a million dollars—on a new building. In response to such lingering obstacles to his personal plan, Henry showed the depth of his and Bache's political acumen by adopting an unhurried, pragmatic, even cynical attitude toward the regents' intentions. "I am advised not to be too urgent as to the prosecution of my plans," he disclosed to Harriet. "The Regents will soon become weary of the affair and then I can direct the institution in the way I may think best for the interest of science and the good of mankind." Convinced of the virtue of his plan, Henry intended to circumvent what he considered a flawed congressional bill. And he was optimistic about his prospects. Indeed, he found his trip to Washington a heady experience, perhaps a reminder of his often daily interactions during his younger years with prominent legislators and officials in Albany. He told Harriet about hobnobbing with representatives, senators, and other government leaders and about excursions to the site of the proposed Smithsonian Building—"it is very beautifully situated and will make a very pleasant residence." Gratified with his stay, he insisted that Harriet accompany him on an intended return trip—although only after she attended to a personal problem. "I think you would be pleased with a visit to Washington," he counseled, "but you must get your teeth fixed before you come on."[27]

The direction in which Henry now intended to take the institution differed in some respects from what he had suggested three months earlier. To achieve the goal of increasing knowledge, he no longer advocated assembling a small company of the nation's leading scientific researchers. Instead, he now favored using the institution's resources to encourage original research *around the world* by publishing papers that passed rigorous review and by rewarding the authors financially. Henry's talk about internationalizing the pool of Smithsonian contributors would not translate during future years into any significant representation of overseas researchers. The talk, however, enabled Henry not only to signal the Smithsonian's autonomy from the parochial interests of the federal government but also to tout the Smithsonian, and hence American science, as being of world stature. The institution would issue deserving papers in a multidisciplinary series titled "Smithsonian Contributions to Knowledge." In addition, following the lead of the British Association for the Advancement of Science, the institution would commission special large-scale studies in fields such as—and here Henry cited two examples close to his heart—meteorology and terrestrial magnetism. Toward the goal of diffusing knowledge, he still favored having the institution distribute widely a multidisciplinary series of review journals—or what he now called reports, again seemingly following the lead of the BAAS. Also, as a concession to the regents, he granted that the institution should sponsor lectures in Washington—"for the edification of members of Congress." He refused to concede, however, that the Institution should become a major museum or library, and he so advised Jewett, the regent's librarian designate.[28]

Although flush with optimism, Henry understood the problems impending with the secretaryship. Indeed, his Princeton colleagues, who were reluctant to lose him, made certain that he understood. Theologian Charles Hodge warned that the job would involve anxieties and responsibilities detrimental to Henry's supposedly already fragile health. He also reminded his friend that he lacked administrative experience. Equally worrisome to Hodge, Henry would become vulnerable to the caprice of party politics and "the harassing questionings of coarse & incompetent" congressmen. Another Princeton professor, political economist Matthew Hope, reinforced Hodge's admonitions. "Politicians and public men even if they cannot directly affect your position," he cautioned, "can torment you if they choose. To a man of your high noble spirit this will be chafing beyond endurance." Similarly, Hope forewarned his colleague that as head of "a great public Institution" he would be expected to perform at a level that justifies his high salary to "a jealous public." (Before the month was out, the editor of the recently established *Scientific American* would be protesting that Henry's salary of $3,500—"this fat office of eleven dollars a day, and not much to do"—was largesse bestowed on him by collusive, elitist regents.) Whereas his colleagues warned him of the long-term problems inherent in the office, Henry himself was also well aware of more immediate hurdles. His most pressing challenge, he told the president of Union College, was to prevent Smithson's endowment from being wasted "on a pile of brick and morter filled with objects of curiosity intended for the embellishment of Washington and the amusement of those who visit that city."[29]

A week before Christmas, while still in Washington, Henry posted a letter to the trustees of the College of New Jersey. He announced that, after fourteen years of service, he was resigning his professorship because of his obligation to "a higher duty." He added, however, that he intended to finish his senior course on natural philosophy by teaching at intervals through the end of the academic year (an arrangement that he would later extend into the next year). He closed with a personal reflection: "The mere though[t] of the seperation has been a source of much grief to myself and my family. The happiest and most peaceful days of my past life have been spent in Princeton." Ten days earlier Matthew Hope had warned Henry that his future as a publicly prominent Smithsonian official did not, unfortunately, portend happy and peaceful days. He cautioned that "thousands of men having some scientific pretensions will take the liberty of taxing your time without conscience" and that "the time thus expended will not have the redeeming circumstances attending that devoted to your classes." Hope predicted that Henry would come to feel "pangs of regret attending the perfect waste of time upon persons who really have no claim upon you but whom you cannot repulse." Decades later, well after Henry's death, daughter Mary marked in the margin of Hope's letter next to his pessimistic prediction: "How prophetic."[30]

Epilogue: The Nestor of American Science

Mary Henry based her confirmation of Hope's prediction on firsthand experience. She, along with her two younger sisters, never married but continued to live with her parents after the family moved to Washington. As early as two years into her father's secretaryship, the teenager might already have noticed that her father was, as he told Smithsonian librarian Jewett, "worked & harassed almost beyond endurance." After two decades of her father's tenure, she would have known without doubt that he found his secretarial duties to be time-consuming and frustrating. "I cannot but think," he told Bache's wife in 1866, "that both the Professor [Bache] and myself would have been better off had we continued in the line of teaching and of original research. I am sure I could have done more in the way of reputation with less anxiety and perplexity of mind."[1] But by the closing years of his life, Henry's frustration would turn out to be only half of his Smithsonian experience. Satisfaction would be the other half.

Henry acknowledged the positive aspect when, for example, in the anonymous 1871 autobiographical sketch, he commented about himself and his stewardship of Smithson's bequest: "He has . . . by constant perseverance in one line of policy, brought the Institution into a condition of financial prosperity and wide reputation. Indeed it is scarcely too much to say that no institution of a scientific character, established by the benevolence of an individual, has done more to render the name of its founder generally and favourably known throughout the civilized world." Henry had never wavered in his "line of policy": to buttress the nation's lagging scientific enterprise by providing Smithsonian support to the most deserving American researchers. He had persisted in his campaign, as he once described it, "to pour fresh material on the apex of the pyramid of science, and thus to enlarge its base." The successes that Henry achieved in the campaign—including the institution's "financial prosperity and wide reputation"—reflected his successes in administering the diverse Smithsonian enterprise. On the one hand, this ex-professor who was accustomed to associating with students and scholars had to hone his interpersonal skills as he hobnobbed almost daily with politicians, government officials, and military officers. On the other hand, he had to develop his administrative skills as he coped with the Smithsonian's organizational complexities, financial challenges, and building problems. In late 1849, three years after accepting the office, he told John Torrey that his personal schedule is "pretty fully occupied" in carrying out the duties of an architect, a financier, an accountant, a director of researches, an editor, and so forth.[2]

In accord with his vision for the institution, Henry had moved quickly to publish a multidisciplinary series of research monographs. In 1848 he piloted into print the first volume in the Smithsonian Contributions to Knowledge, a handsomely bound series that would become the flagship of the institution's small but solid fleet of publications. The hefty inaugural volume featured a groundbreaking archeological study of Native American mounds. Fourteen years later, he launched a second series, Smithsonian Miscellaneous Collections, in each volume of which he assembled assorted articles, reports, and tables. Besides creating these two publications for practitioners, he also captained throughout his Smithsonian years the compilation of an "Annual Report." The report covered administrative matters along with occasional scholarly topics, especially in extensive appendices. It also chronicled various Smithsonian projects and even included translations of key foreign papers. Stressing accessibility and cooperation, Henry insisted that the institution distribute all of these publications free of charge and without copyright restrictions both at home and abroad.[3]

Wanting to guarantee the widest possible distribution of the Contributions volumes, Henry also organized, beginning in 1849, a network for the exchange of scholarly publications. The hiring in 1850 of another assistant secretary, Spencer Baird, ensured the project's success. Eventually relying on Baird to manage what became known as the International Exchange Service, Henry aggressively expanded the network within the United States and around the world. By enlisting individual agents and groups such as the Royal Society of London, he and Baird fashioned the exchange service into a valued, high-volume clearinghouse for scientific, literary, and government publications. In a similar vein Henry sparked a bibliographic project that resulted in the Royal Society's *Catalogue of Scientific Papers,* an innovative transcontinental index issued over the years around 1870.[4]

Henry's research vision had extended beyond publications to encompass Smithsonian sponsorship of original investigations, including large-scale geophysical inquiries. Thus, as early as 1848 the regents approved a plan for the institution to help coordinate a North American chain of volunteers reporting local meteorological conditions; the intention was to reveal continental weather patterns, thus clarifying the science of meteorology, particularly the mechanics of storms. The project caught the public fancy when, in the mid-1850s, Henry began using daily telegraph dispatches to determine national weather patterns, which he charted with color-coded markers on a prominent map in the lobby of the Smithsonian Building. In the early 1870s he arranged for the U.S. War Department's Signal Office to subsume the successful project as part of the government's new, permanent weather bureau. Further indulging his lifelong interest in meteorology, he also personally wrote from 1855 through 1859 a detailed, five-part synopsis of the science. The series, extending to 268 pages, was collectively the longest publication of his career. Although he titled the synopsis "Meteorology in Its Connection with Agriculture"—and published it in successive volumes of the *Agricultural Report of the Commissioner of Patents*—he gave preferential coverage to topics he had mastered during his Albany and Princeton years: terrestrial and atmospheric physics, particularly electricity.[5]

As he neared the end of his career, Henry justifiably felt satisfaction about the Smithsonian effort to foster original research—an effort that did not stop with meteorology but extended to areas such as ethnology and exploration and that often involved extensive networks of correspondents feeding information to Washington. But, from the moment he accepted the secretaryship, he had been forced to compromise his research and publication agenda. In particular, he had had to accept an allocation of Smithsonian resources in which only half of the operating budget went to the increase of knowledge while the other half went to its diffusion—that is, to initiatives involving a library and museum. Assistant Secretary Jewett, whose position as Smithsonian librarian became full-time in 1849, worked determinedly to assemble a first-rate collection of books. Beginning in 1850, Baird, a zoologist who had been a professor at Dickenson College, worked equally hard as assistant secretary responsible primarily for the institution's expanding natural history collections. Baird's tasks expanded significantly in 1857 and 1858 when Congress finally consigned to the Smithsonian the natural history specimens and related artifacts from the Wilkes and other expeditions, thus grafting to the Institution what leaders previously had designated the United States National Museum. Unwilling to accept further drain on the Smithsonian operating budget, Henry saw to it that Congress appropriated separate federal funds for administering and displaying the collections. He also began to sponsor several series of guest lectures, which were well received. Indeed, never merely resigned to the institution's library, museum, and lecture missions, he actively supported them as long as they remained limited in scope. During 1854, however, when he perceived Jewett to be overextending his mandate as librarian, Henry reacted aggressively. He obtained Jewett's dismissal, but at the price of an ensuing, fierce public confrontation and congressional inquiry. (In the midst of the fray, he mused that "had I before entertained any doubt of the Presbyterian doctrine of the depravity of the human heart, I should have entirely been convinced of it before this time.") Later, in 1866, he finagled the transfer of the bulk of the Smithsonian book collection to the Library of Congress. In a similar spirit, to lighten the institution's museum burden even further, he not only convinced Congress by the 1870s to underwrite more of the cost of administering the natural history collections but also persuaded a Washington banker, William W. Corcoran, to fund a gallery to house the art collection.[6]

Incidentally, Henry felt that Samuel Morse took unfair advantage of the distracting clash with Jewett to publish, early in 1855, a humiliating refutation of Henry's claims about discovering the scientific principles behind the electromagnetic telegraph. Henry had made the assertions particularly in an 1849 deposition for court, as an expert witness called by a telegraph entrepreneur contesting Morse's patents. Morse worried that the originality of his telegraphic contributions (and thus the legitimacy of his patents) was being challenged by Henry's construal of events, which was gaining currency at home and abroad. Learning of Morse's broadside, Henry was furious. He wrote to brother-in-law Stephen Alexander: "Morse, the telegrapher[,] has published a most wanton[,] foolish[,] and libellous pamphlet against me. It was . . . intended to put me down in the eyes of the public. Morse is a fool in this, as well as something worse; having seen by the papers that I was attacked and supposing that I would be weakened, concluded, coward as he is, to jump on me. He will find the seat[,] however, if my life and health are spared, somewhat precarious." Although enraged, Henry waited for the Jewett controversy to subside before responding by asking the Smithsonian Regents to adjudicate his and Morse's opposing contentions. This distinguished but partisan board of governmental officials and private citizens vindicated Henry in a meticulously documented, widely cited report. The report probably contributed to the general understanding in the United States during the second half of the nineteenth century that, whereas Morse was "the inventor of a machine," Henry held greater distinction, being "the discoverer of a principle" behind the electromagnetic telegraph and, in fact, the creator of the first experimental model. As early as 1857, aggravated by Henry as well as by "pirates" trying to infringe on his patents, Morse would complain, "*Justice to the inventor,* is no part of the institutions of the country." Indeed, following Henry's death, his presumed ascendancy over Morse would dominate many obituaries and memorial articles that appeared in newspapers around the nation. Eulogists would label him "Father of the Telegraph" and "John the Baptist of the telegraph."[7]

The new Smithsonian Building—whose cornerstone slipped into place on May 1, 1847, with President Polk officiating and thousands of citizens watching—also represented part of the compromise that Secretary Henry had to accept. Initially, he groused about sinking limited dollars into a needlessly grandiose structure—one for which architect James Renwick Jr. indulged in a fanciful Romanesque or, more particularly, a Norman motif. Henry scoffingly spoke of "the erection of the norman cenotaph over one half the buried funds of the Smithsonian legacy." And he surprised visitors, as Mary Henry observed, by declaring that all essential operations of the institution "could be carried on between the four walls of his little office and that the rest of the building was only a clog on its usefulness." Eventually Henry resigned himself to the building's extravagances as he found opportunities to draw on his own architectural talents to redesign to his liking various interior chambers. Only slowly would he realize, however, that Renwick's monumental Smithsonian Building was becoming, as the nineteenth century progressed, synonymous with the Smithsonian Institution.[8] The striking red sandstone edifice—its conspicuous galleries filled with tempting exhibits—was not only legitimating the enterprise, it was defining it. As manifested in stone, mortar, and glass, the Smithsonian Institution was serving ostensibly as a federal museum. Although Henry would spend three decades zealously defending the institution's research mission—the increase rather than the diffusion of knowledge—the Smithsonian Building itself would contribute to thwarting his intention.

The museum mission would expand even further when, after Henry's death, Baird inherited the secretaryship. Already during Henry's waning years, the energetic naturalist had assumed increasing responsibility for the daily operations of the Smithsonian. "For many years Professor Baird has had charge of the details of management of that concern," the *New York Daily Tribune* commented in announcing Baird's appointment as Henry's successor, "and its admirable working system is largely due to his sense of order and capacity for administration." After the close of the 1876 Centennial Exposition in Philadelphia, for example, Baird arranged to display the popular national and international exhibits "illustrating the technical Arts and Industries" in a new Smithsonian building that he helped fashion—a new building that opened in 1881 adjacent to the original building and that assumed the name of National Museum (later the Arts and Industries Building). To be sure, under the leadership of Baird and his successors, the Smithsonian would continue to foster research. But even the research would increasingly be distinguished by ties to the expanding collections in proliferating Smithsonian museums and galleries.[9]

For Henry the original Smithsonian Building also served as a home. In 1855, after living in various temporary quarters, he and his family moved into a well-appointed, eight-room, second-story apartment in the recently redesigned east wing. Joseph, Harriet, and the children could stand in their dignified parlor and greet not only Washington's most prominent families but also famous scientists and scholars from around the nation and the world. From vantage points offered by the Smithsonian Building's expansive arched windows and high towers, the Henrys literally watched history unfold before them in the Mall, bounded on the east by the Capitol and on the northwest by the White House. The incessant troop movements during the Civil War—infantry, cavalry, and artillery—intrigued the family with some of the, no doubt, most fascinating sights.

The Civil War years certainly brought Joseph the most emotionally trying moments of his life. His oldest child William, who became an assistant at the Smithsonian after attending Princeton, contracted cholera and died in late 1862 when he was only thirty-one years old. The tragedy must have painfully reminded the Henrys how they had fled from Albany to Galway three decades earlier to protect their toddler "Bub" from a previous outbreak of the disease. (In her diary, Mary Henry recorded that the last utterances of her dying, delirious brother included the cry: "Oh they are chaining him. They are chaining Father.") Another disaster occurred early in 1865, when a demoralizing fire gutted the upper central section and main towers of the Smithsonian Building. The blaze—from which Henry and his clerk William Rhees only "escaped very narrowly"—destroyed valuable scientific apparatus, irreplaceable portraits of Native Americans, patron Smithson's personal effects, and Henry's record-laden office. The string of calamities culminated at the end of the war in the assassination of Henry's supporter and occasional confidant, President Abraham Lincoln. (Henry had backed Lincoln's action to free the slaves, although he still doubted that ex-slaves could cope and compete in the broader, white society; indeed, in 1869, still advocating the relocation of African Americans to Liberia, he became Life Director of the American Colonization Society.) On the day of Lincoln's "funeral obsequies," Secretary Henry and a few other Smithsonian officers joined a select coterie of mourners in the East Room of the White House for the somber ceremony. Three months after the assassination, in a letter to Harriet, Henry found solace in familiar Christian precepts as he assayed the personal toll of the sequence of calamities:

> Our feelings are much more under our control than they appear to be, we can give way to them, and suffer ourselves to become miserable, or resist them and remain tranquil amidst a sea of troubles. After all what are the gleams of sunshine with which our path through life is heightened at intervals, compared with the peaceful pleasure of the life to come. The events of the last two years have lessened my own ties to life except the strongest of all the cords, my love to my wife and children, for their sake and for the good I may be permitted to do I am willing to remain and toil as long as God sees fit to sustain my life. Let us then be patient and endeavor to realize the fact that affliction has fallen upon us for our good.[10]

Meanwhile, over the years, even with his demanding Smithsonian schedule and busy personal life, Henry had always found time to campaign behind the scenes for American science as a mainstay of the Lazzaroni. This small, informal network of scientific leaders held appreciable sway over the nation's science policies and scientific appointments for more than two decades, before fading at about the time of the Civil War. He also found time to support fledgling American professional and learned societies, as the nation's scientific enterprise gained in strength and size. Having overcome his earlier doubts about the viability in the United States of a truly national, comprehensive scientific society, he took part in founding the American Association for the Advancement of Science in the late 1840s. He went on to serve as president of the popular organization, using the office to revisit the problem of scientific fraud and quackery. (He cautioned that, in evaluating highly abstruse scientific claims, expert opinion must take precedence over public opinion. That is, "the votes must be weighed, not counted.") And though initially wary of the formation in 1863 of the more exclusive National Academy of Sciences, he succeeded Bache as president in 1868 following his dear friend's death. After that, until Henry's own death in 1878, the academy convened in the Smithsonian Building under Henry's leadership for its annual spring meeting (followed by occasional fall meetings in other cities). Retaining his interest in teaching—and clinging to the now flagging psychological theory of mental faculties and the associated mental-discipline approach to instruction—he also served in 1854 as president of the recently chartered American Association for the Advancement of Education.[11] (During the previous year, he had rejected entreaties to assume a more significant academic presidency—of Princeton College; eventually, he did become a Princeton trustee.) And in 1871 he helped organize and then headed the Washington Philosophical Society, an energetic assembly of local scientific and intellectual leaders. A year earlier he had renewed his personal contacts with overseas scientific friends and organizations when he returned to Europe for a four-and-a-half-month excursion.

Washington officials also recruited Henry to fill government posts involving science and technology. During the Civil War he joined with Bache on a special "Permanent Scientific Commission" to evaluate suggestions for improving military technology. The commissioners, for example, gathered atop the Smithsonian Building's tallest tower to fire differently colored aerial rockets, in coded sequences, to test communication with a distant fort. At the nation's Centennial Exhibition of 1876, he also lent his support as a member of the board of judges, helping appraise the displays of the latest scientific and technological marvels, including friend Alexander Graham Bell's new telephone. In a more permanent assignment that began in 1852—one that President Millard Fillmore urged on him—he served as a charter member of the U.S. Light-House Board, putting to use technical knowledge that he had first acquired in Scotland. The new board displayed its scientific acumen by quickly replacing the old parabolic reflecting mirrors in the nation's lighthouses with Augustin Fresnel's efficient, modern lenses. Moreover, Henry conducted his own elaborate experiments on signaling with sound and light, and he cut federal lighthouse fuel expenditures drastically by encouraging the substitution of lard-oil for sperm-oil.[12]

As Chairman of the Light-House Board, an office he assumed in 1871, he enjoyed making extended trips to sites along the east coast where he combined research and relaxation. It was at a favorite test site on Staten Island, New York, during early December of 1877, twelve days short of his eightieth birthday, that he became seriously ill.[13] His ailment—nephritis, commonly called Bright's Disease, a malady of the kidneys—persisted through the winter and spring. Though his condition had deteriorated by the time the National Academy of Sciences convened for its regular meeting at the Smithsonian Building in mid-April, he insisted on being carried from the family's upstairs apartment to preside at the opening session. Able to remain only briefly before his strength ebbed, he allowed another officer to read his written welcome. Toward the end of the short text, he offered to resign the presidency. The membership refused the offer. Moreover, his closest friends used the occasion of the meeting to announce an endowment that would annually generate $2,400 to supplement Henry's modest salary and, after his passing, provide financial protection for his immediate family. The principal of $40,000, which the friends had discreetly raised the previous winter, would at the death of the last family member revert to the NAS where, as the "Joseph Henry Fund," it would support research. The NAS membership also surprised their leader by arranging for the Western Union Telegraph Company to run a special line to his parlor, over which he conversed by telephone for a few moments with a waiting colleague in Philadelphia. These gestures of support and goodwill prompted Henry, in the comments that he sent to be read at the closing session a few days later, to reply:

> Please accept my warmest thanks for the kind expressions of sympathy you have extended to me during this period of my illness, and for your personal partiality in refusing to accept my resignation as president of the Academy. I shall be thankful if a beneficent Providence extends my life during another year and grants me the privilege of greeting you again in a twelvemonth from this time—as successful laborers in the fields of science.
>
> I can truly say that I entertain for each member of the Academy a fraternal sympathy, and rejoice at every step he makes in the development of new truths.
>
> With my best wishes for your safe return to your homes, and for a rich harvest of scientific results in the ensuing year, I now bid you an affectionate farewell.

Providence did not grant Henry the privilege of meeting again with the full academy. Even as his health waned, however, he continued privately to receive colleagues and friends in the family's east wing apartment. Astonomer Maria Mitchell numbered among his final visitors during the spring of 1878. In her diary she recorded her conversations with the, at times, wistful Secretary. Henry began by remarking that, facing imminent death, he felt comforted by the NAS endowment. He assured Mitchell that he did not fear dying and, although having in earlier decades questioned immortality, now leaned toward belief in an afterlife. Interrupting the conversation, Henry opened a desk drawer and retrieved an old but evidently precious book: George Gregory's *Popular Lectures on Experimental Philosophy, Astronomy, and Chemistry.* He did not mention to Mitchell that, when first sailing to Europe four decades earlier, he had inscribed the volume to his now deceased son, William. Rather he recounted that he had chanced on the book as a teenager while bedridden with an illness and that the book persuaded him to devote his life to science. "I asked him if a life of science was a good life," Mitchell closed her diary entry, "& he said he felt that it was."

He died a few weeks later, on 13 May, in his second-floor bedroom, attended by Harriet and his daughters.

Henry's death triggered an outpouring of sentiment that was unprecedented for an American scientist. First came the extensive newspaper coverage of his passing and the Washington funeral with all the pomp of a state ceremony. Later followed the special congressional session and the handsome memorial volume. Homages would peak (but not end) in 1883 with the unveiling of his bronze statue, swelling sixteen feet above the Smithsonian lawn.

Asa Gray, who worried that the unveiling ceremony would be ostentatious, had tried to persuade Mary Henry, a driving force behind the event, to moderate the pageantry. "I have," he lamented to Spencer Baird, Henry's successor as Secretary, "reasoned with her—even to the point of vexing her much upon her schemes for inauguration of the statue of her dear father—with military display, great gathering, and orations—told her such ideas were out of place and vulgar."[14] Contrary to Gray's plea, the pageant went forth. The crowd of between five thousand and ten thousand people that massed on the Smithsonian grounds for the April unveiling witnessed a grand spectacle, complete with politicians, Sousa marches, an on-the-scene telegraph report, and moving orations. On the day after the statue's dedication, the headline in the *Washington Post* heralded, "Prof. Henry in Bronze."

The statue and other commemorations were not merely in recognition of Henry's Smithsonian contributions. To be sure, as Smithsonian Secretary during his last three decades, he had excited an intensity of public notice unusual for an American scientist. By 1878 Henry had achieved the status of "Nestor of American science." In Homer's epic telling of the siege of Troy, Nestor endured as an old Greek chief who provided wise counsel. Eulogists pointed out that Henry had assumed critical positions of leadership in the scientific community during the later decades of his long life—one that during the Washington decades alone spanned nine presidential administrations. One obituary opened with the declaration: "The Nestor of American science, the venerated chief of the Smithsonian Institution . . . has done more than any other who has ever lived in this country—with, perhaps, the single exception of Benjamin Franklin—to organize American science, to promote its interests at home, and to cause it to be esteemed and honored abroad." Although Henry had a handful of critics who would have considered this hyperbole, most Americans agreed with the accolade, lauding the long-lived scientist's performance as Smithsonian Secretary.[15]

But Henry's administrative and organizational achievements—and his longevity—only partially account for the lavish tributes, memorials, and monuments. Ultimately, the acclaim stemmed from his earlier accomplishments as an active natural philosopher. That is, as much as the Smithsonian secretaryship heightened and solidified his national reputation, it was his earlier scientific contributions—or, at least, his fellow citizen's impressions of those contributions—that grounded and sustained his name. Initiatives during the Albany and Princeton years, particularly in electricity, magnetism, and telegraphy, provided him credibility and authority as he flourished in the Washington limelight. Being "Secretary Henry" expanded his reputation, but being "Professor Henry" established it.

In reflecting on the significance of Henry's technical contributions, his compatriots typically classified him uppermost in the hierarchy of the world's physical scientists. According to the *Philadelphia Public Ledger and Daily Transcript,* he "stands in the front rank of the great men of science of his age, his work being of the nature and value that places his name side by side with those of Franklin, Ampère, Arago, Sir Thomas Young, Sir Humphrey Davy, Faraday, Sir William Thomson, and their eminent co-laborers in the same field—men whose peer Professor Henry was." But not all the eulogists were so commendatory. A smattering of American newspersons expressed qualms about the ex-professor's scientific achievements. One hundred miles south of Washington, in Richmond, Virginia, the editors of the *State* rejected as inflated the splashy line with which the *Washington Evening Star* announced Henry's death. Responding to the proclamation, "The great scientist of America is dead!" the *State*'s editors grumbled: "This is in the usual exaggerative style of the day, which turns everything that is 'latest' into 'greatest.'" Similarly, but of more consequence, most Europeans tempered their tributes to the American and his legacy. Recall that Charles Tomlinson, in an obituary for the *London Academy,* characterized Henry as "a useful subaltern, well fitted to obey orders, but not capable of originating the plan of a campaign."[16]

These contrasting reactions—the Americans' enthusiasm and the Europeans' restraint—reflect a variety of circumstances. One is that the European eulogists, being more accustomed to the attainments of many active scientific researchers, were using a more demanding standard of evaluation than the Americans. Even at the crest of Henry's electromagnetic investigations, decades earlier, officials of the Royal Society had deemed his record wanting and withheld the Copley Medal. Most American eulogists, however, were judging Henry by the criterion of their own nation's still developing scientific community—one that, particularly in the recondite realm of physics, was only barely beginning to approach institutional maturity and achieve a sense of professional identity.[17] By that less-evolved standard, these commentators gauged Henry's contributions to be truly exceptional. Also, unlike their overseas counterparts, many American eulogists had known the natural philosopher personally—or knew someone who had. Exceptionally high appraisals of Henry's character seemed at times to lead these acquaintances to equally elevated appraisals of his scientific achievements.

A related factor is that the Europeans based their appraisals mainly on Henry's *documented* research record. And that record, although notable, was patchy. Henry often had failed to report his findings promptly and in widely accessible writings or talks. In addition, he often had failed to grasp—or, at least, report—the fuller import of his usually detailed laboratory gleanings; his pattern was to focus undividedly on a line of investigation only after another researcher announced results involving a similar or identical line. For example, two of his most important early papers appeared not only belatedly but also in an American periodical of constrained circulation, Silliman's *Journal.* The first was the 1831 paper on "intensity" and "quantity" electromagnets, which grew out of findings that he had reported locally to the Albany Institute and which he rushed into print only after Gerrit Moll's announcement of his enhanced electromagnet. The second was the 1832 paper on mutual induction that he hurried into press only after Faraday's celebrated revelation that he had teased electricity from magnetism. And each of these papers contained portentous auxiliary ideas that Henry mentioned only in passing—the design of an electromagnetic telegraph and the concept of self-induction—ideas that he would return to in public forums only after other investigators encroached on them. Indeed, in response to Faraday's independent announcement of self-induction in late 1834, Henry hurriedly resorted to a Philadelphia meeting of the American Philosophical Society to reassert his apparently overlooked claim of priority. Even after this latest incident involving priority, he would continue to issue his main findings on self-induction and related topics in the APS *Transactions,* which suffered from tardy publication and limited circulation.

In contrast to their overseas counterparts, Henry's American eulogists enjoyed easier access to his research reports—Henry having usually lodged his reports in either domestic journals or the proceedings of American societies. They also had greater familiarity with his more revealing, sometimes self-aggrandizing, informal commentaries. These writings included not only the reassessments of his research that he published anonymously in the 1847 encyclopedia article and the 1871 biographical sketch but also the midcentury depositions on the telegraph's development that he provided to various tribunals. So accessible were versions of the biographical sketch, with its list of his twenty-two most significant investigations, that eulogists often tapped it as an informational source, probably unaware that Henry himself had penned it.[18] The Americans also drew on the glowing accounts of Henry's scientific exploits as retold by the many adoring graduates of his Albany and Princeton courses. Eulogists in the United States were thus more aware of Henry's accomplishments involving, for example, strong electromagnets, the reciprocating engine, the telegraph, and self-induction. Ironically, because of Henry's own lifelong reluctance to blazon his contributions to the 1831–32 identification of mutual induction, most eulogists—American and European—initially overlooked this area of achievement. Only gradually would his compatriots realize, and then proclaim in occasionally imperious voices, that he had paralleled Faraday in this pivotal finding of nineteenth-century physics.

Most American political leaders sincerely believed that Henry's scientific achievements matched those of the internationally preeminent researchers, but they also probably realized that trumpeting his attainments served an ulterior purpose: to elevate the nation's cultural image. Accordingly, President Rutherford B. Hayes, senators, representatives, and other politically savvy Americans all would have recognized the symbolic significance of Henry's government-backed tributes: the capital funeral, the congressional session, the memorial volume, and the statue. During an era of aspiration and ambition when many Americans still harbored feelings of insecurity and sensitivity about their country's international cultural standing, the public remembrances signaled to the nation and the world not only that the United States had produced a scientist of the highest global stature but also that its leaders and citizens valued him and his work. Through these gestures of respect, the United States government "honors itself," various newspapers pointed out. General William Tecumseh Sherman, a Smithsonian regent, made a similar remark a month after Henry's death. In a Princeton commencement address, he implored his audience to bolster the nation's reputation by invoking the accomplishments of a select circle of extraordinary Americans, including Princeton's own Professor Henry: "On all grand occasions I beg you to emblazon the names of Franklin, Agassiz and Henry, side by side with our great statesmen—Webster, Clay, and Lincoln—and with our great soldiers—Washington, Jackson, and Grant. 'By their fruits ye shall know them.' A country which can in a single century record such names need not be ashamed, but may, with proud front, claim a place among the most honored nations of earth."[19] Thus, for varied reasons, through obituaries and other remembrances, Americans unflinchingly boosted favorite son Joseph Henry to the pinnacle of world science.

On the one-hundred-and-fiftieth anniversary of the Smithsonian's founding and the two-hundredth anniversary of Henry's birth, his bronze likeness still guards his final home and workplace—the original Smithsonian Building, now known as "the Castle." In 1934 workers moved the statue from the west front of the building to a central location near the main door—the door on which the black crepe had announced his death in 1878. Then, in 1965, a crane rotated Henry's figure so that it now faces away from the Castle, surveying a metropolis of Smithsonian museums and galleries on the expansive "National Mall."[20]

Of the thousands of tourists who daily pass or enter the Castle, few do more than glance at the imposing memorial, although many sightseers include the statue in snapshots or video tapes of the red sandstone building, now an official National Historic Landmark. On a recent summer's day, a family crossing the Mall made a slight detour to inspect the conspicuous monument, triggering a rare but typical conversation about the statue. With no one in the family breaking stride, the father leaned toward the inscription at the base and emotionlessly muttered "Joseph Henry." "Who is that?" asked an older son. "I don't have a clue," the father replied and the family charged along toward the nearby National Air and Space Museum. Caroline H. Dall, an intimate of the Henry family, had been wrong when she foretold the permanency of the statue's personal impact. "There will never be a time," she wrote following the 1883 dedication, "when the eye will not rest with pleasure upon the folds of the drapery, when it may not receive a sacred inspiration from the calm and uplifted gaze which so fully recalls the living form of our vanished friend."[21]

Today, more than a century after its unveiling, the statue has lost its particular meaning to nearly every passerby on the Mall. In Caroline Dall's day, the bronze figure emotionally engaged its beholders, persons who knew or at least knew of the natural philosopher. Few if any visitors through the late nineteenth century would need to ask, "Who is that?" Nor would they find puzzling the horseshoe-shaped electromagnet set in subtle relief on the side of the lectern holding a book on which Henry's left hand rests. But an evolution was already in process. From the moment of its unveiling, the statue was becoming more than an image of a revered scientist; it was on its way to becoming an icon—an autonomous bronze casting with a significance of its own. That is, the statue was losing its immediate and visceral connections to Henry. It was surrendering its status as a memorial to a particular person. Instead, it was evolving into an icon that served a more general and impersonal purpose: helping to garnish and legitimate the Smithsonian enterprise itself.

A stately figure draped in an academic gown and gazing solemnly into the distance connotes the authority, dignity, and mystique of ancient knowledge and, thus, provides a stamp of approval to the Castle, itself the symbolic embodiment of the increasingly expanding Smithsonian complex. Now typically relegated to the lower corner of a tourist's photograph or gift-shop postcard, the statue has become an official seal affixed to ephemeral documents—meaningless in itself but laden with symbolic power. The progression from the particular man, to an idealized image, to an impersonalized icon has reached completion. Behold "Prof. Henry in Bronze."

Notes

There are two, closely related archival depositories in Washington, D.C., that target Henry's career. First, the Joseph Henry Collection in the Smithsonian Institution Archives is an extensive collection of mainly original documents contained in record unit 7001. Second, the Joseph Henry Papers Project, under the auspices of the Smithsonian Archives, is a computer-indexed collection of copies of selected documents from the Henry Collection and other sources around the world.

Throughout these notes, the former depository is cited as HSA, with items usually identified by shelf box number. Because the main series of letters to and from Henry is arranged chronologically in boxes 7 through 12, individual box numbers will not be given for these letters. Also, because what previously were designated the Mary A. Henry Papers and the Harriet Henry Papers have been included in the Joseph Henry Collection, separate mention will not be given of these subcollections. All HSA box numbers refer to the locations of materials after the 1994 reprocessing of the collection.

The latter depository—the Joseph Henry Papers Project—is cited in the notes as HPP, with items identified by either computer-database control number, microfilm batch, or photocopy batch, along with specification of the original depository. The staff of the Henry Papers Project also administers the Alexander Graham Bell-Joseph Henry Library, a comprehensive collection that includes Henry's private books and pamphlets; this collection is cited as Bell-Henry Library.

The staff of the Henry Papers Project is in the midst of publishing key documents in the annotated *Papers of Joseph Henry*. To date, editors Nathan Reingold and then Marc Rothenberg, with the help of their associates, have completed seven volumes (Washington: Smithsonian Institution Press, 1972–1996). Throughout the following notes, if a document from HSA, HPP, or elsewhere has been published in this series, the citation usually will be to *Henry Papers* followed by volume number and page. References to particular pages in the volumes of the *Henry Papers* assume inclusion of the accompanying editorial annotations, which often specify secondary sources. The staffs of both depositories also maintain computer-based bibliographies of books and articles concerning Henry.

Throughout this book, quotations from archival and other primary documents retain, when possible, the original spellings and punctuation.

Prologue

1. The descriptions of Henry's death, funeral, and the events' reverberations rely primarily on the multitude of contemporary newspaper articles collected in the "Henry Memorial. Scrap Book. 1878," HSA, box 47—from here on cited as "Scrap Book"—and the smaller groups included in HSA, boxes 43, 46, 59, and 96. Though many clippings in the informal "Scrap Book" and the smaller collections lack dates and sources, the articles constitute an exhaustive sampling of the coverage. See also "Part I: Obsequies of Joseph Henry," in *Memorial of Joseph Henry*, Smithsonian Institution, Smithsonian Miscellaneous Collections, no. 356 (Washington, D.C., 1881), 5–34; from here on, this volume is cited as *Memorial*. The "local resident" was Horatio King, reporting to a newspaper in his home state of Maine; besides his newspaper comments, see also "Remarks of Hon. Horatio King," in "Henry Memorial. Original Papers. 1878," HSA, box 47. A copy of William T. Sherman's address at "Princeton College, N.J., June 19th 1878" appears not only in "Scrap Book" but also in Manuscript Division, Library of Congress (HPP, microfilm batch M-118). See also William Q. Force, Diary, 13 and 14 May 1878, Manuscript Division, Library of Congress (HPP, nos. 14716 and 14717).
2. "Introduction," in *Memorial*, 1–3; "Part II: Memorial Exercises at the Capitol," in *Memorial*, 35–122, esp. 49, 75; see also "Obsequies," 7–12, 27–34. For copies of the British telegrams, 16 Jan. 1879, see HSA, box 57.
3. "Part III: Memorial Proceedings of Societies," in *Memorial*, 123–508; see also "Introduction," 3–4. For a copy of the "Memorial Exercises" as reported in the *Congressional Record*, 4 March 1879, see HSA, box 46. William B. Taylor, "Memorandum," Smithsonian Institution, 16 Feb. 1888, HSA, box 46. "Joseph Henry Memorial," *Public Ledger and Daily Transcript*, Philadelphia, 15 April 1881, copy in William A. Henry's "Rhetoric" notebook, HSA, box 59.
4. Files titled "Letters from friends after Joseph Henry's death" and "Miscellaneous correspondence with daughters after Joseph Henry's death," HSA, box 57. File titled "Henry Condolence Letters," HSA, box 46. "Henry Memorial. Original Papers. 1878," HSA, box 47.
5. Sherman, "Address," in *Memorial*, 119–20. "Appendix: Proceedings in Congress Relative to a Monument to Joseph Henry," in *Memorial*, 511–14. "Report of the Executive Committee of the Board of Regents on the Henry Statue," in *Annual Report of the Board of Regents of the Smithsonian Institution, 1883* (Washington: Government Printing Office, 1885), xvi–xxxvii.

6. For the letter from Gray to Spencer Baird (11 April 1881), various 1881 letters from Story, program and invitations to the unveiling, and newspaper accounts (including the *Post* article), see the files titled "Statue by William W. Story," HSA, box 46. Caroline H. Dall, "Prof. Henry's Statue Unveiled," correspondence to the *Republican*, dated 20 April 1883, copy in William A. Henry's "Rhetoric" notebook, HSA, box 59. See also Story to Miss Henry, 12 Aug. 1886, HSA, box 39; John Maclean to Caroline Maclean, 6 Dec. 1880, William J. Rhees to John Maclean, 18 Dec. 1880, Maclean Papers, Princeton University Library (HPP, nos. 10130 and 10131).
7. In 1897 a bronze cast of Henry joined an imposing circle of statues in the rotunda of the Library of Congress—alongside Moses, Newton, Saint Paul, Columbus, Fulton, Gibbon, Herodotus, James Kent, Michelangelo, Solon, Plato, Bacon, Shakespeare, Homer, and Beethoven. Nathan Reingold et. al., eds., "Introduction," in *The Papers of Joseph Henry,* vol. 1, *December 1797–October 1832: The Albany Years* (Washington: Smithsonian Institution Press, 1972), xvi–xviii; from here on cited as *Henry Papers* 1.
8. James C. Welling, "The Life and Character of Joseph Henry," in *Memorial,* 186–87. [Henry], "Henry, Joseph," in *The Biblical Repertory and Princeton Review: Index Volume from 1825 to 1868* (Philadelphia: Peter Walker, 1871), 198; see below, chap. 1, n. 2, for evidence of Henry's authorship of this anonymous biographical sketch.

1. An Unsettled Beginning

1. For the preceding descriptions of Albany, see Cuyler Reynolds, comp., *Albany Chronicles: A History of the City Arranged Chronologically* (Albany: J. B. Lyon, 1906), 354–57, 379–97. See also Robert V. Bruce, *The Launching of Modern American Science* (Ithaca: Cornell University Press, 1987), 15–16, and John T. McClintock, "Albany and Its Early Nineteenth Century Schools" (Research report prepared for Harvard Graduate School of Education, summer 1967), 1–12 and app. A.
2. One brief document contains Henry's own description of his elders: "Note on the Life of James Henry My Grand Father—From his daughter E. Selkirk," in file labeled "Miscellaneous," HSA, box 39. Early in 1864 Henry gave daughter Mary "a more connected account of his life than ever before"; see Diary of Mary A. Henry, 4 Feb. 1864, HSA, box 51. For a copy of this lengthy diary entry, see MS that begins "Feb. 4th Thursday," in folder 8, HSA, box 56 (typed version in HPP, no. 5850). An aged and ailing Henry also informally related his family history to chief Smithsonian clerk William J. Rhees, who summarized the information in a brief note with the heading "Mar 16, 78. Interview with Prof H"; the original is in the Rhees Collection at the Huntington Library, item RH 4233, and a copy exists in HPP,

no. 2612 (microfilm batch M-24a). This and later information from the Rhees Collection is published by permission of The Huntington Library, San Marino, Calif. Internal and external evidence also suggests that Henry was the anonymous author of the published biographical sketch, "Henry, Joseph," in *The Biblical Repertory and Princeton Review: Index Volume from 1825 to 1868*, 194–200. The internal evidence is that the 1871 sketch contains information to which only Henry had access, such as close paraphrases of the preceding "Note on the Life of James Henry." The external evidence is threefold. First, there is the annotation "by himself," probably added by Henry's longstanding clerk Rhees, written near Henry's name in the title of a republished copy among Henry's papers (see "Sketch of the Life and Contributions to Science of Prof. Joseph Henry, LL.D." n.p., n.d., copy in "Scrap Book"); similarly, there is another copy labeled "Autobiography of Joseph Henry" (in HSA, box 43). Second, there is brother-in-law Stephen Alexander's letter to Asa Gray stating that he has "good reason to think" that at least part of the article "had undergone the revision of the Professor himself"; see Alexander to Gray, 4 Dec. 1878, Archives of the Gray Herbarium, Harvard University (HPP, no. 7333). Gray repeated this conjecture as fact in his Henry memorial (Gray, "Biographical Memorial," in *Memorial*, 63). Third, there is the note "it is said to have been written by Henry himself" added to a typed copy at Princeton passed on to the Smithsonian (see file labeled "Articles and clippings concerning Henry," HSA, box 29). An updated, edited version appeared as "Sketch of Professor Henry," *Popular Science Monthly* 2 (April 1873): 741–44. A shortened version, with minor revisions, appeared as "Prof. Joseph Henry, LL.D.," *Eclectic Magazine* 21 (March 1875): 376–78. After Henry's death, revised or paraphrased versions of the sketch appeared in various newspapers; see, e.g., "Special Dispatch" from Washington dated 13 May 1878 to the *Evening Telegraph*, Philadelphia, copy in "Scrap Book." A list of accomplishments excerpted from the sketch, along with an identification of Henry as author, appeared in Benjamin Silliman, "Joseph Henry, LL.D.," *American Journal of Science and Arts* 15 (June 1878): 464–65, copy in "Scrap Book." For further details on the Alexander and Henry lineages, see the charts and records—particularly as compiled by Robert G. Lester, R. F. Meredith, and James C. Selkirk—in the genealogical file maintained by HPP; see also the printed "Reminiscences of Mary Lydia Kelly," pp. 1–6, in the miscellaneous information file maintained by HPP. For James Henry's living on the estate of Rensselaer Nicoll, see federal census (records for 1790 and 1800), copies of which are on file in the Colonial Albany Social History Project, New York State Museum, Albany; Stefan Bielinski, of the museum, identified Nicoll for me and called the census records to my attention. In personal email messages to the author of 5 and 6 May 1997, Michael Wolfe and James Selkirk, descendants of Joseph Henry's aunt Elizabeth Selkirk, also provided historical data from various records.

3. Albany tax records (assessment rolls for 1788 and 1799) and federal census (record for 1790), copies of which are on file in the Colonial Albany Social History Project, New York State Museum, Albany; Stefan Bielinski searched these records for Henry citations and called them to my attention. For conjectures about home sites, see: Cuyler Reynolds, "Prof. Joseph Henry and His Life Work," *Albany Argus*, 17 Dec. 1899; and transcript of Alexander Selkirk's anecdotes, in "Prof. Joseph Henry, First Wizard," *Albany Argus*, 17 Dec. 1899. Copies of both articles are included in a file labeled "Henry—Albany Centennial Celebration," HSA, box 46. See also school records, originally from the archives of the Albany Academy, in the file labeled "The Albany Academy," HPP, photocopy batch X-90. See also the small painting labeled "Joseph Henry's birthplace South Pearl St. Albany" in the Henry filing cabinet at the Department of History of Science and Technology, National Museum of American History, Smithsonian Institution; a more polished version of the painting is in the file labeled "Henry Illustrative Material," HSA, box 30. For the childhood anecdote, see Diary of Mary A. Henry, 8 July 1863, HSA, box 51.
4. Gray, "Biographical Memorial," in *Memorial*, 54n. Newcomb, "Biographical Memoir," in *Memorial*, 442. File labeled "Henry Birth Date Substantiation," HSA, box 27. MS on Psalm, ca. late 1860s, in the file labeled "Religion," HSA, box 30; copy also in HPP, no. 5556.
5. File labeled "Henry Birth Date Substantiation." See also "Henry's Baptismal Record," *Henry Papers* 1:3, and Reynolds, "Prof. Joseph Henry and His Life Work." "John H. Wendell to Ann Alexander Henry," *Henry Papers* 1:14. Diary of Mary A. Henry, 4 Feb. 1864. Albany tax records (assessment roll for 1799), copies of which are on file in the Colonial Albany Social History Project, New York State Museum, Albany. The Montesquieu volume (a new edition printed for A. Donaldson) is in the Bell-Henry Library.
6. Before his interview of 16 March 1878 (see note 2 above), Rhees recorded the details of two other conversations in which Henry initiated discussions of his biography. In the course of dictating a letter on about 15 June 1868, Henry gave Rhees "quite a full account of his early life" including the general point: "He saw the evil effects of drinking & made a vow never to drink which he kept faithfully"; a copy survives as HPP, no. 2613. Rhees then prepared a more polished version of his notes (HPP no. 2614), which he later amended slightly in light of a second conversation, "a long and interesting account of his early life," which was triggered about 1 Dec. 1876 in response to a letter Henry received from a woman claiming to be a relative. Rhees afterward wrote down "the principal points," including Henry's disclosure of

his father's intemperance, and these survive as HPP, no. 3969. Originals of all three accounts are in the Huntington's Rhees collection, RH 3292 and 4232; microfilm copies exist in HPP, batch M-24a. Throughout this chapter, unless otherwise indicated, the narrative of Henry's earliest decades relies on: his anonymous biographical "Sketch," his "Note" on his grandfather, his reminiscences to his daughter Mary in 1864, Rhees 1878 interview, and genealogical records (see note 2 above), or the accounts that he personally narrated to Rhees (as listed in this note). For Henry hiring Rhees, see Henry to Alexander Dallas Bache, 11 July 1853, Bache Papers, Smithsonian Archives, record unit 7053 (HPP, no. 6475).

7. Researchers in recent decades have catalogued these patterns through studies of adults who, as children, had an alcoholic parent. This and the next three paragraphs rely on the following relevant sources. Bernadette Matthews and Michael Halbrook, "Adult Children of Alcoholics: Implications for Career Development," *Journal of Career Development* 16 (Summer 1990): 24–27. Sharon Wegscheider, *Another Chance: Hope & Health for the Alcholic Family* (Palo Alto, Calif.: Science and Behavior Books, 1981). Barbara L. Wood, *Children of Alcoholism: The Struggle for Self and Intimacy in Adult Life* (New York: New York University Press, 1987). Frank J. Sulloway and Hans Eysenck, "Rebels with a Cause," *Economist* 329 (20 Nov. 1993): 99–100. Frank J. Sulloway, *Born to Rebel: Birth Order, Family Dynamics, and Creative Lives* (New York: Pantheon Books, 1996); for factoring "parental loss" into birth order, see pp. 136–45, 190–91. Diary of Mary A. Henry, 2 Jan. 1865. Of course, the interpretation in the text assumes that late-twentieth-century findings about adult children of alcoholics have bearing on similar adult children in the late eighteenth and early nineteenth centuries. As later sections will show, however, the assumption finds support in significant parallels in the personality profiles of Joseph Henry and his modern counterparts.
8. *Henry Papers* 1:249 n.3. Henry to James Henry, 26 Nov. 1833, in Nathan Reingold et al., eds., *The Papers of Joseph Henry*, vol. 2, *November 1832–December 1835: The Princeton Years* (Washington: Smithsonian Institution Press, 1975), 126–28; from here on cited as *Henry Papers* 2. Theodore L. Cuyler to Miss Henry, 20 March 1888, file labeled "Addresses on Putting up Tablet in Princeton—Personal Recollections," HSA, box 54. In a published version of his recollections a few years earlier, Cuyler seemingly was even more protective of the acclaimed scientist and his family, not specifying the Albany connection when recounting that Joseph illustrated his Princeton pleas for sobriety with "a most pathetic incident." See Cuyler, "Professor Henry—The Christian Philosopher" (n.p., n.d., but probably *New York Evangelist*, May 1878), copy in HSA, box 46. See also Cuyler to "very dear friends" [Henry's daughters], 17 May 1878, in file labeled "Henry—Condolence Letters," HSA, box 46.

9. Only when near death, in the interview with Rhees of 16 March 1878, did Henry list his age at the time of his father's demise as 12—a faithful figure given Henry's misconception that he was born in 1799 rather than 1797. He stuck to the dissembled version of his early years not only in the autobiograpical "Sketch" that first appeared in print around 1871 but also in his 1864 reminiscences with daughter Mary and his 1868 and 1876 conversations with Rhees (see notes 2 and 6 above). The version also appears in the anonymous "Sketch of Professor Henry," *Popular Science Monthly* 2 (April 1873): 741–43. He also apparently related this version to Charles Lanman, a friend since 1848, who eventually published it in his book *Haphazard Personalities; Chiefly of Noted Americans* (Boston: Lee and Shepard, 1886), 7; Lanman presented a preliminary version in "Professor Henry in Private," *New York Evening Post,* 24 May 1878; copy of latter in "Scrap Book." Subsequent obituaries, memorials, and biographies typically perpetuated Henry's contorted version.
10. "To Parents," *Henry Papers* 1:3–4; for original copy, see Joseph Henry to Father and Mother, 26 July 1808, HSA, box 7. The outside of the letter carries the address, "Mr. William Henry, Albany," along with later annotations describing Henry's circumstances as a student.
11. "To Parents," *Henry Papers* 1:3–4 n. 1. Diary of Mary A. Henry, 14 Feb. 1863 and 26 May 1865; quoted in Paul H. Oehser, "Joseph Henry Builds an Institution," *Journal of the Washington Academy of Sciences* 37 (15 Nov. 1947):378 n. 2. Gray, "Biographical Memorial," 71. Joseph to William A. Henry, 27 Jan. 1847, in Marc Rothenberg et al., eds., *The Papers of Joseph Henry,* vol. 7, *January 1847–December 1849: The Smithsonian Years* (Washington: Smithsonian Institution Press, 1996), 30–32; from here on cited as *Henry Papers* 7. For Henry later mentioning going to work in the store at age eleven, see Marc Rothenberg et al., eds., *The Papers of Joseph Henry,* vol. 6, *January 1844–December 1846: The Princeton Years* (Washington: Smithsonian Institution Press, 1975), 286 n. 6; from here on cited as *Henry Papers* 6.
12. In addition to Henry's accounts of his Galway years included in the sources described above in notes 2, 6, and 8, see the recollections of close family friend Welling, "The Life and Character of Joseph Henry," 177–79. As for an uncle who might have died, John Alexander, Ann's twin in Galway, lived until 1841, although step-uncle Robert Alexander died in November 1815.
13. Reynolds, *Albany Chronicles*, 400–428. Diary of Mary A. Henry, 7 Jan. 1863.
14. Thurlow Weed, "How Prof. Henry Missed Being an Actor," *Journal of the Telegraph* 11 (1 June 1878): 164; this reminiscence, which appeared originally in the *New York Tribune*, 22 May 1878, was also reprinted in *New York Commercial Advertiser*, 22 May 1878. For copies, see "Scrap Book"; see also "Professor Henry Almost an Actor," copy in folder labeled "Various newspaper clippings regarding Henry," HSA, box 27. Reynolds, *Albany Chronicles*, 437. See also

T. Romeyn Beck to Martin Van Buren, 10 March 1826, *Henry Papers* 1:120 n.8. Henry C. Cameron, "Reminiscences," in *Memorial*, 167. Alexander to Asa Gray, 4 Dec. 1878.

15. The Gregory book (the first of two vols., printed for R. Phillips by R. Taylor and Co.), with Henry's inscription of 7 Feb. 1837, survives in the Bell-Henry Library; see pp. 1–9, 223. See also, *Henry Papers* 1:xxii. The Scotsman was Robert Boyle. For Henry's "zeal" for reading, see Franklin B. Hough to Spencer F. Baird, 15 May 1878, Huntington Library, RH 4237 (HPP, no. 3923).
16. Weed, "How Prof. Henry Missed Being an Actor"; Beck became principal in 1817. See also Weed's recollections in "Boys Who Became Famous Men," a note that appeared in a Presbyterian periodical, ca. 1886; copy in folder labeled "Various newspaper clippings regarding Henry," HSA, box 27. Franklin B. Hough to Spencer F. Baird, 15 May 1878; the trustee was Gideon Hawley, who assumed this position in 1818. Reynolds, *Albany Chronicles*, 428.
17. Henry mentions studying under Hamilton in "Sketch," 2, and hiring a tutor in HPP, no. 3969. Also, Welling has him studying under "a peripatetic teacher of English grammar"; see Welling, "The Life and Character," 182–83. For Hamilton's summer teaching excursions to northeastern towns, see James Hamilton, *The History, Principles, Practice, and Results of the Hamiltonian System,* new ed. (London: W. Aylott and Co., n.d.), 13.
18. Selkirk, in "Prof. Joseph Henry, First Wizard." (Uncle James Selkirk was the father of silversmith William Selkirk, with whom Joseph apprenticed.) Diary of Mary A. Henry, 4 Feb. 1864. Cuyler to Miss Henry, 20 March 1888; Benjamin Silliman Sr. first published the *American Journal of Science* in 1819.
19. Beck to Martin Van Buren, 10 March 1826, *Henry Papers* 1:117–18. School record card (n.d.) in the file labeled "The Albany Academy," HPP; in the same file, see also what seems to be a record of Henry's absences from Sept. and Oct. 1821. "List of Students," in *Celebration of the Semi-Centennial Anniversary of the Albany Academy* (Albany: J. Munsell, 1863), 109. See also Henry's 25-cent receipt for lending the academy a "casting stove," 23 Oct. 1820, in folder labeled "Albany Academy Miscellaneous Accounts," HSA, box 28.

2. From Student to Surveyor

1. "Historical Sketch," in *The Celebration of the Seventy-Fifth Anniversary of the Founding of the Albany Academy* (Albany: Charles Van Benthuysen & Sons, 1889), 7–23, esp. p. 13. Orlando Meads, "Historical Discourse," in *Celebration of the Semi-Centennial Anniversary of the Albany Academy*, 18–27. *Henry Papers* 1:4–54. William L. Lassiter, *Philip Hooker and the Old Albany Academy* (n.p., n.d.), 1–10; Lassiter apparently published this pamphlet in association with the New York State Museum, Albany, ca. mid-twentieth century.

2. Parkes, *Chemical Catechism,* 6th English ed. with rev. (New York: Collins and Co., 1816), i–ix, 2–25, 510–12; personal copy in Bell-Henry Library. See also *Henry Papers* 1:51–52, n. 10.
3. Henry, MSS, "Stereographic projection of the Sphere," Albany, 1 Sept. 1821, and "Astronomical Problems Solved by Oblique Spherics . . . ," Albany, 22 Nov. 1821, in HSA, box 17. See also *Henry Papers* 1:54–57. Vince, *Elements of Astronomy*, 1st Amer. ed. (Philadelphia: Kimber and Conrad, 1811); personal copy in Bell-Henry Library.
4. Henry, MS, "Mathematical Problems," 27 June 1822, HSA, box 39. See also *Henry Papers* 1:60–61.
5. Reynolds, "Prof. Joseph Henry and His Life Work." Selkirk, in "Prof. Joseph Henry, First Wizard."
6. Beck to Van Buren, 10 March 1826. Meads, 1872, quoted in William B. Taylor, "The Scientific Work of Joseph Henry," 1878, rpt. in "Part III: Memorial Proceedings of Societies," in *Memorial*, 375, note A. Rhees's conversations with Henry of 1876, as described above in chap. 1, n. 6. For March's teaching of medicine and for Beck's *Elements*, which he wrote with his brother John, see Reynolds, *Albany Chronicles*, 434, 440. For Henry's continuing interests in anatomy, see *Henry Papers* 1:256, 258.
7. Henry to T. Romeyn Beck, 28 Nov. 1825, *Henry Papers* 1:114.
8. Henry to John Maclean, 28 June 1832, *Henry Papers* 1:436. Amos Eaton to Benjamin Silliman Sr., 20 Oct. 1829, Gratz Collection, Historical Society of Pennsylvania (HPP, microfilm batch M-90). See also *Henry Papers* 1:xxiii–xxiv, 7 n. 17, 8 n. 26, 19 n.18, 64 nn. 5, 6, and pp. 65–72. R. W. B. Lewis, *The Jameses: A Family Narrative* (New York: Farrar, Straus and Giroux, 1991), 18, 33. C. Hartley Grattan, *The Three Jameses: A Family of Minds* (1932; rpt. New York: New York University Press, 1962), 24–26.
9. Henry's reminiscences to his daughter Mary, Diary of Mary A. Henry, 4 Feb. 1864 (see above, chap. 1, n. 2). For Van Rensselaer recommendations, see letters in *Henry Papers* 1:122, 131, 436.
10. Minutes, Albany Institute, *Henry Papers* 1:65–77, 92–97. Nathan Reingold, "The New York State Roots of Joseph Henry's National Career," *New York History* (April 1973): 140.
11. Henry, "On the Chemical and Mechanical Effects of Steam," *Henry Papers* 1:78–92, 95–96. See also, *Scientific Writings of Joseph Henry*, 2 vols. (Washington: Smithsonian Institution, 1886), 1:1–2.
12. Reynolds, *Albany Chronicles*, 428–42, 449–58.
13. For background on the survey and its personnel, see *Henry Papers* 1:92–93 n. 1, 96–97 n. 2, 98–99 n. 3, 96–105. Reynolds, *Albany Chronicles*, 440, 461. Henry to James Henry, 25 [May 1833], *Henry Papers* 2:73–76. See also Beck to Van Buren, 10 March 1826. Amos Eaton to Benjamin Silliman Sr., 20 Oct. 1829. Henry to [Jabez D. Hammond], [mid-June 1849], *Henry Papers* 7:558–61.

14. Henry's eleven notebooks for the westward survey and three notebooks for the eastward survey, July–Dec. 1825, are in the box labeled "Books of Levels," HSA, box 20. William Campbell to Henry, 3 Oct. 1825, Henry to Jabez D. Hammond, ca. 15 Oct. 1825, Henry to Campbell, 31 Oct. 1825, and Henry to Campbell, 16 Nov. 1825, *Henry Papers* 1:108–13.
15. H.W.E. [Henry W. Elliott], "Joseph Henry: Character and Habits of the Late Secretary of the Smithsonian Institution," *Cleveland Herald*, ca. 19 May 1878, in "Scrap Book." *Henry Papers* 1:89 n. 48.
16. Cuyler to Miss Henry, 20 March 1888. Cuyler, "Professor Henry—The Christian Philosopher."
17. Henry to Beck, 28 Nov. 1825, Campbell to Henry, 12 Dec. 1825, and Henry to Stephen Alexander, 19 Dec. 1825, *Henry Papers* 1:113–17. For David H. Burr's compilation of the *Atlas*, see *Henry Papers* 1:250 n. 1. Henry "Topographical Sketch" (Oct. 1829), rpt. in *Scientific Writings* 1:8–36, esp. 36n.
18. Henry's account sheet, whose heading begins "The State of New York . . ." and which bears the date 28 Feb. 1826, is in the file labeled "Albany Academy Accounts for Philosophical Apparatus," HSA, box 28. Henry to Beck, 28 Nov. 1825, and Henry to Stephen Alexander, 19 Dec. 1825.
19. Henry to John Maclean, 28 June 1832, *Henry Papers* 1:435–37. For background on U.S. Army's and West Point's initiatives on civil engineering, see *Henry Papers* 1:118–20 n. 7.
20. Beck to Van Buren, 10 March 1826, Hammond to James Barbour, 16 March 1826, Van Buren to Barbour, 17 March 1826, [Isaac Roberdear] to Barbour, ca. 17 March 1826, and Hammond to John W. Yates, 28 April 1826, *Henry Papers* 1:117–20, 122–24, 131–32. See also Reynolds, *Albany Chronicles*, 441.
21. Minutes, Academy Trustees, 14, 24, and 28 April 1826, *Henry Papers* 1:128–31, 132–33. See also school records, originally from the archives of the Albany Academy, in the file labeled "The Albany Academy," HPP, photocopy batch X-90, especially item no. 7472. Rhees's conversations with Henry of 1868 and 1876, as described above in chap. 1, n. 6.
22. Hammond to Yates, 28 April 1826. Rhees's conversation with Henry of 1876, as described above in chap. 1, n. 6. Bruce, *Launching of Modern American Science*, 82, 104–05, 111.

3. Teaching at Albany Academy

1. Unless otherwise indicated, this and the next six paragraphs rely on the following sources. "Fitch's Canal Tour Journal," "Notes of a Tour from the Hudson to Lake Erie in May and June of 1826," and Henry to George W. Clinton, 7 June 1826, *Henry Papers* 1:133–55. George W. Clinton, "Journal of a Tour from Albany to Lake Erie by the Erie Canal in 1826," *Buffalo Historical Society Publications* 14 (1910): 277–305. Samuel Rezneck, "Joseph Henry

Learns Geology on the Erie Canal in 1826," *New York History* 50 (Jan. 1969): 28–42. Samuel Rezneck, "A Traveling School of Science on the Erie Canal in 1826," *New York History* 40 (July 1959): 255–69.

2. George W. Clinton to the daughters of Joseph Henry, 24 May 1878, in the file titled "Henry—Condolence Letters," HSA, box 46. Amos Eaton to Anna Eaton, 7 May 1826, Gratz Collection, Historical Society of Pennsylvania (HPP, microfilm batch M-90).
3. L. C. Beck to Joseph Henry, 9 May 1826, HSA, box 7.
4. Excerpt, Amos Eaton to Benjamin Silliman, 26 Feb. 1828, *Henry Papers* 1:205–6, esp. n. 3.
5. Unless otherwise indicated, this and the next five paragraphs rely on "Henry's Journal of a Trip to West Point and New York," *Henry Papers* 1:155–61.
6. Lewis C. Beck to Henry, 15 April 1827, "Lewis C. Beck and Joseph Henry: Description of a Scale of Chemical Equivalents," "Excerpt from Lewis C. Beck's Autobiography," Henry to [Lewis C. Beck], 21 Sept. 1827, and Henry to John Torrey, 4 Oct. 1827, *Henry Papers* 1:191–200. See also Excerpt, Minutes, Albany Institute, 25 Nov. 1828, *Henry Papers* 1:212.
7. W. James King, *The Development of Electrical Technology in the 19th Century: 1. The Electrochemical Cell and the Electromagnet*, United States National Museum Bulletin 228 (Washington, D.C., 1962), 233–56. For Oersted's initial theory and Faraday's representation of it, see also L Pearce Williams, *Michael Faraday: A Biography* (New York: Simon and Schuster, 1965), 138–40, 154.
8. Charles C. Sellers, *Mr. Peale's Museum* (New York: Norton, 1980), 2, 302, 330. Henry, draft of commemorative address on "the life and labors" of Peale, ca. 1875–76, copy in file labeled "Eulogy of Charles Willson Peale," HSA, box 26. For the effect of Henry's New York experiences on his Smithsonian directorship, see Reingold, "The New York State Roots of Joseph Henry's National Career," 133–44, esp. p.140.
9. Unless otherwise indicated, this and the next six paragraphs rely on the following source, including footnotes: "Inaugural Address to Albany Academy, September 11, 1826," *Henry Papers* 1:163–79.
10. Unless otherwise indicated, this and the next three paragraphs rely on the following sources, including editors' footnotes. Minutes, Academy Trustees, *Henry Papers* 1:188–90. Henry to the Academy Trustees, [ca. June or July 1829], *Henry Papers* 1:224–29. The Statutes of the Albany Academy, *Henry Papers* 1:232–48. Henry to the Academy Trustees, 12 Nov. 1830, *Henry Papers* 1:297–301. Henry to Charles Davies, 15 Feb. 1831, *Henry Papers* 1:325–27. Diary of Mary A. Henry, 18 July 1867.
11. Excerpt, Minutes, Academy Trustees, 27 Jan. 1832, *Henry Papers* 1:397–400. For Henry's extrcurricular interactions, see *Henry Papers* 1:351–53, 355–56, 357–58, 428–32.

12. Reminiscences of George W. Carpenter, 15 Nov. 1893, included in Carpenter to Mary Henry, 15 Nov. 1893, HSA, box 23. Smyth to Mary Henry, 23 Aug. 1897 and 25 Oct. 1897, file titled "Recollection of Joseph Henry," HSA, box 27 (HPP, no. 5370 and 5368). Bradford, "Commemorative Address," in *Celebration of the Semi-Centennial Anniversary of the Albany Academy*, 48. "Recollections of incidents given by Prof. Henry in his College Course of Lectures in Princeton N.J.," included in J. J. Halsey to William J. Rhees, 21 May 1883, in file titled "Recollection of Joseph Henry," HSA, box 27.
13. Excerpt, Minutes, Academy Trustees, [11 April 1828], *Henry Papers* 1:208.
14. This and the next four paragraphs rely on the following sources, including footnotes. "Henry's Notes on a Trip to Yale," 6 May 1830, *Henry Papers* 1:274–77. "Henry's Descriptive Catalogue of Philosophical Apparatus Purchased for the Albany Academy," [18 Dec. 1830], *Henry Papers* 1:303–12. "Henry's Introductory Lecture on Chemistry," [Jan.–March 1832], *Henry Papers* 1:380–97.
15. H. N. Smyth to Mary Henry, 23 Aug. 1897, in the file titled "Recollections of Joseph Henry," HSA, box 27.
16. "Henry's Descriptive Catalogue of Philosophical Apparatus," pp. 305–6, 309, esp. 306 n. 4. King, *Development of Electrical Technology in the 19th Century*, 240.
17. "Lecture Notes on Magnetism and Electromagnetism," *Henry Papers* 1:465–69.
18. Henry to [Beck], 21 Sept. 1827, *Henry Papers* 1:196–99.
19. *Henry Papers* 1:xxiii–xxiv, 182, 208–10, 214, 220–21. Reingold, "The New York State Roots of Joseph Henry's National Career,"139–40. "Officers of the Albany Institute," *Transactions of the Albany Institute* 1 (1830): 86, 152.
20. Williams, *Michael Faraday,* 153–56. [Faraday], "Historical Sketch of Electro-magnetism," *Annals of Philosophy* 18 (Sept. 1821): 195–200; 18 (Oct. 1821): 274–90; 19 (Feb. 1822): 107–21; see esp. p. 279. Apparently, Henry was still unaware that Faraday identified himself as author of the sketch in 1823. See Faraday, *Experimental Researches in Electricity* (London: Taylor Printers, 1844), 2:161; see also Faraday to Richard Phillips, ca. June 1823, in Frank A. J. L. James, ed., *The Correspondence of Michael Faraday* (London: Institution of Electrical Engineers, 1991), vol. 1, *1811–December 1831, Letters 1–524,* letter 200.
21. Henry to John Torrey, 4 Oct. 1827, *Henry Papers* 1:199.
22. Minutes, Albany Institute, 10 Oct. 1827, *Henry Papers* 1:201. Unless otherwise indicated, this and the next three paragraphs rely on Henry, "On Some Modifications of the Electro-Magnetic Apparatus" (Oct. 1827), rpt. in *Scientific Writings* 1:3–7. See also King, *Development of Electrical Technology in the 19th Century*, 256–60.
23. Recollections of Mary A. Henry, MS, 28 Dec. 1897, initiated by George C. Maynard and transcribed by John Boyle, copy in the Henry filing cabinet at the Department of History of Science and Technology, National Museum of American History, Smithsonian Institution. Mary Henry, "Joseph Henry,"

typescript ("Article prepared by me and published in a History of New Jersey"), pp. 2–3, in folder 16, HSA, box 56. Henry, "Statement in Relation to the History of the Electro-Magnetic Telegraph" (1857), rpt. in *Scientific Writings* 2:432. For a certified copy of Dana's lecture notes from the New York Athenaeum, Jan. and Feb. 1827, see folio titled "James Freeman Dana," Houghton Library, Harvard University, Ms Am 595. For edited excerpts from Dana's notes, particularly his key second lecture, see Samuel F. B. Morse, "The Electro-Magnetic Telegraph. A Defence Against the Injurious Deductions Drawn from the Deposition of Prof. Joseph Henry," *Shaffner's Telegraphy Companion, Devoted to the Science and Art of the Morse American Telegraph* 2 (Jan. 1855): 67–73. For Henry's personal critique of Morse's interpretation of Dana's notes, see the MS that begins, "It should be recollected that in my testimony," folder 6, HSA, box 29. See also Samuel I. Prime, *The Life of Samuel F. B. Morse* (New York: D. Appleton, 1875), 164–68; Carleton Mabee, *The American Leonardo: A Life of Samuel F. B. Morse* (New York: Octagon Books, 1969), 119, 393 n. 29. In a personal letter to the author of 6 Nov. 1995, Megan Hahn (Manuscript Department, New-York Historical Society) reported that a search of Athenaeum records neither confirmed nor denied Henry's attendance at Dana's lectures. Sturgeon, "Account of an improved Electro-magnetic Apparatus," *Annals of Philosophy* 12 (July–Dec. 1826): 357–61; copy in Bell-Henry Library.

4. Master of Electromagnets

1. King, *Development of Electrical Technology,* 256–60. Paul F. Mottelay, *Bibliographical History of Electricity & Magnetism* (London: Charles Griffin & Company, 1922), 412–15.
2. [Faraday], "Historical Sketch of Electro-magnetism," 287–88. Ampère, *Exposé des nouvelles découvertes sur l'électricité et le magnétisme, de MM. Oersted, Argo, Ampère, H. Davy, Biot, Erman, Schweiger, De la Rive, etc.* (Paris, 1822). Green, *Electro-Magnetism: Being an Arrangement of the Principal Facts Hitherto Discovered in that Science* (Philadelphia: Judah Dobson, 1827), 80–93, 207–10; personal copies, one inscribed by Henry and the other by John Torrey, in Bell-Henry Library. For Green's biography, see *Henry Papers* 1:280, 322–25 including n. 1. For evidence of Green borrowing from Faraday's "Historical Sketch," cf. Green's p. 33 and Faraday's p. 277. Faraday, *Chemical Manipulation* (London: W. Phillips, 1827); Henry's copy in Bell-Henry Library.
3. Excerpt, Minutes, Albany Institute, 1 and 15 April 1829, *Henry Papers* 1:216, 217. Henry, "On the Application of the Principle of the Galvanic Multiplier to Electro-Magnetic Apparatus, and Also to the Development of Great Magnetic Power in Soft Iron, with a Small Galvanic Element" (Jan. 1831), rpt. in *Scientific Writings* 1:37–39.

4. Harriet Henry to Stephen Alexander, 7 May 1830, and Joseph Henry to Harriet Henry, [18] May 1830, *Henry Papers* 1:277–81. (This latter letter perhaps should be dated after May 27; cf. Joseph to Harriet Henry, 27 May 1830, in file labeled "1825-May 1830," HSA, box 57.
5. Henry to Silliman, 10 Dec. 1830, Silliman to Henry, 17 Dec. 1830, and Henry to Silliman, [2]8 Dec. 1830, *Henry Papers* 1:301–3, 316–17.
6. Moll, "Electro-Magnetic Experiments," rpt. in *American Journal of Science and Arts* 19 (Jan. 1831): 329–37. Unless otherwise indicated, the remainder of this second section of this chapter relies on the following source. Henry, "On the Application of the Principle of the Galvanic Multiplier to Electro-Magnetic Apparatus, and Also to the Development of Great Magnetic Power in Soft Iron, with a Small Galvanic Element" (Jan. 1831), rpt. in *Scientific Writings* 1:37–49.
7. Silliman to Henry, 6 Jan. 1831, Isaac W. Jackson to Henry, 16 Feb. [1831], *Henry Papers* 1:318–19, 327–28. Silliman, *Elements of Chemistry, in the Order of the Lectures Given in Yale College* (New Haven: Hezekiah Howe, 1831), 2:680–83.
8. Folio titled "James Freeman Dana," Houghton Library. Charles Weiner, "Joseph Henry's Lectures on Natural Philosophy: Teaching and Research in Physics, 1832–1847" (Ph.D. diss., Case Institute of Technology, 1965), 194–95. A. Alexander Hodge, notebook on natural philosophy, 1840–41, Princeton University Library, copy in HPP, microfilm batch M-139; this and later quotations from cognate notebooks are published with permission of the Department of Rare Books and Special Collections, Princeton University Library. Joseph J. Halsey, notebook on natural philosophy, vol. 2, 1841–42, HSA, box 31, copy in HPP, microfilm batch M-170. William J. Gibson, notebook on natural philosophy, 1842–44, Princeton University Archives, copy in HPP, microfilm batch M-135. Cf. Henry's later recollections regarding his improvements in electromagnet design, in Henry, "Statement in Relation to the History of the Electro-Magnetic Telegraph," 429–30. Alfred M. Mayer, "Henry as a Discoverer," in *Memorial*, 481–82; cf. Cameron, "Reminiscences," in *Memorial*, 173, and Thomas Coulson, *Joseph Henry: His Life and Work* (Princeton: Princeton University Press, 1950), 49–50.
9. Excerpt, Minutes, Albany Institute, 20 Dec. 1830, *Henry Papers* 1:312–13.
10. David Gooding, "'He Who Proves, Discovers': John Herschel, William Pepys and the Faraday Effect," *Notes and Records of the Royal Society of London* 39 (March 1985): 229–44. For Henry's adaptation of Hare's theory, see L. Pearce Williams, "The Simultaneous Discovery of Electro-Magnetic Induction by Michael Faraday and Joseph Henry," *Bullétin de la Société des Amis D'André-Marie Ampère* 22 (Jan. 1965):17–18.
11. For one of the earliest commentaries on Ohm's relevance to Henry, see Taylor, "The Scientific Work of Joseph Henry," 227.

12. Green, *Electro-Magnetism*, 160–61, 180–81. Cf. Faraday, *Chemical Manipulation*, 460–61. [Faraday], "Historical Sketch of Electro-magnetism," 288–89. King, *Development of Electrical Technology*, 238–39. For Henry's early notes on an article by Antoine Becquerel that contributed to Henry's views on the capabilities of the two types of batteries relative to different lengths of wire, see the notebook that begins "For the composition of an unchangable cement . . . ," pp. 21–23, HSA, box 17.
13. Green, *Electro-Magnetism*, 37–39; Henry also owned Barlow, *An Essay on Magnetic Attractions, and on the Laws of Terrestrial and Electro Magnetism* (London: J. Mawman, 1824); copy in Bell-Henry Library. Henry, "Statement in Relation to the History of the Electro-Magnetic Telegraph," 434–37. Taylor, "The Scientific Work of Joseph Henry," 380–81 (note C). William B. Taylor, "Henry and the Telegraph," in *Annual Report of the Board of Regents of the Smithsonian Institution, 1878* (Washington, D.C.: Government Printing Office, 1879), 289–90, 334–36 (note E). Newland, "To the Editor of the Saratogian," *Saratogian*, ca. May 1878, copy in "Henry Memorial. Original Papers. 1878," HSA, box 47. For Carpenter's recollections, see Reynolds, "Prof. Joseph Henry and His Life Work." See also "Henry's Telegraph: Result of an Effort to Identify the Room," *Times-Union*, ca. 1899, copy included in a file labeled "Henry Albany Centennial Celebration," HSA, box 46.
14. Hall to Henry, 19 Jan. 1856, rpt. in Henry, "Statement in Relation to the History of the Electro-Magnetic Telegraph," 437, app. B.
15. Henry to Silliman, [2]8 Dec. 1830, 28 March 1831, *Henry Papers* 1:316–17, 331–33. Henry (with Ten Eyck), "An Account of a Large Electro-Magnet, Made for the Laboratory of Yale College" (April 1831), rpt. in *Scientific Writings* 1:50–53. Silliman to Henry, 5 March 1832, *Henry Papers* 1:407–9.
16. For the correspondence involving other Americans requesting information, see *Henry Papers* 1:347, 380, 400–402, 416–17; see also 320–21, 345–46. For the correspondence between Cleaveland and Henry, see *Henry Papers* 1:373, 375–76, 377–79, 402–4, 413, 420–26, 432–33. Henry to Green, 3 Feb. 1831, *Henry Papers* 1:322–23. Henry to John R. Henry, 9 July 1831, *Henry Papers* 1:348–49.
17. For the correspondence involving the difficulty and expense of building magnets, see *Henry Papers* 1:319–20, 331–33, 333–35, 340–41, 400–402, 402–3, 420–26. For the silk insulation, see Joseph S. Ames, "Certain Aspects of Henry's Experiments on Electromagnetic Induction," *Science* 75 (22 Jan. 1932): 87–92; see particularly Henry to Edward Hitchcock, 27 Jan. 1832, and Henry to Cleaveland, 8 May 1832, *Henry Papers* 1:401, 421.
18. Henry to Charles Davies, 15 Feb. 1831. This and the next two paragraphs rely on the following source: Henry, "On a Reciprocating Motion Produced by Magnetic Attractions and Repulsion" (July 1831), rpt. in *Scientific Writings* 1:54–57. See also King, *Development of Electrical Technology*, 260–61. See also

Frederick A. P. Barnard, "Henry (Joseph)," *Johnson's New Universal Cyclopædia* (New York: A. J. Johnson & Son, 1876–78), 2:878–79. Benjamin Silliman Jr. reported that Henry had personally revised Barnard's article; see Silliman, "Joseph Henry, LL.D.," *American Journal of Science and Arts*, 3d ser., 15 (June 1878): 464.

19. Henry to Thomas R. Pynchon, 16 June 1868, Letterpress 1867–77, p. 70, HSA, box 38. Decades later, former pupil H. N. Smyth reminisced about Henry dazzling the class with a miniature railroad that he built about this time using a locomotive powered presumably by a variant of his electromagnetic engine. Henry supposedly propelled a car around a small circle of track "to the amazement & delight of the pupils!" There is no corroborating evidence that Henry did this. Perhaps the student was referring to a model made by Thomas Davenport, who had been inspired by Henry's work. Smyth to Mary Henry, 23 Aug. 1897 and 25 Oct. 1897. For Davenport, see King, *Development of Electrical Technology,* 263–64.
20. Green to Henry, 13 Feb. 1834, Henry to Green, 17 Feb. 1834, Van Rensselaer to Henry, 29 June 1835, Henry to Silliman, 10 Sept. 1835, in *Henry Papers*, 2:158–66, 416–17, 445–51. See also Silliman to Henry, 21 July 1831, *Henry Papers* 1:354–55. Arthur P. Molella, "The Electric Motor, the Telegraph, and Joseph Henry's Theory of Technological Progress," *Proceedings of the Institute of Electrical and Electronics Engineers* 64 (Sept. 1976): 1273–78. Arthur P. Molella and Nathan Reingold, "Theorists and Ingenious Mechanics: Joseph Henry Defines Science," *Science Studies* 3 (Oct. 1973): 323–51. King, *Development of Electrical Technology,* 260–71. See also Coulson, *Joseph Henry*, 69–72.
21. Henry to Samuel B. Dod, 4 Dec. 1876, rpt. in Dod, "Discourse Memorial" (1878), in *Memorial,* 149–50. Henry, "Communication from Prof. Henry . . . Relative to a Publication of Prof. Morse," in *Annual Report of the Board of Regents of the Smithsonian Institution, 1857* (Washington, D.C.: William A. Harris, Printer, 1858), 86.
22. Molella, "The Electric Motor, the Telegraph, and Joseph Henry's Theory of Technological Progress," 1274–76.
23. Allen Penfeld and Timothy Taft jointly to Henry, 30 May and 27 June 1831, Henry to Mr. Rogers, [4 Nov.] 1831, *Henry Papers* 1:340–41, 346, 364–72. *Henry Papers* 2:416–17 n. 1. See also editor's n. 17 in Mary A. Henry, "America's Part in the Discovery of Magneto-Electricity—A Study of the Work of Faraday and Henry—V," *Electrical Engineer* 13 (20 Jan. 1892): 54. See also Coulson, *Joseph Henry*, 66–67.
24. Henry to Silliman, 28 March 1832, *Henry Papers* 1:409–12.
25. Fragment of newspaper (n.d., n.p.), possibly from Albany or Princeton area, ca. mid to late 1830s, folder labeled "Memoir preface III," HSA, box 55. Lewis W. Washington to Henry, 22 Nov. 1835, *Henry Papers* 2:475–76.

5. The Hunt for Electromagnetic Induction

1. See George Carpenter's recollections in Reynolds, "Prof. Joseph Henry and His Life Work." Lassiter, *Philip Hooker and the Old Albany Academy*, 9–10. Henry to John R. Henry, 9 July 1831; for earliest European reports, but doubts about French republication, see *Henry Papers* 1:348 n. 4. Eaton to Henry, 16 Aug. 1832, *Henry Papers* 1:442. See also Henry to [Thomas Cooper], 26 Nov. 1833, *Henry Papers* 2:130–32.
2. Sturgeon, "On Electromagnets," *Philosophical Magazine*, n.s., 11 (March 1832): 199; quoted in Taylor, "The Scientific Work of Joseph Henry," 230. Peter M. Roget, "Electro-Magnetism," *Library of Useful Knowledge*, no. 117 (n.p., n.d.): 55; an editorial "Postscript" in this installment, p. 100, is dated 12 Dec. 1831. For the correspondence involving other Americans, see above, chap. 4, n. 16. Silliman to Henry, 12 March 1831, *Henry Papers* 1:330. Shepard, note appended to "An Account of a Large Electro-Magnet, Made for the Laboratory of Yale College," 50–51 n. Webster and Hare, letters to the editor in section titled "Galvano-magnetism," *American Journal of Science and Arts* 20 (July 1831): 143–47. Joslin, "Electro-Magnetic Apparatus," *American Journal of Science and Arts* 21 (January 1832): 86-90. For the instrument maker, James Chilton, see *Henry Papers* 2:296, 311, 316–17, 442.
3. Moll to Faraday, 7 June and 13 Nov. 1831, *Correspondence of Michael Faraday*, 1: letters 497, 519. Both letters are also reprinted in L. Pearce Williams, ed., *Selected Correspondence of Michael Faraday*, vol. 1, *1812–1848* (Cambridge: Cambridge University Press, 1971), 197–98, 204–7. The 7 June letter is also reprinted in *Henry Papers* 1:341–45.
4. Sydney Ross, "The Search for Electromagnetic Induction, 1820–1831," *Notes and Records of the Royal Society of London* 20 (1965): 183–219. Samuel Devons, "The Search for Electromagnetic Induction," in *Physics History from AAPT Journals* (College Park, Md.: American Association of Physics Teachers, 1988), 37–43.
5. Faraday, 29 Aug. 1831, *Faraday's Diary*, vol. 1, *Sept., 1820–June 11, 1832*, Thomas Martin, ed. (London: G. Bell and Sons, 1932), 367–68. Faraday, "First Series" (24 Nov. 1831), reprinted in *Experimental Researches in Electricity* (London: Richard and John Taylor, 1839), 1:7–16; these pages refer to sect. 2 of the paper, "On the Evolution of Electricity from Magnetism."
6. One of the first to call attention to Faraday's seeming debt to Henry was Taylor, "The Scientific Work of Joseph Henry," 222 n and 390, note D. For a detailed but biased examination of the debt, see also Mary A. Henry, "America's Part in the Discovery of Magneto-Electricity—A Study of the Work of Faraday and Henry—IV," *Electrical Engineer* 13 (10 Feb. 1892): 134–36. For the contrasting view that "there is no evidence, explicit or implicit, that Faraday was 'repeating' Henry's experiments or even used a magnet similar in design to Henry's," see *Henry Papers* 2:131 n. 6.

7. Faraday, 6–10 Sept. 1821, 28 Nov. 1825, *Faraday's Diary* 1:54–57, 279. Ryan D. Tweney and David Gooding, eds., *Michael Faraday's "Chemical Notes, Hints, Suggestions and Objects of Pursuit" of 1822* (London, Peter Peregrinus Ltd., 1991), 72–73; the editors raise the possibility that some of the entries in these 1822 "Notes" extend into the mid-1820s. See also Ryan D. Tweney, "Faraday's Discovery of Induction: A Cognitive Approach," in *Faraday Rediscovered: Essays on the Life and Work of Michael Faraday, 1791–1867*, eds. David Gooding and Frank A. J. L. James (New York: Stockton Press, 1985), 192–99, 204–6.
8. For the British synopses, see "14. Powerful Electro-Magnets" and "3. Powerful Electro-Magnet," *Journal of the Royal Institution* 1 (May 1831): 609–10, and 2 (Aug. 1831):182. See also *Henry Papers* 1:348 n. 4. For Faraday's role in the *Journal*, see James, ed., "Introduction," in *Correspondence of Michael Faraday*, 1:xxxii; see also letters 486, 457, 468, 513. Moll to Faraday, 11 March 1831, *Correspondence of Michael Faraday*, 1: letter 487. (In hindsight, Moll's failure to duplicate Henry's results seems to reflect his inattention to details of Henry's design; this perhaps reflected his reliance on Faraday's second-hand account.) Henry to De La Rive, 24 Nov. 1841, in Nathan Reingold et al., eds., *The Papers of Joseph Henry,* vol. 5, *January 1841–December 1843: The Princeton Years* (Washington: Smithsonian Institution Press, 1985), 119–23; from here on cited as *Henry Papers* 5. Henry to Thomas R. Pynchon, 16 June 1868, p. 72. See also Henry to De La Rive, 8 May 1846, *Henry Papers* 6:420–23.
9. Faraday to Peter M. Roget, 4 July 1831, *Correspondence of Michael Faraday*, 1: letter 501; for Faraday's other concerns, see Frank James, "The Military Context of Chemistry: The Case of Michael Faraday," *Bulletin for the History of Chemistry* 11 (1991): 36–40. Shepard, note appended to "An Account of a Large Electro-Magnet, Made for the Laboratory of Yale College," 50–51 n; one of the first to call attention to Shepard's experiment was Taylor, "The Scientific Work of Joseph Henry," 233–34. "3. Powerful Electro-Magnet," 182. For background on Shepard, see Silliman to Henry, 5 April 1831; see also Shepard, "Notice of the Mine of Spathic Iron . . . ," *American Journal of Science and Arts* 19 (Jan. 1831), 311–25. By 1841 Shepard could write, "By means of several ingenious machines for exalting the effects of magneto-electricity, all the well known effects of voltaic currents may be produced, such as shocks, sparks, decompositions, &c."; see Shepard, *Syllabus to Lectures on Chemistry* (New Haven: B. & W. Noyes, 1841), 104–6. In 1858 Shepard published an article in an Albany newspaper that disputed Henry's contribution to the Morse telegraph; see Henry to Alexander Dallas Bache, 11 Sept. 1858, Bache Papers (HPP, no. 6540).
10. At the point in his text where he reports that all the electrical effects could be strengthened using iron-core electromagnets as arranged by Moll, Henry, and Ten Eyck, Faraday mentions neither the multiple-coil design nor Henry's contribution to it. At that point, however, he does refer by number to an ear-

lier paragraph in his paper describing an iron-core electromagnet wrapped with multiple coils. Faraday, "On the Evolution of Electricity from Magnetism," 9, 15; see also p. 32.

11. Tweney and Gooding, eds., *Michael Faraday's "Chemical Notes,"* 53, 73; next to an explicit sketch of a bar magnet in a helix connected to wire looped around a magnetized needle, Faraday jotted: "Magnets in copper coils connected with other coils and galvanometer." See also, Faraday, 28 Dec. 1824, *Faraday's Diary* 1:178, and Faraday, "Electro-magnetic Current (under the Influence of a Magnet)" (July 1825), reprinted in *Experimental Researches* 2:162–63. Faraday's Royal Institution talk is quoted in Henry, "On the Production of Currents and Sparks of Electricity from Magnetism" (July 1832), reprinted in *Scientific Writings* 1:74. For the importance of the 1822 experiment, see Tweney, "Faraday's Discovery of Induction," 197–99, 205. See also Ross, "The Search for Electromagnetic Induction," 184–85, 198–208, 215 n. 2, and Silvanus P. Thompson, *Michael Faraday: His Life and Work* (London: Cassell and Co., 1901), 89–91.
12. Commentators have long recognized that the electromagnet's strength helped Faraday; see, e.g., Williams, *Michael Faraday,* 169.
13. Jean Hachette to Michael Faraday, 9 July 1832, in Frank A. J. L. James, ed., *The Correspondence of Michael Faraday,* vol. 2, *1832–December 1840, Letters 525–1333* (London: Institution of Electrical Engineers, 1993), letter 598. For Pouillet's possible appropriation of Henry's coil design, see Taylor, "The Scientific Work of Joseph Henry," 226n, and "Henry and the Telegraph," 333–34, note D. See also Pouillet, *Éléments de Physique*, 4th ed. (Paris: Béchet Jeune, 1844), 648–49; copy in Bell-Henry Library. For Henry's notes on Pouillet's 1832 announcement, see HSA, box 18. Ritchie, "Experimental Researches in Electro-Magnetism and Magneto-Electricity," *Philosophical Transactions of the Royal Society of London* (1833): 317, quoted in *Henry Papers* 2:162 n. 4, 195 n. 48. Robert Hare to De La Rive, 11 Aug. 1841, Henry to De La Rive, 24 Nov. 1841, *Henry Papers* 5:79–82, 119–23. See also Henry's annotation in De La Rive, *Treatise on Electricity* (London: Longman, Brown, Green, and Longmans, 1853), 1:293; copy in Bell-Henry Library. Henry to Thomas R. Pynchon, 16 June 1868, p. 72. Henry to Page, 31 Oct. 1838, Page to Henry, 8 Nov. 1838, in Nathan Reingold et al., eds., *The Papers of Joseph Henry,* vol. 4, *January 1838–December 1840: The Princeton Years* (Washington: Smithsonian Institution Press, 1981), 146–47, 157–58; from here on cited as *Henry Papers* 4. See also Nathan Reingold, "Joseph Henry on the Scientific Life: An AAAS Presidential Address of 1850," in *Science American Style*, ed. Reingold (New Brunswick, N.J.: Rutgers University Press, 1991), 164.
14. For Sturgeon's 1839 and Brewster's 1837 acknowledgments of Henry's engine, see *Henry Papers* 2:446 n. 2. For Joule's 1839 acknowledgment of Henry's engine, see Taylor, "Henry and the Telegraph," 290n. Brewster's

lengthy article for the *Encyclopædia Britannica* was reprinted as *A Treatise on Magnetism* (Edinburgh: Adam and Charles Black, 1837); see pp. 130–31 of copy in Bell-Henry Library. For the instrument maker, Francis Watkins, see Robert Hare to Henry, 21 April 1838, *Henry Papers* 4:36, esp. n. 2.

15. For Faraday's priority dispute with the Italians, see Faraday to William Jerdan, 27 March 1832, and Faraday to Forbes, 16 April 1832, *Correspondence of Michael Faraday* 2: letters 560, 570. See also Faraday's replies reprinted in *Experimental Researches* 2:164–200. See also Williams, *Michael Faraday*, 200–202. For his dispute with Ampère, see Forbes to Faraday, 2 May 1832, Ampère to Faraday, 13 April 1833, Faraday to Ampère, 4 May 1833, and Ampère to Faraday, 17 Feb. 1835, *Correspondence of Michael Faraday* 2: letters 576, 654, 660, 765. See also Ross, "The Search for Electromagnetic Induction," 208–13. See also José Romo and Manuel B. Doncel, "Faraday's Initial Mistake Concerning the Direction of Induced Currents, and the Manuscript of Series I of his Researches," *Archive for History of Exact Science* 47 (Aug. 1994): 298–302.
16. Faraday, 24 Nov. 1831, *Faraday's Diary* 1:389. For the likelihood that the submission of Faraday's manuscript to the Royal Society occurred on 21 Nov. 1831 and the formal reading actually took place in stages on 24 Nov., 8 Dec., and 15 Dec. 1831, see Romo and Doncel, "Faraday's Initial Mistake Concerning the Direction of Induced Currents," 292, 311–13. Henry to Cleaveland, 16 Nov. 1831, *Henry Papers* 1:375–76. Henry, "On the Production of Currents and Sparks of Electricity from Magnetism," 73–79; unless otherwise indicated, this paragraph and the remainder of the chapter rely on this source. Weiner, "Joseph Henry's Lectures on Natural Philosophy," 200–201. William Gledhill, notebook on natural philosophy, 1842–43, Princeton University Library, copy in HPP, microfilm batch M-136. Henry probably obtained the iron core for the 101-pound magnet from the same furnace and foundry operators that he went on to assist with their extraction machine; see Penfield and Taft to Henry, 30 May 1831.
17. Henry to Silliman, 28 March 1831; see also Silliman to Henry, 26 April 1832, *Henry Papers* 1:418–19.
18. Henry to John Maclean, 28 June 1832, *Henry Papers* 1:435–37.
19. Roget, "Electro-Magnetism," 99. Roget attached a "Note" about Faraday's discovery to the "Electro-Magnetism" component of his four-part survey of electricity and magnetism and followed it with an editorial "Postscript" dated 12 Dec. 1831; a bound copy of this and other installments of the *Library of Useful Knowledge*, with a few annotations by Henry, survives in the Bell-Henry Library. Roget's note is also quoted in Coulson, *Joseph Henry*, 76–77. For Volta-electric induction, see Romo and Doncel, "Faraday's Initial Mistake Concerning the Direction of Induced Currents," 291–85. For Henry's undated but later notes on the full, published version of Faraday's results, see the notebook that begins "See for Mr Espy how. . . ," HSA, box 17.

20. Henry to John Maclean, 28 June 1832.
21. Silliman to Henry, 30 June 1832, *Henry Papers* 1:437–39. Forbes to Faraday, 13 April 1832, *Correspondence of Michael Faraday*, 2: letter 569. Decades later, Princeton student Cuyler recalled Henry lamenting that he missed making his finding before Forbes because he lacked necessary chemicals, presumably for the battery in the experiment. "I should have got the start of Forbes," Cuyler remembered Henry saying, "if I had not been obliged to wait for some chemicals from the city of New York." In an even later reminiscence, however, Cuyler was uncertain if the anecdote about the chemicals applied to Forbes or Faraday. Cuyler, "Professor Henry—The Christian Philosopher"; Cuyler to Miss Henry, 20 March 1888.
22. Saxton, "Notice of Electro-Magnetic Experiments," *American Journal of Science and Arts* 22 (July 1832): 409–15; Saxton's letters, addressed to Isaiah Lukens, were forwarded to Silliman by Alexander Dallas Bache. See also Henry, "Memoir of Joseph Saxton" (1874), *National Academy of Sciences: Biographical Memoirs* 1(1877): 289–316.

6. Staking a Claim

1. Henry's oldest daughter, Mary, fixated through her later years on establishing her deceased father's priority in the discovery of not only electromagnetic induction but also the telegraph. See, e.g., Mary A. Henry, "America's Part in the Discovery of Magneto-Electricity—A Study of the Work of Faraday and Henry," a six-part series, *Electrical Engineer* 13 (13 Jan.–9 March 1892).

   Mary Henry buttressed one of her most detailed arguments with a recollection from George Carpenter, Henry's pupil in 1826 and later an academy tutor and assistant to Henry. Carpenter recalled that Henry began his investigation of electromagnetic induction by evoking sparks. He remembered Henry saying that it was after going to bed one night that he figured out how to draw a spark from a magnet. Supposedly, the next morning, he rushed to his laboratory at the academy and executed the experiment successfully. Carpenter further recalled Henry mentioning that, "some time after," Faraday repeated the experiment and publicly announced it. "Mr. Henry regretted," Carpenter added, "that through his own negligence he had been deprived of the honor awarded to Faraday." Reminiscences of George W. Carpenter, 15 Nov. 1893, included in Carpenter to Mary Henry, 15 Nov. 1893, HSA, box 23.

   Mary Henry felt that Carpenter's recollection helped explain a small, intriguing sheet of her father's notes and drawings that she discovered among his earliest records; comparing watermarks on similar documents, she estimated the sheet to date to around 1827. The notes and drawings present ways of drawing sparks or shocks by breaking the contact of a coil-wrapped armature on a permanent horseshoe magnet while using different arrangements of external wires and a helix. One caption to a sketch reads: "A helix on the leg of the magnet produces very strong shock." Concerned about priority, Mary Henry argued that her father did draw sparks from permanent magnets early in his Albany researches and that his results later alerted him to look for similar effects in activating or deactivating electromagnets. For Mary Henry's comments and her fathers notes and drawings, see the file titled "Electricity and magnetism notes III," HSA, box 23.

Carpenter's recollection and Mary Henry's interpretation of the notes are almost certainly inaccurate. In fact, the sheet of notes probably describes Henry's experiments in 1835 at Peale's museum in Philadelphia—experiments with a large horseshoe magnet to test and extend a colleague's findings announced in 1833 in *Silliman's Journal.* For example, the first of Henry's sketches on the sheet duplicates the colleague's basic finding, and the last sketch duplicates one of Henry's supplemental finding that he described in *Silliman's Journal* in 1835. Carpenter, who was recalling events from sixty years earlier, probably blurred memories. In particular, he probably attributed to Henry's Albany years a Princeton story that he later heard from Henry. In his laboratory notebook for January 1835—four months before the experiments at Peale's museum—Henry recorded that an important idea occurred to him before rising from bed one morning. The idea was that he could attach "copper handles" to the extremities of circuits and experience shocks when grasping the handles or create sparks between them (discussed in chap. 11, below). In the notes that Mary Henry discovered, Henry similarly relied on spherical handles to produce shocks and sparks. Joseph to Harriet Henry, 4 May 1835, and "Record of Experiments," entries for 7 Jan. 1835 and between 9 May and 12 May 1835, *Henry Papers* 2:328, 386–89, 391–94. Henry, "Appendix to the Above—Action of a Spiral Conductor" (July 1835), reprinted in *Scientific Writings* 1:91.

For another defense of Henry's priority over Faraday, see Ames, "Certain Aspects of Henry's Experiments on Electromagnetic Induction," 87–92. Although Henry biographer Thomas Coulson challenged Mary Henry's interpretation of Carpenter's recollection, Coulson carried a moderated version of her campaign into the mid-twentieth century; see Coulson, *Joseph Henry*, 85–89. For a rebuttal to Coulson, see Ross, "The Search for Electromagnetic Induction," 214–15.

2. Faraday, "First Series" (24 Nov. 1831), reprinted in *Experimental Researches* 1:2, 16–24; these pages refer to the introduction and to sect. 3, "New Electrical State or Condition of Matter." Henry Van Vleck Rankin, notebook on natural philosophy, 1842–43, Princeton University Library, copy in HPP, microfilm batch M-135. Williams compares the theoretical orientations of Faraday and Henry, in "Simultaneous Discovery," pp. 12–20. For analysis of

Faraday's Ampèrian leanings and deviations, see Romo and Doncel, "Faraday's Initial Mistake Concerning the Direction of Induced Currents," 291–35. See also L. Pearce Williams, "Why Ampère Did Not Discover Electromagnetic Induction," *American Journal of Physics* 54 (April 1986): 306–11; "Faraday and Ampère: A Critical Dialogue," in *Faraday Rediscovered*, 83–104. See also Friedrich Steinle, "Work, Finish, Publish: The Formation of the Second Series of Faraday's Experimental Researches in Electricity," *Physis* (forthcoming); see esp. sect. 2. For a revisionist discussion of the role of experimentation in Faraday's early work, see Friedrich Steinle, "Looking for a 'Simple Case': Faraday and Electromagnetic Rotation," *History of Science* 33 (1995): 179–202.

3. Williams argues that Faraday and Henry, though holding "diametrically opposed theories," were led to their discoveries of electromagnetic induction by theoretical concerns; see Williams, "Simultaneous Discovery," pp. 12–20. See also Williams, *Michael Faraday*, 169–83, and Steinle, "Work, Finish, Publish," esp. sect. 2. For a summary of factors that possibly preconditioned Faraday to think about transient effects, see Tweney, "Faraday's Discovery of Induction," 204–6. See also Geoffrey Cantor, *Michael Faraday: Sandemanian and Scientist* (New York: St. Martin's Press, 1991), 226–44, and Ross, "The Search for Electromagnetic Induction," 196–209. For a general discussion of the role of experiment, see David Gooding, "'In Nature's School': Faraday as an Experimentalist," in *Faraday Rediscovered*, 105–35. For Henry's stance toward contemporary theoretical notions, see "Introduction," *Henry Papers* 2:xxviii–xxix; see also pp. 258–59 n. 2. Albert E. Moyer, "Benjamin Franklin: 'Let the Experiment be made,'" in *Physics History from AAPT Journals*, ed. Melba N. Phillips (College Park, Md.: American Association of Physics Teachers, 1985), 1–10. John L. Heilbron, "Franklin's Physics," *Physics Today* 29 (July 1976), 32–37.

4. Albert Van Helden and Thomas L. Hankins, eds., *Instruments*, vol. 9 of *Osiris* (Chicago: University of Chicago Press, 1994), 1–6; see also Jed Z. Buchwald, ed., *Scientific Practice: Theories and Stories of Doing Physics* (Chicago: University of Chicago Press, 1995). In a personal letter to the author of 20 Sept. 1994, Frank A. J. L. James called attention to a distinction between scientific "instruments" and "apparatus." In particular, there is a shading of meaning between the terms: "instrument" commonly connotes that the hardware and requisite techniques for using it are somewhat settled, shared, and replicable, while "apparatus" refers to a more ad hoc and singular collective of hardware and requisite techniques. Contrary to the author, however, James suggests that while Henry's electromagnet may have achieved the status of a "black-boxed" instrument by the 1830s, Faraday's ring electromagnet remained a piece of apparatus constructed solely for the purpose of his particular investigation.

5. James, ed., "Introduction," in *Correspondence of Michael Faraday*, 1:xxvii–xxxviii. Williams, *Michael Faraday*, 28–29, 329. In a personal letter to the author of 11 May 1995, Frank A. J. L. James identified an early source of the Faraday maxim; see John H. Gladstone, *Michael Faraday*, 1st ed. (London: 1872), 122. See also Bern Dibner, "The Beginning of Electricity," in *Technology in Western Civilization*, eds. Melvin Kranzberg and Carroll W. Pursell (New York: Oxford University Press, 1967), 1:446–49.
6. "Excerpt, Minutes, Albany Academy," 3 April 1828, 25 Nov. 1828, 18 March 1829, 8 April 1830, *Henry Papers* 1:207–8, 212, 213–14, 269–70. See also item no. 6, "Henry's Descriptive Catalogue of Philosophical Apparatus Purchased for the Albany Academy."
7. Ross, "The Search for Electromagnetic Induction," 186–89, 208–13. Faraday to Richard Phillips, 23 Sept. 1831, *Correspondence of Michael Faraday* 1: letter 515. See also Thompson, *Michael Faraday*, 109–10; Bence Jones, *Life and Letters of Faraday* (London: Longmans, Green, and Co., 1870), 1:299, 2:3; and Mary A. Henry, "America's Part in the Discovery of Magneto-Electricity," 134.
8. Frederick S. Giger, notebook on natural philosophy, 1840–41, Princeton University Library, copy in HPP, microfilm batch M-157. Halsey, notebook on natural philosophy, vol. 2, 1841–42. Henry C. Cameron, notebook on natural philosophy, 1846–47, vol. 2, Princeton University Library, copy in HPP, microfilm batch M-138.
9. Faraday, "Ninth Series: On the Influence by Induction of an Electric Current on Itself" (18 Dec. 1834), reprinted in *Experimental Researches* 1:322–43. Henry to John Vaughan, 17 Nov. 1834, and Henry to Bache, 17 Dec. 1834, *Henry Papers* 2:279–80, 297–304, including editor's notes.
10. This paragraph and the next rely on the following sources. Henry, "Facts in Reference to the Spark, Etc., from a Long Conductor Uniting the Poles of a Galvanic Battery" (March 1835), "Appendix to the Above—Action of a Spiral Conductor" (July 1835), "Contributions to Electricity and Magnetism. No. II. On the Influence of a Spiral Conductor in Increasing the Intensity of Electricity from a Galvanic Arrangement of a Single Pair, etc." (6 Feb. 1835), reprinted in *Scientific Writings* 1:87–100. Henry to Bache, 9 Feb. 1835 and 18 May 1835, *Henry Papers* 2:346–49, 403–8, including editor's notes. For Henry's additional reflections on priority see: Henry to Silliman, 15 May 1835, *Henry Papers* 2:399–400; Henry to Alexander, 15 May 1835, *Henry Papers* 2:400–403; Henry to Forbes, 7 June 1836, *Henry Papers* 3:72–76. For Henry's return to self-induction, see "Record of Experiments," 15 Aug. 1834, *Henry Papers* 2:220–22, including editor's notes.
11. Henry, "Memoir of Joseph Saxton," 296. Cameron, notebook on natural philosophy, vol. 2, 1846–47. Gledhill, notebook on natural philosophy, 1842–43. Weiner, "Joseph Henry's Lectures on Natural Philosophy," 168, 199–202. "Diary of Jonathan Homer Lane," 10 June 1848, *Henry Papers* 7:339–41.

Mary Henry reported not only frequently hearing her father express "disappointment" in Faraday preceding him in the announcement of mutual induction but also hearing him profess his priority in the discovery; see Mary A. Henry, "America's Part in the Discovery of Magneto-Electricity—A Study of the Work of Faraday and Henry—V," *Electrical Engineer* 13 (17 Feb. 1892): 153.

12. John Vaughan to Henry, 5 Aug. 1835, *Henry Papers* 2:429–30, including editor's notes. James, ed., "Introduction," in *Correspondence of Michael Faraday*, 2:xxv–xxxviii. See also Jones, *Life and Letters of Faraday*, 2:44–46; Thompson, *Michael Faraday*, 149–52.
13. Henry, "Statement of Professor Henry in Relation to the History of the Electro-Magnetic Telegraph," in *Annual Report of the Board of Regents of the Smithsonian Institution, 1857*, 99–106. Henry, "Contributions to Electricity and Magnetism. No. II," 92–93. Mary Henry puts full credence in her father's later claim of interrupting his telegraphic work because of his self-inductive studies; she uses the claim to help buttress her priority argument, dating her father's discovery of self-induction to 1829. See Mary A. Henry, "America's Part in the Discovery of Magneto-Electricity—A Study of the Work of Faraday and Henry—VI," *Electrical Engineer* 13 (9 March 1892): 249. See also Coulson, *Joseph Henry*, 89–91.
14. "Magnetism," in *Encyclopædia Americana: Supplementary Volume* (Philadelphia: Lea and Blanchard, 1847), 14:412–26. On a front page of his personal copy in the Bell-Henry Library, Henry has inscribed: "This volume contains a number of articles written by myself. J.H." For his handwritten draft of the article, with deletions, see the file labeled "Electricity and magnetism notes I," HSA, box 23. Henry Vethake to Henry, 7 June 1845, Henry to James Alexander, 14 March 1846, *Henry Papers* 6:296–99, 393–95.
15. Gibson, notebook on natural philosophy, 1842–44. See also Weiner, "Joseph Henry's Lectures on Natural Philosophy," 171–77.
16. "Henry, Joseph," in *The Biblical Repertory and Princeton Review: Index Volume from 1825 to 1868*, 194–200. See above, chap. 1, n. 2, for evidence of Henry's authorship and for periodicals that reprinted the biographical sketch. For the *Review*'s influence, see Henry to Bache, 15 March 1845, *Henry Papers* 6:251, esp. n. 5.
17. Gledhill, notebook on natural philosophy, 1842–43. Gibson, notebook on natural philosophy, 1842–44. A. Alexander Hodge, notebook on natural philosophy, 1840–41. W. C. Cattell, notebook on natural philosophy, 1847–48, HSA, box 31, copy in HPP, microfilm batch M-135. For commentary on Henry's lectures on electricity and magnetism, see Weiner, "Joseph Henry's Lectures on Natural Philosophy," 170–206.
18. Cattell, notebook on natural philosophy, 1847–48. John Miller, notebook on natural philosophy, 1836, Princeton University Library, copy in HPP, microfilm batch M-134. Halsey, notebook on natural philosophy, vol. 2,

1841–42. Giger, notebook on natural philosophy, 1840–41. Cameron, notebook on natural philosophy, 1846–47.

19. Henry to Oersted, 27 April 1841, Henry to Berzelius, 29 April 1841, Henry to A. De La Rive, 24 Nov. 1841, *Henry Papers* 5:27–30, 119–23. Henry to De La Rive, 8 May 1846, *Henry Papers* 6:420–23. See also Coulson, *Joseph Henry*, 146–48, 293.
20. Henry to anonymous, 29 April 1844, *Henry Papers* 6:93. [Crosby S. Noyes, ed.], 27 Jan. 1872, *Washington Evening Star,* copy with Noyes's dedication in HSA, box 23; copy also inserted in Diary of Mary A. Henry, HSA, box 51.
21. Mayer, "Henry as a Discoverer," in *Memorial*, 490. Taylor, "The Scientific Work of Joseph Henry," 233–37. Mary A. Henry, "America's Part in the Discovery of Magneto-Electricity—A Study of the Work of Faraday and Henry." For more on the campaign to establish Henry's priority over Faraday, see also n. 1 above. For the depiction of Henry in a typical, leading, recent textbook, see David Halliday, Robert Resnick, Jearl Walker, *Fundamentals of Physics*, 4th ed. (New York: John Wiley and Sons, 1993), 874, 900–901.
22. For the International Electrical Congress, see the documents in the folder labeled "Materials regarding the establishment of the henry," HSA, box 27; see also "Magnetic Symbols and Units," *The Electrical World* 26 (3 Aug. 1895): 125.

7. Joseph and Harriet

1. Alexander S. Alexander to Henry, 11 Feb. 1827, *Henry Papers* 1:181. Obituary note on Alexander Alexander, *Albany Gazette*, 7 Sept. 1809, copy in folder labeled "Miscellaneous family papers," HSA, box 58. Henry to [William H. Alexander], 31 May 1845, *Henry Papers* 6:285–86. Henry to Stephen Alexander, 19 Dec. 1825, *Henry Papers* 1:115–17. Harriet Alexander to Stephen Alexander, [March? 1829], *Henry Papers* 1:215–16. *Charles A. Young, Memoir of Stephen Alexander, 1806–1883*, read before the National Academy (Washington: Judd & Detweiler Printers, 1884), 251–59. For genealogical details on the Alexanders, see the charts and records as compiled by R. F. Meredith in the genealogical file maintained by HPP.
2. This and the next nine paragraphs rely on the letters variously between Henry and Harriet, Maria, and Stephen Alexander during the period from 15 Nov. 1828 through 23 April 1830, *Henry Papers* 1:211–12, 215–16, 222–24, 230–31, 249, 251–54, 254–55, 255–57, 258–60, 261, 261–62, 263, 264–65, 265, 266, 267, 268, 271, 271–72, 272–73. See also, Harriet to Stephen Alexander, 27 Sept. [should be Jan.] 1829, in file labeled "1825–May 1830," HSA, box 57. *Henry Papers* 6:443 n. 4. For the record of Harriet and Joseph's marriage, see C. A. Church to Kathleen W. Dorman, 11 Jan. 1984, copy in the genealogical file maintained by HPP.
3. This and the next four paragraphs rely on "Henry's Notes On a Trip to Yale,"

6 May 1830, and on the letters variously between Joseph Henry, Harriet Henry, and Stephen Alexander during the period from 7 May 1830 through 30 Aug. 1830, *Henry Papers* 1:274–89, and on 3 May 1847, *Henry Papers* 7:97–100. The paragraphs also rely on Joseph to Harriet Henry, 27 May 1830, in file labeled "1825–May 1830," HSA, box 57. See also *Henry Papers* 1:252, 267, 268. Another source is Diary of Mary A. Henry, 28 Nov. 1867. See also Stephen Van Rensselaer Ryan to Joseph Henry, 8 May 1832, *Henry Papers* 1:427; for the births of the Henrys' children, see the charts and records compiled by R. F. Meredith in the genealogical file maintained by HPP. Rhees's conversations with Henry of 1876, as described above in chap. 1, n. 6.

4. This and the next five paragraphs rely on the documents and letters from 27 April 1825 through 27 Aug. 1832, *Henry Papers* 1:96–97, 196–99, 280–81, 288–89, 289–91, 291–93, 331–33, 335–37, 338–39, 356–60, 402–4, 404–7, 409–12, 414–16, 428–32, 443–45, 465–69. Sabine, "Observations on the Magnetism of the Earth . . . in a letter from Capt. Edward Sabine, to Professor Renwick," *American Journal of Science and Arts* 14 (July 1828): 145–56. Henry, "On a Disturbance of the Earth's Magnetism, In Connection with the Appearance of an Aurora Borealis, as Observed at Albany, April 19th, 1831" (April 1832), reprinted in *Scientific Writings* 1:58–72. Stephen Alexander to Asa Gray, 4 Dec. 1878. Green, *Electro-Magnetism*, 198–205. John Cawood, "Terrestrial Magnetism and the Development of International Collaboration in the Early Nineteenth Century," *Annals of Science* 34 (1977): 568–70. See also Coulson, *Joseph Henry*, 72–74.
5. Faraday, 19 April 1831, *Faraday's Diary* 1:320. James, ed., "Introduction," in *Correspondence of Michael Faraday*, 1:xxxvi.
6. This and the next paragraph rely on the documents and letters from 9 Sept. 1825 through 28 June 1832, *Henry Papers* 1:105–7, 251–54, 264–65, 397–400, 409–12, 435–37; see also pp. 4–5, n. 1. Henry to William Kelly, 30 Dec. 1833, *Henry Papers* 2:142–47. Henry to Alexander Dallas Bache, [19–27] March 1839, *Henry Papers* 4:186–92. Taylor, "The Scientific Work of Joseph Henry," 212, 258; see also Taylor's accompanying "List of the Scientific Papers of Joseph Henry," 365. James R. Fleming, *Meteorology in America, 1800–1870* (Baltimore: Johns Hopkins University Press, 1990), 19–21, 48; Fleming quotes Cawood, "Terrestrial Magnetism," 581, 586.

8. The Call to Princeton

1. Henry to T. Romeyn Beck, 28 Nov. 1825.
2. Rhees's conversations with Henry of 1868 and 1876, as described above in chap. 1, n. 6. *Henry Papers* 1:69 n. 15, 214 n. 6.
3. This and the next two paragraphs rely on letters between Henry and Silliman during the period from 25 Jan. 1831 through 5 Mar. 1832, *Henry Papers*

1:321–22, 331–33, 354–55, 404–7, 407–9. Editor of Journal [Silliman], note appended to "An Account of a Large Electro-Magnet, Made for the Laboratory of Yale College," 50 n. See also Rhees's conversations with Henry of 1868 and 1876.

4. Stephen Van Rensselaer Ryan to Henry, 24 May 1827, and Henry to Bennett H. Henderson, 25 July 1827, *Henry Papers* 1:186–87, 188. For a possible West Point vacancy beginning in late 1826, see career notes on Jonathan Prescott, *Henry Papers* 1:157 n. 7.
5. Lanman, *Haphazard Personalities*, 11–12, as excerpted in "Professor Henry in Private," *New York Evening Post,* 24 May 1878. Rhees's conversation with Henry of 1876. See also Cuyler to Miss Henry, 20 March 1888, and Cuyler, "Professor Henry—The Christian Philosopher."
6. This and the next five paragraphs rely on the documents and letters from 18 June 1832 through 23 Oct. 1832, *Henry Papers* 1:433–34, 435–37, 439–40, 441, 443, 455–64. Rhees's conversation with Henry of 1876. See also John Maclean to Miss [Mary] Henry, 29 Nov. 1879, HSA (HPP, no. 5697), and Harriet to Joseph Henry, 6 April 1833, in file labeled "1831–1834," HSA, box 57.
7. This paragraph relies on the letters variously between Joseph Henry, Stephen Alexander, James Henry, and Harriet Henry during the period from 27 Aug. 1832 through 4 Sept. 1832, *Henry Papers* 1:443–55.
8. Franklin B. Hough to Spencer F. Baird, 15 May 1878, Huntington Library, RH 4237 (HPP, no. 3923). Lanman, *Haphazard Personalities*, 12–13, excerpted in "Professor Henry in Private," *New York Evening Post,* 24 May 1878. Rhees's conversation with Henry of 1876. Cf. Welling, "The Life and Character," 184. Henry's reminiscences to his daughter Mary, Diary of Mary A. Henry, 4 Feb. 1864 (see above, chap. 1, n. 2); see also the apparent diary extract, 14 Jan. 1877, in folder 8, HSA, box 56 (typed version in HPP, no. 5851). See also *Henry Papers* 1:339 n. 2.
9. Dunlap diary, 8 Oct. 1832, quoted in Coulson, *Joseph Henry*, 94.

9. Settling In

1. Joseph Henry to James Henry, 9 Feb. 1833, *Henry Papers* 2:45–49. Harriet Henry to [Nancy Henry], 5 Dec. 1832, *Henry Papers* 2:17–18. See also "Introduction," *Henry Papers* 2:xiv.
2. John Maclean, *History of the College of New Jersey, 1746–1854*, 2 vols. in one, American Education: Its Men, Ideas, and Institutions (1877; rpt., New York: Arno Press, 1969), 2:284–94. Thomas J. Wertenbaker, *Princeton, 1746–1896* (Princeton: Princeton University Press, 1946), 215–16, 221–22, 250, 256. Stanley M. Guralnick, *Science and the Ante-Bellum American College* (Philadelphia: American Philosophical Society, 1975), 42–43. *General Catalogue of Princeton*

University, 1746–1906 (Princeton, N.J.: Princeton University, 1908), 12, 30–31, 52–53. *Catalogue of the Officers and Students of the College of New-Jersey; for 1832–1833* (Princeton, N.J.: J. T. Robinson & Co., 1832). Henry to Silliman, 18 April 1833, *Henry Papers* 2:63–66. Henry to Stephen Van Rensselaer, 23 Oct. 1834, *Henry Papers* 2:270–71.

3. Mark A. Noll, ed., "Introduction," in *The Princeton Theology, 1812–1921* (Grand Rapids, Mich.: Baker Book House, 1983), 11–40. Wertenbaker, *Princeton*, 232–33, 239. See also "Introduction," *Henry Papers* 2:xviii. For Stephen Alexander's Princeton affiliations, see various letters from Henry, *Henry Papers* 2:13, 31, 56, 60, 68–69, 149; see also "Stephen Alexander," in *The Princeton Book: A Series of Sketches . . . by Officers and Graduates of the College* (Boston: Houghton, Osgood, 1879), 138–39.
4. This and the next paragraph rely on V. Lansing Collins, *Princeton, Past and Present*, rev. ed. (Princeton: Princeton University Press, 1946), esp. sects. 1–4, 69, 81–83. Henry L. Savage, ed., *Nassau Hall, 1756–1956* (Princeton: Princeton University, 1956), 140–42. See also: Guralnick, *Science and the Ante-Bellum American College*, 67.; Wertenbaker, *Princeton*, 243; and various letters from Henry, *Henry Papers* 2:9, 45–49, 55–57, 71, 81–82.
5. This and the next paragraph rely on various letters from Harriet and Joseph Henry in *Henry Papers* 2:3, 6–8, 9–12, 17–18, 45 49, 55, 205–6, 400–401 and *Henry Papers* 3:46, 100–102 esp. n. 1. See also *Henry Papers* 3:113 n. 4.
6. Reynolds, "Prof. Joseph Henry and His Life Work." For Mary Anna, see various letters from Harriet and Joseph Henry, *Henry Papers* 2:167–68, 205–6, 334, 344; for Helen Louisa, see *Henry Papers* 3:1042, 1050; and for Caroline, see *Henry Papers* 4:249, 257. Joseph to James Henry, [10–11 Oct. 1839] and 23 Dec. 1840, *Henry Papers* 4:272–74, 451–52. Clement Clarke Moore's "A Visit From St. Nicholas" first appeared, anonymously, in 1823.
7. "Died" [death notice for Ann Henry], *Daily Albany Argus*, 6 April 1835, p. 2, col. 6. Various letters from Joseph and Harriet Henry, between 13 Nov. 1832 and 23 April 1835, *Henry Papers* 2:7, 10–11, 12–13, 17–18, 22, 27, 31, 36, 51, 76, 120, 151, 171, 335, 337–38, 354–55, 367–69, 371, 373, 378–79. See also Joseph to Harriet Henry, 15 Dec. 1832, in file labeled "1831–1834," HSA, box 57.
8. See various letters from Harriet and Joseph in *Henry Papers* 2:17, 43, 71, 73, 76, 78, 83, 121, 168. Joseph to James Henry, 28 June 1844, 30 July 1844, 13 Jan. 1845, *Henry Papers* 6:119–21, 125–28, 165–68.
9. Joseph to James Henry, 1 Dec. 1832, 30 Jan. 1833, *Henry Papers* 2:12–16, 41–44. For Henry's appraisal of his Princeton years, see Rhees's conversations with Henry of 1876, as described above in chap. 1, n. 6.
10. Various letter from Henry in *Henry Papers* 2:7, 9, 19, 63. See also Silliman to Henry, 23 April 1833, *Henry Papers* 2:66–68. Diary of Mary A. Henry, 4 Feb. 1864.

11. Charles Weiner, "Joseph Henry and the Relations between Teaching and Research," *American Journal of Physics* 34 (July 1966): 1096–97. Weiner, "Joseph Henry's Lectures on Natural Philosophy: Teaching and Research in Physics, 1832–1847" (Ph.D. diss., Case Institute of Technology, 1965), 68–79. Barbara Myers Swartz, "Joseph Henry—America's Premier Physics Teacher," *Physics Teacher* 16 (Sept. 1978): 350. *Henry Papers* 2:xvii–xix, 202 n. 1, 280 n. 2, 353 n. 1. *Henry Papers* 6:237–40 n. 8. See also Henry, "Closing Remarks for Natural Philosophy," and Henry to Charles F. McCay, 25 Aug. 1846, in *Henry Papers* 6:412–15, 486–88. The remainder of this section on Henry's teaching continues to rely on Weiner's two studies.
12. Wertenbaker, *Princeton,* 249–50. Henry to Torrey, 14 Sept. 1840, *Henry Papers* 4:433.
13. Wertenbaker, *Princeton,* 236–37. Maclean, *History*, 285–94. Edward Hall, *Reminiscences of Princeton College* (Princeton: Princeton University Press, 1914), 10–11. Henry, "Closing Remarks for Natural Philosophy," 412–13. Swartz, "Joseph Henry—America's Premier Physics Teacher," 350.
14. Henry, "Syllabus, &c.," with annotations, p. 9, in folder labeled "Somotology book and notebook," HSA, box 16. Halsey, notebook on natural philosophy, 1841–42. James F. Davison, notebook on natural philosophy, 1846–47, Princeton University Library, copy in HPP, microfilm batch M-138. Notice that in the aside on the human body, Henry was insinuating that science not only justifies a belief in white racial superiority but also creates racial superiority.
15. Hall, *Reminiscences*, 10–11. Various letter from Joseph Henry in *Henry Papers* 2:52, 127, 365. Henry, "Introductory Remarks for Natural Philosophy," and Henry to Peter Bullions, 5 Aug. 1846, in *Henry Papers* 6:426–30, 457–62.
16. Henry to Silliman, 1 Oct. 1834 and 10 Sept. 1835, *Henry Papers* 2:261–63, 445–51. Henry to Torrey, 20 Sept. 1834, *Henry Papers* 2:253.
17. "Professor Henry's Magnet," *New-York Standard*, 1 Oct. 1833, p. 2, reprinted in *Henry Papers* 2:122–23 n. 2. MS that begins, "In exhibiting the lifting power of the great magnet," in folder labeled "Electricity and magnetism notes," HSA, box 16. "Professor Henry's Big Magnet," *Scientific American* (11 Dec. 1880): 370. "The Lesson of a Working Life," *New York Evening Post,* 14 May 1878; copy in "Scrap Book" and in HSA, box 46. Mary Henry, "Joseph Henry," typescript, p. 6. Excerpt, Diary of John R. Buhler, 21 Feb. [1846], and "Record of Experiments," 21 Feb. 1846, *Henry Papers* 6:376–77, 377–78. "Record of Experiments," 6 June 1840, *Henry Papers* 4:408–10. For other students' memories of the magnet, see James K. Kelly to Misses Henry, 25 Feb. 1901, and Parke Godwin to Miss Henry, 6 Sept. 1897, file labeled "Recollections of Joseph Henry," HSA, box 27.
18. In one of his earliest, most trustworthy reconstructions of the timing of his introduction of the relay, an 1849 court deposition, Henry expressed confidence

that he had the idea before visiting Wheatstone in 1837 but was unsure when he first detailed it to his class. In the deposition, an inadvertent substitution of the adverb "previously" for "subsequently" probably confused his account; he seemed to mean to say that, while he might have informed his classes before the 1837 visit, he was certain that he had regularly told the classes after the visit. In the draft of an 1869 letter, he said he showed his class "as early as 1834." Henry, "Deposition of Joseph Henry, in the Case of Morse *vs.* O'Reilly," reprinted in *Annual Report of the Board of Regents of the Smithsonian Institution, 1857,* 111–12. Henry to anonymous, Jan. 1869, Huntington Library (HPP, no. 2893). Henry's European Diary, 1 April 1837, *Henry Papers* 3:219–20. Stephen G. Dodd, "Princeton in the Forties," *Princeton Alumni Weekly* 13 (Dec. 1912): 240. Weiner, "Joseph Henry's Lectures on Natural Philosophy," 197. See also Coulson, *Joseph Henry,* 107–8, 120, 223, 234.

19. "Introductory Lecture," [ca. 11 Nov. 1837], *Henry Papers* 3:519–21. Henry, "Introductory Lecture," n.d., quoted in Weiner, "Joseph Henry and the Relations between Teaching and Research," 1096–97; see also Weiner, "Joseph Henry's Lectures on Natural Philosophy," 63.
20. Edward Courtenay to Henry, 3 Jan. 1833, and Joseph to James Henry, 27 Jan. 1834, *Henry Papers* 2:32–33, 152–54. Guralnick, *Science and the Ante-Bellum American College,* 67. Henry to Bache, 9 Feb. 1835, and Henry to Torrey, [27] April 1835, *Henry Papers* 2:346–49, 381–82.
21. Henry to [no name], 16 July 1844, quoted in Weiner, "Joseph Henry and the Relations between Teaching and Research," 1096. Various letters from Henry in *Henry Papers* 5:134, 449, and *Henry Papers* 4:38–39. Wiley & Putnam to Henry, 30 Jan. 1844, Joseph to James Henry, [Jan. 1844], Lea & Blanchard to Henry, 27 April 1844, *Henry Papers* 6:12–13, 15–16, 92–93. Hall, *Reminiscences,* 11.
22. Lewis R. Gibbes to Henry, 1 March 1845, *Henry Papers* 6:236–44. Various letters from Henry in *Henry Papers* 6:137–39, 280–84, 486–88. Henry, "Syllabus, &c." Henry, "Syllabus of a Course of Lectures on Physics," in *Annual Report of the Board of Regents of the Smithsonian Institution, 1856* (Washington, D.C.: A. O. P. Nicholson, Printer, 1857), 187–91.
23. *Henry Papers* 2:33 n. 4, 87–88 n. 3, 365 n. 5, 431 n. 8; *Henry Papers* 3:112 n. 5, 121 n. 3. *Henry Papers* 6:426 n. 1. Various letters from Henry in *Henry Papers* 3:84, 95, 101. "Journal of Job M. Allen," 25 Jan. [1837], in *Henry Papers* 3:136. From the Senior Class of the College of New Jersey, 25 June 1841, Henry to Silliman, 24 Feb. 1842, *Henry Papers* 5:43–45, 149. Diary of Mary A. Henry, 13 Dec. 1862. Maclean to Henry, 25 July 1836, *Henry Papers* 3:80–82. Maclean, *History,* 300, 314; Wertenbaker, *Princeton,* 235.
24. John D. Colt, notebook on natural philosophy, 1842–43, Princeton University Library, copy in HPP, microfilm batch M-158. *Henry Papers* 6:323 n. 13. Various letters from Henry in *Henry Papers* 2:55, 97.

10. Professor Henry

1. This and the next three paragraphs rely on the following sources. Theodore L. Cuyler to Miss Henry, 20 March 1888, file labeled "Addresses on Putting up Tablet in Princeton—Personal Recollections," HSA, box 54; for a similar, published version, see Cuyler, "Professor Henry—The Christian Philosopher" (n.p., n.d., but probably *New York Evangelist,* May 1878), copy in HSA, box 46. William E. Schenck, MS, n.d., file labeled "Addresses on Putting up Tablet in Princeton—Personal Recollections," HSA, box 54. Dodd, "Princeton in the Forties," 238–40. Halsey to William Rhees, 21 May 1883, file labeled "Recollections of Joseph Henry," HSA, box 27.
2. This paragraph, unless otherwise indicated, relies on the following sources. W. M. Whitehead, notebook on natural philosophy, 1836–37, Princeton University Library, copy in HPP, microfilm batch M-134. Grier to Mary Henry, 7 Dec. 1893, A. V. C. Schenck to Grier, 10 Oct. 1893, and John P. Stockton to Mary Henry, 13 Nov. 1895, in the file labeled "Recollections of Joseph Henry," HSA, box 27. Unsigned article probably by Grier, "Joseph Henry: Science and Religion," n.p., n.d., copy in William A. Henry's "Rhetoric" notebook, HSA, box 59. See also P. C. Van Wyck's letter to the *New York Times,* quoted in Silliman, "Joseph Henry, LL.D.," 465, and Weiner, "Joseph Henry's Lectures on Natural Philosophy," 197.
3. The grounded telegraph wire, routed from Philosophical Hall to the study in the house via the Library, probably went up for the first time during winter term 1842–43; a similar wire for out-of-doors induction experiments (see chap. 11), routed via Nassau Hall, went up sooner, in October 1842. Whereas an 1849 court deposition contains one of the earliest, most trustworthy, public reconstructions of the timing of Henry's introduction of the ground circuit—sometime after the German announcement in 1837—his 1876 account of his researches that he sent to Samuel Dod contains a contradictory version. The account that Dod received and published has Henry introducing the grounded telegraph in 1836 using a line routed from the hall to the house via the Library; however, the draft of the account reveals that the date 1836 is a substitution for an original, crossed-out date of 1838—a revision made by either Henry or, perhaps, clerk Rhees. Though the date of 1838 is still probably premature, the two-year difference is significant in that the German announcement came in 1837. The draft also has a crossed-out passage indicating that Harriet Henry operated one end of the grounded telegraph, presumably in the new house occupied in late 1838; the similarity of this deleted description to the students' description of the lecture-hall demonstration during the winter term 1842–43 suggests that Henry first arranged a full-scale grounded telegraph during that term. Similarities between Henry's "Record of Experiments" and the original and published versions of the Dod letter reinforce this conclusion; the "Record of Experiments" further indicates that Henry

probably first tested the ground circuit in October 1842 in conjunction with his induction experiments. Henry, "Deposition of Joseph Henry, in the Case of Morse *vs.* O'Reilly," 111–12. Cf. Henry to Dod, 4 Dec. 1876, reprinted in Dod, "Discourse Memorial" (1878), in *Memorial,* 149–50; the draft is located in the file labeled "Letter about Scientific Investigations of Joseph Henry while at Princeton," HSA, box 26. Henry, "Record of Experiments," 10 June 1842, 6 Oct. 1842, and 16 Oct. 1843, *Henry Papers* 5:224–25, 276–77, 411–12. See also Coulson, *Joseph Henry,* 107, 223.

4. P. C. Van Wyck's letter to the New York *Times*, quoted in Silliman, "Joseph Henry, LL.D.," 465–66. Dodd, "Princeton in the Forties," 240. *Henry Papers* 4:452 n. 5. Joseph to Harriet Henry, [23] Jan. [1839], Henry to Bache, [25] Jan. 1839, Henry to Wheatstone, [Dec. 1839], *Henry Papers* 4:172–74, 322–23.
5. Francis Levison to Henry, 29 Dec. 1846, *Henry Papers* 6:619–21. Henry Alexander to Joseph Henry, 11 Sept. 1847, *Henry Papers* 7:177–79.
6. Unless otherwise indicated, this and the next two paragraphs rely on the following sources. Diary of John R. Buhler, excerpts between 21 Feb. and 3 June [1846], *Henry Papers* 6:376–77, 387–88, 391–92, 395–96, 424–25, 425–26, 430, 431. Hall, *Reminiscences,* 10. See also Buhler, notebook on natural philosophy, 1845, Louisiana and Lower Mississippi Collections of the Louisiana State University Libraries (HPP, microfilm batch M-449); the later pages of the notebook contain "Poems and Scraps!"
7. Joseph to Harriet Henry, [28 Sept.] 1836, *Henry Papers* 3:111–12. Cf. Henry, "The Philosophy of Education" (1854), in *A Scientist in American Life: Essays and Lectures of Joseph Henry,* ed. Arthur Molella et al. (Washington: Smithsonian Institution Press, 1980), 73.
8. Joseph to Harriet Henry, 12 Oct. 1836, *Henry Papers* 3:113–18. Diary of Mary A. Henry, 4 July 1862. Henry to Bache, 17 Dec. 1834, Henry to Torrey, 20 Dec. 1834 and 26 Aug. 1835, *Henry Papers* 2:297–304, 304–7, 438–40.
9. "Henry, Joseph," in *The Biblical Repertory and Princeton Review: Index Volume from 1825 to 1868,* 200. See above, chap. 1, n. 2, for evidence of Henry's authorship and for periodicals which reprinted the biographical sketch. Rhees's conversations with Henry of 15 June 1868, as described above in chap. 1, n. 6. *Henry Papers* 6:475–76 n. 8.
10. Various letter from Henry in *Henry Papers* 2:84, 86, 206, and *Henry Papers* 3:101–2, 103–4, 111. Joseph to [James] Henry, April 1842, *Henry Papers* 5:160–61. *Henry Papers* 1:117–18 n. 1.
11. Newland to Harriet Henry, 16 May 1878, in file labeled "Henry—Condolence Letters," HSA, box 46; see also various letter from Henry in *Henry Papers* 2:467, and *Henry Papers* 3:10, 104. Richard Dewitt to Henry, 22 Dec. 1834, and various letters from Joseph to James Henry between 13 Jan. and 23 Feb. 1835, *Henry Papers* 2:310, 334, 338, 345, 355–56. Joseph to [James] Henry, April 1842.

12. Anonymous student to Henry, ca. 1843, *Henry Papers* 5:479; for original, see file labeled "Undated L–Z," HSA, box 12. See also Savage, ed., *Nassau Hall,* 136–37. Henry to Torrey, 17 Aug. 1836, *Henry Papers* 3:83; see also Wertenbaker, *Princeton*, 241–42, and Savage, ed., *Nassau Hall,* 144.
13. Henry to William Gledhill, 12 Sept. 1844, Thomas Sparrow to Henry, 13 Feb. 1845, Henry to [Sparrow], 15 Aug. 1846, Henry to anonymous, 15 March 1845, *Henry Papers* 6:131–32, 218–21, 478–82, 246–50.
14. Joseph to James Henry, 17 March 1835, *Henry Papers* 2:367. *Henry Papers* 2:175 n. 6, and *Henry Papers* 3:371 n. 8. See also Henry to Torrey, 18 May 1838, *Henry Papers* 4:52–53, and Henry to John Foster, 25 Oct. 1845, *Henry Papers* 6:320–24. This and the next two paragraphs also draw on the following sources: Wertenbaker, *Princeton*, 217–19, 250–53; Guralnick, *Science and the Ante-Bellum American College*, 42–43; Maclean, *History*, 297–309; and Collins, *Princeton, Past and Present*, sections 81–83.
15. Various letters from Henry, *Henry Papers* 2:270–71, 276. "Points which May be Noticed in the Circular of the Whig Society," [Aug. 1836], *Henry Papers* 3:93–95. Excerpt, Minutes, Trustees, College of New Jersey, 15 April 1835, *Henry Papers* 2:375–76.
16. Maclean to Henry, 25 July 1836, "American Whig Society Circular," 30 Aug. 1836, and Excerpt, Minutes, Trustees, College of New Jersey, 29 Sept. 1836, *Henry Papers* 3:80–82, 88–92, 113. *Henry Papers* 2:84 n. 9. Various letters from Henry to Torrey, *Henry Papers* 2:363, 371–72, 376–78, 439. Excerpt, Minutes, Trustees, College of New Jersey, 15 April 1835. Henry's Drawing of Philosophical Hall, [April 1835], *Henry Papers* 2:382–84.
17. "Princeton, N.J.," newspaper article ca. early 1860s, file labeled "Henry Illustrative Material," HSA, box 30; see also Henry C. Cameron, "Reminiscences," in *Memorial*, 171–72. Henry to [Robley Dunglison], 10 Feb. 1844, *Henry Papers* 6:27.
18. Various letters from Harriet and Joseph Henry in *Henry Papers* 2:158, 168, 210, 356, 481, and *Henry Papers* 3:7, 14, 40, 42–43.
19. Maclean to Henry, 25 July 1836. Silliman to Henry, 15 March 1836, *Henry Papers* 3:24–25.

11. New Experiments

1. This and the next two paragraphs rely on various letters from Joseph to James and Harriet Henry, during the period from 30 May 1834 through 16 Sept. 1834, *Henry Papers* 2:200, 234–36, 250.
2. "Record of Experiments," 15 Aug. 1834, *Henry Papers* 2:215–22. Various letters from Henry, *Henry Papers* 2:63–66, 70, 142–47. Henry to Edward C. Herrick, 9 April 1849, *Henry Papers* 7:504–5. "Galvanic Experiments on the Body of the Murderer Le Blanc," 6 Sept. 1833, *Henry Papers* 2:90–96.

Charles Hodge to Hugh L. Hodge, 4 Sept. 1834 and 12 Oct. 1834, *Henry Papers* 2:240–42, 266–67.

3. "Record of Experiments," 15 Aug. 1834, *Henry Papers* 2:215–20. Henry, "Record of Experiments," entries between 15 Aug. and 26 Sept. 1834, *Henry Papers* 2:220–22, 225–34, 242–49, 253, 258–60.
4. Henry, "Record of Experiments," 31 Dec. 1834 and 1 Jan. 1835, *Henry Papers* 2:317–19, 321–23.
5. Henry to Bache, 17 Dec. 1834 and 12 Jan. 1835, *Henry Papers* 2:297–304, 329–31. Henry to Torrey, 20 Dec. 1834, *Henry Papers* 2:304–7. Henry, "Contributions to Electricity and Magnetism. No. I. Description of a Galvanic Battery for Producing Electricity of Different Intensities" (read Jan. 1835), "Contributions to Electricity and Magnetism. No. II. On the Influence of a Spiral Conductor in Increasing the Intensity of Electricity from a Galvanic Arrangement of a Single Pair, Etc." (read Feb. 1835), reprinted in *Scientific Writings* 1:80–86, 92–100. Faraday, "Preface" (March 1839), in *Experimental Researches* 1:iii–vi. James, ed., "Introduction," in *Correspondence of Michael Faraday*, 2:xxx–xxxi.
6. Henry to Torrey, 23 Feb. 1835, Henry to Bache, 18 May 1835, *Henry Papers* 2:356–60, 403–8.
7. Henry to Bache, 12 Jan. 1835, *Henry Papers* 2:329–31. Henry, "Record of Experiments," 7 Jan. 1835 and 13 May [1835], *Henry Papers* 2:328, 394–97. Henry, "Contributions to Electricity and Magnetism. No. II," 95. Henry, "Notice of Electrical Researches—the Lateral Discharge" (1837), reprinted in *Scientific Writings* 1:101–4.
8. *Henry Papers* 2:xxiv–xxv, 403 n. 1, 405 n. 11, 451 n. 13. Henry to Bache, 18 May 1835. Henry Notebook Entry, 14 June 1835, *Henry Papers* 2:415. Joseph to James Henry, [2 April 1836], *Henry Papers* 3:42–43. Henry, "Record of Experiments," 16 March 1836 and 20 March [1836], *Henry Papers* 3:, 27–28, 31–32.
9. Henry, "Record of Experiments," various entries from 19 March through 14 May 1836, *Henry Papers* 3:30–65. Henry to Bache, 10 May 1836, *Henry Papers* 3:60–62. Henry to Torrey, 30 Oct. 1838, Henry to Gray, 1 Nov. 1838, Henry to William Redfield, 17 Dec. 1838, *Henry Papers* 4:145–46, 152–54, 164–65.
10. Various letters from Henry, from 6 May 1838 through 6 May 1839, *Henry Papers* 4:44–45, 54, 115–17, 145–46, 150–51, 152–54, 215. For the general circumstances surrounding the preparation and publication of the third article in Henry's APS series, see the various letters to and from Henry and his entries in "Record of Experiments," from 6 April through 17 Dec. 1838, *Henry Papers* 4:17–165.
11. Various letters to and from Henry, during the period from 12 Feb. 1839 to 21 April 1840, *Henry Papers* 4:176, 209, 215, 221, 222, 231–42, 248, 252, 261–64, 283, 296–303, 311, 354, and during the period from 9 to 29 April 1841,

Henry Papers 5:18–19, 26, 27, 29. Nomination of Henry for Copley Medal, *Henry Papers* 4:295–96. "Introduction," *Henry Papers* 4:xiv.

12. This third section of the chapter—concerning Henry's third article in his series—relies on, unless otherwise indicated, "Induced Currents from Ordinary Electricity" (read Nov. 1838), and "Contributions to Electricity and Magnetism. No. III. On Electro-Dynamic Induction" (read Nov. 1838), reprinted in *Scientific Writings* 1:106–7, 108–45. See also "Introduction," *Henry Papers* 4:xi–xiii.
13. Henry, "Contributions to Electricity and Magnetism. No. II," 98–100. *Henry Papers* 2:329–31 n. 5. Halsey, notebook on natural philosophy, vol. 2, 1841–42; Gledhill, notebook on natural philosophy, 1842–43. [Henry], "Magnetism," in *Encyclopædia Americana: Supplementary Volume*, 414–15. Faraday, 12 Nov. 1839, *Faraday's Diary* 4:14.
14. Henry to Gray, 1 Nov. 1838, Sturgeon to Henry, 23 Sept. 1839, and the exchange of letters between Henry, Silliman, and Bache, between 2 and 18 Nov. 1839, *Henry Papers* 4:152–54, 261–64, 296–306. Henry, "Contributions to Electricity and Magnetism. No. IV. On Electro-Dynamic Induction. Continued" (read June and Nov. 1840), reprinted in *Scientific Writings* 1:159. See also Cantor, *Michael Faraday: Sandemanian and Scientist,* 144–45.
15. Harold I. Sharlin, "Applications of Electricity," in *Technology in Western Civilization,* ed. Melvin Kranzberg and Carroll W. Pursell, Jr. (New York: Oxford University Press, 1967), 1:581. George Shiers, "The Induction Coil," *Scientific American* 224 (May 1971): 80–82. See also Coulson, *Joseph Henry,* 132–34.
16. Henry, "Letter on Electrical Induction" (Oct. 1839), reprinted in *Scientific Writings* 1:147–48; "Contributions to Electricity and Magnetism. No. IV," 149–88. "Introduction," *Henry Papers* 4:xiii–xiv. "Record of Experiments," entries for 13 May, 14 Oct., and 24 Oct. 1839, *Henry Papers* 4:216–17, 275–77, 284–85; see also entries for 14 May, 16 June, and 31 Oct. 1840, pp. 382–83, 414–16, 447–48. Various letters from Henry to Bache, from [20] May 1839 to 17 Dec. 1840, *Henry Papers* 4:218–21, 280–82, 440–42, 450–51. Henry to Torrey, 13 May and 4 Sept. 1839, *Henry Papers* 4:227–29, 255–57. Faraday, "Relation of the Electric and Magnetic Forces," sect. 21 of "Fourteenth Series" (21 June 1838), reprinted in *Experimental Researches* 1:545–52. Ames, "Certain Aspects of Henry's Experiments on Electromagnetic Induction," 91–92. See also Coulson, *Joseph Henry*, 135–37.
17. Henry, "Contributions to Electricity and Magnetism. No. IV," 161. Henry to Silliman, 17 Aug. 1839, *Henry Papers* 4:247–49.
18. This fourth section of the chapter—concerning Henry's large-distance electrostatic effects—relies on, unless otherwise indicated, Henry, "Contributions to Electricity and Magnetism. No. V. On Induction from Ordinary Electricity; and on the Oscillatory Discharge" (read June 1842), reprinted in *Scientific Writings* 1:200–203. Henry, "Record of Experiments," entries between 6 May and 19 July 1842, *Henry Papers* 5:163–250; see also Henry to Bache,

24 June 1842, pp. 229-30. Lewis Gibbes to Henry, 11 March 1843, HSA, quoted in *Henry Papers* 5:468 n. 6.

19. Exchange of letters between Hare and Henry, 26 May 1842, 14 and 18 Oct. 1843, *Henry Papers* 5:189-90, 408–10, 414–16. Henry to Stephen Alexander, 6 May 1838, "Record of Experiments," 7 May [1838], *Henry Papers* 4:44–47. "Introduction," *Henry Papers* 6:xviii–xix. Weiner, "Joseph Henry's Lectures on Natural Philosophy," 171–75. Williams, *Michael Faraday*, 148–51. P. M. Harman, *Energy, Force, and Matter: The Conceptual Development of Nineteenth-Century Physics* (Cambridge: Cambridge University Press, 1982), 31.
20. Henry, "Record of Experiments," entries between 6 Oct. and 8 Oct. 1842, *Henry Papers* 5:276–79. [Paragraph on oral communication of Henry for meeting of 21 Oct. 1842], *Proceedings of the American Philosophical Society*, 2 (1841–43): 229. Henry to Samuel B. Dod, 4 Dec. 1876, reprinted in Dod, "Discourse Memorial" (1878), in *Memorial,* 150–52. See also Henry, "Record of Experiments," entries between 16 Oct. and 19 Oct. 1843, *Henry Papers* 5:411–14, 416–20. Gibson, notebook on natural philosophy, 1842–44 ; see also Halsey, notebook on natural philosophy, vol. 2, 1841–42. Albert G. Gluckman, "Joseph Henry's 1842 and 1843 Out-of-Doors Electrical Transmission Signal Experiments," *Journal of the Washington Academy of Sciences* 82 (Sept. 1992): 111–31.
21. Henry, "Record of Experiments," entries between 28 May and 6 June 1842, *Henry Papers* 5:194–217. Gibson, notebook on natural philosophy, 1842–44; see also Gledhill, notebook on natural philosophy, 1842–43. Henry to Thomas R. Pynchon, 16 June 1868, pp. 76–77. See also: Henry, "Contributions to Electricity and Magnetism. No. III," 144; Henry Notebook Entry, [1834], *Henry Papers* 2:320–21; *Henry Papers* 4:438 n. 1; and "Introduction," *Henry Papers* 5:xviii–xxvii. See also Taylor, "The Scientific Work of Joseph Henry," 250, 255, 257, 396–97 note F; Coulson, *Joseph Henry*, 139–41, 146; Nathan Reingold, "Henry, Joseph," in *Dictionary of Scientific Biography* (New York: Charles Scribner's Sons, 1970–76), 278–79; and Albert G. Gluckman, "The Discovery of Oscillatory Electric Current," *Journal of the Washington Academy of Sciences* 80 (March 1990): 16–25.
22. Henry, "Record of Experiments," 28 May and 10 June 1842, *Henry Papers* 5:194–97, 224–25. "Introduction," *Henry Papers* 5:xxii. Henry, "Meteorology in Its Connection with Agriculture" (1855–59), reprinted in *Scientific Writings* 2:334–35.
23. Henry, "Notice of Electrical Researches—the Lateral Discharge," 101–4. Henry, "Contributions to Electricity and Magnetism. No. III," 139–41. "Introduction," *Henry Papers* 5:xxiii–xxix. Henry, "On the Effects of a Thunder-Storm" (read 5 Nov. 1841), reprinted in *Scientific Writings* 1:193–99. Henry, "Record of Experiments," 15 July 1841, *Henry Papers* 5:66–72. Henry to Forbes, 2 July 1842, Espy to Henry, 20 Aug. 1842, Henry to James Coffin,

9 Sept. 1842, exchange of letters between M. Jacobus and Henry, 18 and 19 July 1843, *Henry Papers* 5:235–36, 257–58, 266–68, 372–74. Exchanges of letters among George M. Dallas, Robert M. Patterson, and Henry, between 6 and 27 Aug. 1847, *Henry Papers* 7:154–73. Henry to anonymous, 15 April 1876, HSA (HPP, no. 17575).

24. Henry, "Record of Experiments,"15 Oct. 1842, *Henry Papers* 5:279–80. Lefroy to Edward Sabine, 25 Oct. 1842, *Henry Papers* 5:281–87. See also Lefroy, "The Scientific Writings of Joseph Henry," *Nature* 38 (31 May 1888): 98–99; copy in HSA, box 42. See also Lefroy to Mary Henry, 24 Feb. 1887, 7 Feb. 1888, 27 April 1888, in file labeled "Miscellaneous correspondence with daughters after Joseph Henry's death," HSA, box 57.
25. Henry to [Trustees, College of New Jersey], [1838], *Henry Papers* 4:166. William to Thomas Kinney, 15 March [1841], *Henry Papers* 5:15–16.
26. Henry to Torrey, 3 Feb. 1840, Joseph to James Henry, 23 Dec. 1840, *Henry Papers* 4:327–29, 451–52, esp. n. 5. Henry, "Record of Experiments," 11 June 1842, *Henry Papers* 5:225–26.
27. Henry to Elias Loomis, 29 Nov. 1841, Joseph to Harriet Henry, [8–10 Aug. 1843], *Henry Papers* 5:125–28, 382–85. Joseph to James Henry, 30 July [–2 Aug.] 1844, *Henry Papers* 6:125–28. Joseph to Harriet Henry, [19 Jan. 1847], *Henry Papers* 7:17–19.
28. [Henry], "Magnetism," in *Encyclopædia Americana: Supplementary Volume*, 414–15. Henry, "On the Atomic Constitution of Matter" (read Nov. 1846), reprinted in *Scientific Writings* 1:255–59. Henry, "Record of Experiments," 19 Jan. 1844, *Henry Papers* 6:7–8. "Diary of Jonathan Homer Lane," 10 June 1848, *Henry Papers* 7:339–41. See also Taylor, "The Scientific Work of Joseph Henry," 257, and Weiner, "Joseph Henry's Lectures on Natural Philosophy," 177–79. For Faraday's speculations in 1832 on the time-dependent, "vibratory," "progressive" nature of magnetic and electrical actions, see Williams, *Michael Faraday*, 181.
29. For a negative view on Henry transmitting and detecting electromagnetic waves, see John A. Wheeler and Herbert S. Bailey Jr., "Joseph Henry (1797–1878): Architect of Organized Science," *American Scientist* 34 (Oct. 1946): 627. For positive views, see: Gluckman, "Joseph Henry's 1842 and 1843 Out-of-Doors Electrical Transmission Signal Experiments," 124–27; William F. Magie, "Joseph Henry: Why the World Joins Princeton in Honoring His Memory," *Princeton Alumni Weekly* 33 (7 Oct. 1932): 62; and Coulson, *Joseph Henry*, 141–46, 337.

12. Reassessments

1. Henry to Pynchon, 16 June 1868. Henry, "Theory of the Discharge of the Leyden Jar" (June 1842), reprinted in *Scientific Writings* 1:216. James, ed.,

"Introduction," in *Correspondence of Michael Faraday* 2:xxx–xxxi. Williams, *Michael Faraday*, 404 n. 11. Henry to Faraday, 22 April 1845, *Henry Papers* 6:270–272.

2. For Henry's "idiosyncratic study of static electricity," see "Introduction," *Henry Papers* 6:xvi–xix. For his main research initiatives in the 1840s, see his *Scientific Writings* 1:191–260. For mesmerism, see Henry, "Record of Experiments," 16 [March] 1843, *Henry Papers* 5:313–16, and excerpt, diary of John R. Buhler, *Henry Papers* 6:395–96. For soap bubbles, see "Record of Experiments," 21 March 1844, *Henry Papers* 6:59–61. Secchi taught briefly in Washington and conferred with Henry at the Smithsonian; see Henry to Bache, 25 June 1852, Bache Papers (HPP, no. 6469).
3. This and the next two paragraphs, unless otherwise indicated, rely on the following sources. Francis Watkins to Henry, [16 Dec.] 1845, Henry to Gray, [mid-Dec. 1845], Henry to Benjamin Peirce, 30 Dec. 1845, and Henry to Henry M. Alexander, 6 Jan. 1846, *Henry Papers* 6:266–67, 340–42, 355–59, 359–63. Henry, "Record of Experiments," 6 April 1838 and 30 Aug. [1839], *Henry Papers* 4:17–18, 253–54. Henry, "Record of Experiments," 1 Nov. 1841, *Henry Papers* 5:106–7. Henry, "Record of Experiments," 27 Dec. and 30 Dec. 1845, *Henry Papers* 6:349–52, 353–55. See also Faraday, "On Simple Voltaic Circles," sect. 1 of "Eighth Series" (5 June 1834), reprinted in *Experimental Researches* 1:284–85.
4. Gooding, "'He Who Proves, Discovers': John Herschel, William Pepys and the Faraday Effect," 234. Cf. "Diary of Jonathan Homer Lane," 10 June 1848, *Henry Papers* 7:339–41.
5. Excerpt, Diary of John R. Buhler, 21 April [1846], *Henry Papers* 6:411–12.
6. Henry to the Editors of the *New York Evening Post*, 28 Feb. 1846, *Henry Papers* 6:386–87. Eben N. Horsford to Henry, 31 Dec. 1846, *Henry Papers* 6:621–23.
7. Henry to Silliman, 22 March 1836, *Henry Papers* 3:33–36. Weiner, "Joseph Henry and the Relations between Teaching and Research," 1096; see also Weiner, "Joseph Henry's Lectures on Natural Philosophy," 44, 53–54.
8. This and the next two paragraphs rely on Henry to Charles F. McCay, 25 Aug. 1846, *Henry Papers* 6:486–88. *Henry Papers* 6:480 n. 7. Henry, "On the Cohesion of Liquids" (April and May 1844) and "On the Capillarity of Metals" (June 1845), reprinted in *Scientific Writings* 1:217–20, 228–30.
9. Lewis R. Gibbes to Henry, 13 March 1844, Henry to [J. J. Sylvester], 26 Feb. 1846, Henry to [Thomas Sparrow], 15 Aug. 1846, Henry to [Samuel Tyler], 4 Dec. 1846, *Henry Papers* 6:53–56, 380–81, 478–82, 560–64. Noll, ed., "Introduction," in *The Princeton Theology*, 21–22. Albert E. Moyer, *A Scientist's Voice in American Culture: Simon Newcomb and the Rhetoric of Scientific Method* (Berkeley: University of California Press, 1992), 34–35, 37–38.
10. Henry, "On the Origin and Classification of the Natural Motors" (Dec.

1844), reprinted in *Scientific Writings* 1:220–23; the original title in the APS *Proceedings* was "Classification and Sources of Mechanical Power." See also Weiner, "Joseph Henry's Lectures on Natural Philosophy," 113–14, and Coulson, *Joseph Henry*, 155–59.

11. Henry, "Contributions to Electricity and Magnetism. No. II," 100. Henry to Bache, 25 Nov. 1835, *Henry Papers* 2:478–82. Henry, "Contributions to Electricity and Magnetism. No. IV," 169. Henry to Robert Hare, 18 Oct. 1843, *Henry Papers* 5:414–16.
12. Henry, Visit to Melrose Abbey to Dr. Brewster, pocket notebook, [19–21 Aug. 1837], *Henry Papers* 3:473–83. *Henry Papers* 5:239–40 n. 13. *Henry Papers* 6:563 n. 15. Henry, "Syllabus of a Course of Lectures on Physics," pp. 188–91; Henry, "Syllabus, &c.," with annotations, pp. 7–8. Gibson, notebook on natural philosophy, 1842–44; see also Halsey, notebook on natural philosophy, 1841–42. See also Moyer, *A Scientist's Voice in American Culture*, 10, 29, 43.
13. This and the concluding paragraph rely on Henry, "On the Atomic Constitution of Matter" (Nov. 1846), reprinted in *Scientific Writings* 1:255–59; the original title in the APS *Proceedings* was "On the Corpuscular Hypothesis of the Constitution of Matter." Weiner, "Joseph Henry's Lectures on Natural Philosophy," 104–13. Henry to Henry James, 22 Aug. 1843, "Record of Experiments," 28 Oct. 1843, *Henry Papers* 5:387–88, 438. See also "Excerpt, Henry Notebook Entry," [ca. 1835], *Henry Papers* 2:491–93, and Coulson, *Joseph Henry*, 159–65.

13. Philadelphia Connections

1. *Henry Papers* 5:68 n. 3; see also "Introduction," *Henry Papers* 2:xiv, xxii. Henry to Vaughan, 13 Oct. 1838, Henry to Bache, 28 Oct. 1839, *Henry Papers* 4:119–20, 289–90; Henry to Bache, 24 June 1842, *Henry Papers* 5:229–30. Various letters from Joseph to Harriet and James Henry, between 11 April 1833 and 4 May 1835, *Henry Papers* 2:57–59, 337–40, 386–89. Henry's Notes on a Trip to Philadelphia, 29 April–2 May 1834, *Henry Papers* 2:181–96. Henry to [Charles Wheatstone], 27 Feb. 1846, *Henry Papers* 6:383. "Boston to Washington, Circa 1830," map supplement to *National Geographic* (July 1994).
2. *Henry Papers* 2:109–11 n. 10. Hugh R. Slotten, *Patronage, Practice, and the Culture of American Science: Alexander Dallas Bache and the U.S. Coast Survey* (New York: Cambridge University Press, 1994), 6–8, 13–24.
3. Joseph to James Henry, 17 Aug. 1833, and Joseph to Harriet Henry, 1 Nov. 1833 and 24 [March 1834], *Henry Papers* 2:87–88, 107–13, 170–74.
4. Joseph to Harriet Henry, 1 Nov. 1833 and 5 Dec. 1834, *Henry Papers* 2:107–13, 284–87. *Henry Papers* 2:108–9 n. 7, 383 n. 5. Slotten, *Patronage, Practice, and the Culture of American Science*, 164.

5. Henry's Notes on a Trip to Philadelphia, 29 April–2 May 1834, p. 185. Joseph to James Henry, 12 Jan. 1842, Joseph Henry to Petty Vaughan, [Feb. 1842], *Henry Papers* 5:141–42, 153–54. Henry to Forbes, 19 Sept. 1836, *Henry Papers* 3:96–98. Slotten, *Patronage, Practice, and the Culture of American Science*, 5–24.
6. Silliman to Vaughan, 9 Oct. 1834, Charles Hodge to Hugh L. Hodge, 4 Sept. 1834 and 12 Oct. 1834, *Henry Papers* 2:264–65, 240–42, 266–67. *Henry Papers* 2:80 n. 4, 240–41 nn. 1, 2.
7. Various letters from Henry, between 20 Dec. 1834 and 23 Jan. 1835, *Henry Papers* 2:305, 329, 332, 339. Nomination for Membership in APS, 7 Nov. 1834, *Henry Papers* 2:277. Torrey to Henry, 29 Dec. 1834 and 16 Jan. 1835, *Henry Papers* 2:311–15, 335–37. Vaughan to Henry, 3 Jan. [1835], *Henry Papers* 2:324. *Henry Papers* 2:326 n. 10.
8. Various letters from Henry, between 18 April 1833 and [1 May 1835], *Henry Papers* 2:63–64, 88, 272, 385. "Record of Experiments," 15 Aug. 1834, 217–18. Henry to Bache, *Henry Papers* 5:229–30.
9. Silliman to Henry, 27 May 1835, *Henry Papers* 2:410–11. "List of Persons to Whom Copies of My Paper Have Been Sent," [ca. 21 Aug.] 1835, *Henry Papers* 2:432–38. *Henry Papers* 2:278 n. 1. Henry to Vaughan, 13 Oct. 1838, Henry to Silliman, 4 Nov. 1839, *Henry Papers* 4:119–20, 299–301. [Henry], Comments on American Philosophical Society Dinner, [6 Jan. 1841], *Henry Papers* 5:3–4. Benjamin Silliman Jr. to Henry, 17 Sept. 1846, *Henry Papers* 6:504–7.
10. *Henry Papers* 1:394 n. 28. Joseph to Harriet Henry, 1 Nov. 1833, p. 113. William Hamilton to Henry, 21 May 1834, Bache to Henry, 13 July 1834 and 15 Feb. [1835], *Henry Papers* 2:198, 207–9, 349–50. Various letters and documents concerning the "Peacemaker," between 3 March and 8 June 1844, *Henry Papers* 6:47, 66–80, 105–18. Bruce Sinclair, *Philadelphia's Philosopher Mechanics* (Baltimore: Johns Hopkins University Press, 1974), 1–19.
11. Espy to Henry, 19 July 1836, and Bache to Henry, 20 July 1836, *Henry Papers* 3:77–79. [Henry], "To the Editor of the Newark Daily Advertiser," [1 Aug. 1838], *Henry Papers* 4:81–83. Bern Dibner, *Benjamin Franklin: Electrician* (Norwalk, Conn.: Burndy Library, 1976), 32–36. Henry, "Record of Experiments," 10 May 1836, *Henry Papers* 3:62–63; *Henry Papers* 5:66–67 n. 1. Excerpt, Henry Notebook Entry, [ca. 1835], *Henry Papers* 2:491–93; Henry to Hare, 18 Oct. 1843, *Henry Papers* 5:414–16; and *Henry Papers* 6:297–98 n. 4. Henry, "Record of Experiments,"11 Oct. 1841 and 17 Oct. 1843, *Henry Papers* 5:96–97, 412–14; see also *Henry Papers* 5:446 n. 3. Andrew Lane to Henry, 12 April [1842], *Henry Papers* 5:159. See also Weiner, "Joseph Henry's Lectures on Natural Philosophy," 171–80.
12. Various letters to and from Henry, between 31 March and 21 Oct. 1843, *Henry Papers* 5:318, 346–47, 352–53, 353–54, 423. Newspaper article beginning "The American Philosophical Society celebrated . . . " (n.p., n.d., but probably *Princeton Whig*, 20 Oct. 1843); clipping inserted in front of Freder-

ick S. Giger, notebook on natural philosophy, 1840–41, Princeton University Library, copy in HPP, microfilm batch M-157. "Henry, Joseph," in *The Biblical Repertory and Princeton Review: Index Volume from 1825 to 1868*, 196; Henry extended the Franklin comparison to his Princeton electrical studies in Henry to Thomas R. Pynchon, 16 June 1868, p. 76.

13. Henry's Notes on a Trip to Philadelphia, 5–7 Dec. 1834, *Henry Papers* 2:290–91. Henry to Torrey, 20 Dec. 1834, *Henry Papers* 2:304–7.
14. Various letters from Henry, between 9 Feb. 1835 and 16 March 1835, *Henry Papers* 2:348–49, 356–60, 364–66, 366–67. Various letters from Torrey to Henry, between 16 Jan. 1835 and 2 [March] 1835, *Henry Papers* 2:335–37, 350–52, 360–61. Henry's Notes on a Trip to Philadelphia, 5–7 Dec. 1834, pp. 290–92; Excerpt, Henry Notebook Entry, [ca. 1835], p. 491. Bache to Henry, 13 Feb. 1840, *Henry Papers* 4:329–31.
15. Joseph to James Henry, 26 Nov. 1833. Minutes, Vaughan Club, 15 July 1840, Joseph to James Henry, 23 Dec. 1840, *Henry Papers* 4:419–22, 451–52. See also Reingold et al., *Henry Papers* 5:238 n. 11, and Rothenberg et al., *Henry Papers* 6:274–75 nn. 7–12.
16. Peter Bullions to Henry, 19 Dec. 1833, 138–41, Joseph to James Henry, 23 Dec. 1833, *Henry Papers* 2:142.
17. This and the next paragraph rely on various letters to Henry, between 12 June 1835 and 9 July 1835, *Henry Papers* 2:413–15, 418–19, 419–21.
18. This and the next paragraph rely on all of the following sources. Excerpt, Minutes, Trustees, College of New Jersey, 9 April 1834, 30 Sept. 1835, *Henry Papers* 2:174–77, 460–62. "Henry, Joseph," in *The Biblical Repertory and Princeton Review: Index Volume from 1825 to 1868*, 196–97. Henry Vethake to Henry, 20 July 1835, John S. Henry to Maclean, 22 July 1835, and Charles Hodge to Henry, 29 July 1835, *Henry Papers* 2:423, 424, 426. Henry to John H. Cocke, 24 July 1835, Joseph to James Henry, [3] Aug. 1835, *Henry Papers* 2:425–26, 428–29.
19. "American Whig Society Circular," 30 Aug. 1836, and Excerpt, Minutes, Trustees, College of New Jersey, 29 Sept. 1836, *Henry Papers* 3:88–92, 113. Henry to Patterson, 18 Feb. 1839, *Henry Papers* 4:180–81. Maclean, *History*, 295–303.
20. John Page to Henry, 23 Sept. 1835, *Henry Papers* 2:457–60. See also Henry's recommendation for William Barton Rogers, 6 July 1835, *Henry Papers* 2:417–18.
21. This and the next paragraph rely on Bache to Henry, 20 July 1836, Joseph to Harriet Henry, 23 July 1836, Patterson to Henry, 14 Aug. 1836, Henry to Patterson, 20 Aug. 1836, and Patterson to Henry, 23 Aug. 1836, *Henry Papers* 2:79, 80, 82–83, 86, 87. *Henry Papers* 2:192 n. 39. Samuel B. Dod, "Princeton and Science," in *The Princeton Book,* pp. 93–94. Torrey to Gray,

28 July 1836, quoted in A. Hunter Dupree, *Asa Gray: 1810–1888* (Cambridge: Harvard University Press, 1959), 37–38.

22. Joseph to James Henry, 22 Jan. 1836, Joseph to Harriet Henry, 20 April 1836, Maclean to Henry, 25 July 1836, *Henry Papers* 2:9–12, 48–51, 80–82. Henry to anonymous, 6 Feb. 1837, Drawing of the *Wellington*, Feb. 1837, *Henry Papers* 3:142–43, 163–64. "Henry, Joseph," in *The Biblical Repertory and Princeton Review: Index Volume from 1825 to 1868*, 197. Excerpt, Minutes, Trustees, College of New Jersey, 29 Sept. 1836, *Henry Papers* 2:112–13. "Introduction," *Henry Papers* 3:xiii–xv.
23. This and the next four paragraphs rely on Henry to Trustees, College of New Jersey, [Oct. 1836], and on the letters to, from, and involving Henry between 20 Dec. 1836 and 22 Feb. 1837, *Henry Papers* 3:131–59. For Henry's tie, through his uncle McCulloch and Principal Beck, to Van Buren, see Beck to Van Buren, 10 March 1826, *Henry Papers* 1:117–20.
24. W. M. Whitehead, notebook on natural philosophy, 1836–37.
25. *Henry Papers* 2:38 n. 8; "Introduction," *Henry Papers* 5:xvi–xvii; *Henry Papers* 6:373 n. 9. Henry to Robert Patterson, 31 Jan. 1843, Joseph to James Henry, 16 Oct. 1843, *Henry Papers* 5:302, 410. Henry to Bache, 20 Oct. 1846, Joseph to Harriet Henry, 17 and [18] Dec. 1846, *Henry Papers* 6:524–26, 590–93, 599–600. Joseph to Harriet Henry, 20[–21] Jan. 1847, *Henry Papers* 7:19–22. Cuyler to Miss Henry, 20 March 1888; Cuyler, "Professor Henry—The Christian Philosopher." Horace G. Hinsdale, *An Historical Discourse, Commemorating the Centenary of the Completed Organization of the First Presbyterian Church, Princeton, New Jersey* (Princeton: Princeton Press, 1888), 55–60. William E. Schenck, *An Historical Account of the First Presbyterian Church of Princeton, N.J.* (Princeton: John T. Robinson, 1850), 77.

14. London, Paris, and Edinburgh

1. This chapter's first two sections—concerning Henry's voyage and initial stay in London—rely on, unless otherwise indicated, the documents and letters from 22 Feb. through 23 May 1837, *Henry Papers* 3:160–347.
2. Henry to Forbes, 30 June 1834, *Henry Papers* 2:203–5; various letters between Henry and Forbes, between 7 June and 22 Sept. 1836, *Henry Papers* 3:72–76, 96–98, 98–100.
3. Henry, "Deposition of Joseph Henry, in the Case of Morse *vs.* O'Reilly"; see also chap. 10, n. 3. Henry to Bache, 17 Dec. 1834, and Bache to Henry, 3 Jan. 1835, *Henry Papers* 2:297–304, 324–26.
4. Henry to Torrey, 31 Dec. 1839, *Henry Papers* 4:320–22. Gibson, notebook on natural philosophy, 1842–44.
5. Henry's Notes on a Trip to Philadelphia, 29 April–2 May 1834, *Henry Papers*

2:181–96, esp. n. 21. Henry to William Kinney, 12 July 1842, *Henry Papers* 5:247–49.

6. Gibson, notebook on natural philosophy, 1842–44. See also Cantor, *Michael Faraday: Sandemanian and Scientist,* 261–62, 280–81.
7. Faraday, "Third Series: Identity of Electricities Derived from Different Sources," (10 Jan. 1833), reprinted in *Experimental Researches* 1:99, 102. [Henry], "Magnetism," in *Encyclopædia Americana: Supplementary Volume,* 422; see also a deleted passage on p. 60 of his handwritten draft of this article, in the file labeled "Electricity and magnetism notes I," HSA, box 23. See also Henry to Thomas R. Pynchon, 16 June 1868, p. 73.
8. This and the next three paragraphs rely on the following sources. *Henry Papers* 3:317–18 n. 8, 425 n. 4; Rothenberg et al., eds, *Henry Papers* 6:77–78 n. 10. Cuyler to Miss Henry, 20 March 1888; Cuyler, "Professor Henry—The Christian Philosopher." (Cuyler offers his own anecdote about Henry's interactions with Wheatstone and Faraday.) Mayer, "Henry as a Discoverer," in *Memorial,* 506. For a retelling of Mayer's account, see Coulson, *Joseph Henry,* 119–20; Coulson notes that as of 1930 King's College still had two coils of copper ribbon attributed to Henry. For an additional apocryphal anecdote, see "The Late Professor Henry," *Commercial,* perhaps a Cincinnati newspaper, n.p., n.d. (circa 1878), copy in "Scrap Book." See also Brian Bowers, "Faraday, Wheatstone and Electrical Engineering," in *Faraday Rediscovered,* 169–71.
9. This chapter's third section—concerning Henry's stay in Paris—relies on, unless otherwise indicated, the documents and letters from 23 May through 5 July 1837, *Henry Papers* 3:345–408.
10. Henry to Stephen Alexander, 28 July 1837, Henry to Charles Coquerel, 1 Oct. 1837, Bache to Petty Vaughan, 5 Dec. 1837, *Henry Papers* 3:421–23, 510–13, 528–30. Charles Thilorer to Faraday, 10 July 1837, *Correspondence of Michael Faraday* 2: letter 1016.
11. Henry, "Observations on the Relative Radiation of the Solar Spots" (1845), reprinted in *Scientific Writings* 1:224–27. "Record of Experiments," 4 Jan. 1845, *Henry Papers* 6:145–48. Alexander to Asa Gray, 4 Dec. 1878, Archives of the Gray Herbarium, Harvard University (HPP, no. 7333). See also Henry, "Record of Experiments," entries between 27 April. and 3 Oct. 1843, *Henry Papers* 5:327–31, 344–45, 392–94.
12. Joseph to Harriet Henry, 16 Aug. 1837, Henry to Lewis R. Gibbes, 27 Dec. 1837, *Henry Papers* 3:465–71, 536–38.
13. Henry to Stephen Alexander, 28 July 1837, Henry to Lewis R. Gibbes, 27 Dec. 1837, and Bills for Philosophical Apparatus [editor's appendix], *Henry Papers* 3:421–23, 536–38, 540–45. Various letters to and from Henry, between 14 March and 9 Aug. 1838, *Henry Papers* 4:9–12, 38–39, 52–53, 73–74, 95–106, between 20 Dec. 1841 and 9 June 1842, *Henry Papers* 5:132, 135, 156–57, 161–62, 222–23, and 10 Feb. 1844, *Henry Papers* 5:28. Joseph to James

Henry, 28 June 1844, *Henry Papers* 6:119–21. Diary of Mary A. Henry, 8–10 Jan. 1863 and 10 April 1865.

14. This chapter's fourth section—concerning Henry's visits to Brussels, Antwerp, and second stay in London—relies on, unless otherwise indicated, the documents and letters from 24 July through 2 Aug. 1837, *Henry Papers* 3:412–35.
15. Henry's European Diary, 24 Mar. [1837], *Henry Papers* 3:312–15.
16. Cuyler to Miss Henry, 20 March 1888.
17. Lewis, *The Jameses*, 33–34. Diary of Mary A. Henry, 14 Nov. 1863. Pledge to Princeton Colonization Society, Feb. 1839, *Henry Papers* 4:183. Joseph to Harriet Henry, 19 July 1841, *Henry Papers* 5:74–75.
18. This chapter's fifth section—concerning Henry's stay in Scotland—relies on, unless otherwise indicated, the documents and letters from 4 Aug. through 27 Aug. 1837, and the letters from Henry on 1 Oct., 2 Oct., and 2 Nov. 1837, *Henry Papers* 3:435–93, 506–10, 513, 518–19.
19. Bache to Basil Hall, 16 July 1837, *Henry Papers* 3:409–10. Diary of Mary A. Henry, 18 Jan. 1861.
20. For Newton memorabilia, see *Henry Papers* 3:189, 262, 299, 439, 479–80, 491–92, 503.
21. "Introduction," *Henry Papers* 3:xix–xx; see also 480–81 n. 35.
22. Henry to John W. Webster, 17 Aug. 1842, *Henry Papers* 5:256. Joseph to James Henry, 8 Feb. 1837, *Henry Papers* 3:145–46. Fragment beginning, "The grandfather of John Hendrie," with closing initials "J.H.," in the file labeled "Miscellaneous," HSA, box 19. See also Joseph to James Henry, [11 Oct.] and [18] Dec. 1839, *Henry Papers* 4:272–74, 310–12, and Henry to Thurlow Weed, 7 [March] 1844, *Henry Papers* 6:51–53.

15. Measuring Up

1. This chapter's first two sections—concerning Henry's stay in Liverpool—rely on, unless otherwise indicated, the documents and letters from 15 Sept. through 1 Oct. 1837, and the letter from Henry to Lewis R. Gibbes, 27 Dec. 1837, *Henry Papers* 3:499–513, 536–38. These two sections also rely on Henry to Bache, 9 Aug. 1838, *Henry Papers* 4:95–106.
2. Joseph to Harriet Henry, [5 July 1837], *Henry Papers* 3:408. Henry to Thomas R. Pynchon, 16 June 1868, p. 75. Sturgeon to Henry, 23 Sept. 1839, *Henry Papers* 4:261–65, esp. n. 6. See also Henry, "Record of Experiments," 28 Oct. 1843, Henry to William O'Shaughnessy, 30 Oct. 1843, *Henry Papers* 5:438–41, esp. n. 5, 445–46, esp. n. 4. See also Coulson, *Joseph Henry*, 125–26. 3. For the steam-boiler controversy, see *Henry Papers* 3:67, 251–52, 340–41, 376–77.
4. Recommendation for William Barton Rogers, 6 July 1835, *Henry Papers* 2:417–18. Henry to Stephen Alexander, 10 April 1837, *Henry Papers* 3:259–66.

5. "Introduction," *Henry Papers* 5:xiii; see also Henry to John Gassiot, 11 July 1842, pp. 246–47.
6. The remainder of this chapter's first section—concerning Henry's closing days in England and return home—relies on, unless otherwise indicated, the documents and letters from 5 Sept. through 2 Nov. 1837, *Henry Papers* 3:495–519. For background of the geomagnetic experiment, see *Henry Papers* 3:410–12, 483–85. For Henry support of Bache's involvement in geomagnetic research, see *Henry Papers* 4:236, 280–81, 315–20.
7. Henry to Faraday, 9 Oct. 1838, James to Henry, 20 April 1839, *Henry Papers* 4:115–17, 208–10. Cuyler to Miss Henry, 20 March 1888.
8. [Henry], "Review of 'Report of the Committee on Naval Affairs, to Whom was Referred the Memorial of Henry Hall Sherwood . . . ,'" *Biblical Repertory and Princeton Review* 10 (July 1838), reprinted in *Henry Papers* 4:75–79. John C. Warren to Henry, 29 Sept. 1838, and Henry to Warren, 16 Oct. 1838, *Henry Papers* 4:114–15, 121. Henry to Torrey, [7 Nov. 1838], Torrey to Henry, 9 Nov. 1838, *Henry Papers* 4:156–57, 158–59. Henry to De La Rive, 24 Nov. 1841, *Henry Papers* 5:119–23.
9. Henry to Benjamin Peirce, 25 Nov. 1843, *Henry Papers* 5:461. Henry to Bache, 16 April 1844, *Henry Papers* 6:75–80. *Henry Papers* 7:149–50 n. 6. For Henry's later involvement, see "Excerpt, Minutes, National Institute," 15 Jan. 1849, *Henry Papers* 7:449–50. See also Bruce, *Launching of Modern American Science*, 252–53, and Sally Kohlstedt, "A Step toward Scientific Self-Identity: The Failure of the National Institute, 1844," *Isis* 62 (1971): 341–45.
10. Henry to La Roy Sunderland, 28 Aug. 1841, exchange of letters between Bache and Henry, 6 and 12 Nov. 1843, *Henry Papers* 5:86–88, 447–50. See also "Introduction," *Henry Papers* 5:xiv. Henry to [Francis Dwight], 3 March 1845, Henry to [Benjamin Silliman Jr.], 13 Aug. 1846, *Henry Papers* 6:244–46, 474–78. See also Bruce, *Launching of Modern American Science*, 11–13, 44–45.
11. Henry to Torrey, [20] and 29 Dec. 1841, John S. Hart to Henry, 18 Mar. 1842, *Henry Papers* 5:132–34, 136–37, 158. Paul Theerman, "Speaking of Nature: Lectures and Writings on the Natural World in Nineteenth-Century America," paper presented at annual meeting of the History of Science Society, New Orleans, 1994. See also *Henry Papers* 3:302 n. 15.
12. Henry to Henslow, 5 July 1842, *Henry Papers* 5:242–45.
13. Dodd, "Princeton in the Forties," 239. Various letters from Henry, between [23] Jan. and [19] Feb. 1839, *Henry Papers* 4:172–80; see also the exchange of letters between Horatio Otis and Henry, 8 Sept. and 3 Oct. 1840, pp. 430–31, 443. Exchange of letters between Mercantile Library Company and Henry, 21 May and 7 June 1842, *Henry Papers* 5:188, 221–22.
14. Henry to John W. Webster, 17 Aug. and 2 Sept. 1842, *Henry Papers* 5:256, 264–66. Hodge, "Lecture, Addressed to the Students of the Theological Seminary" (1829), reprinted in *The Princeton Theology, 1812–1921*, 108–13. Henry,

"Geology and Revelation" (ca. 1840s) and "The Philosophy of Inductive Science" (1866), in *A Scientist in American Life: Essays and Lectures of Joseph Henry*, ed. Arthur Molella et al., pp. 28, 95, 98. Henry to Robert Were Fox, 26 April 1844, *Henry Papers* 6:87–89.

15. Nomination of Henry for Copley Medal, 30 Oct. 1839. Mayer, "Henry as a Discoverer," in *Memorial*, 506. See also Faraday to Bache, 28 July 1840, *Henry Papers* 4:425.
16. Tomlinson, "Professor Henry," *Academy* (London), 25 May 1878, copy in "Scrap Book." Henry, "Contributions to Electricity and Magnetism. No. III. On Electro-Dynamic Induction" (Nov. 1838), reprinted in *Scientific Writings* 1:114.
17. Excerpt, Minutes, Council of the Royal Society, 27 Jan. 1842, *Henry Papers* 5:142–44. *Henry Papers* 3:375 n. 3. Henry to William Ferrel, 31 Jan. 1861, HSA. "Introduction," *Henry Papers* 4:xvi. Francis Watkins to Henry, [16 Dec.] 1845, *Henry Papers* 6:264–67. John Barlow to Henry, 29 June 1849, *Henry Papers* 7:566–67. See also the annotated, chronological list of Henry's "Offices and Honors," copy extracted from *Bulletin of the Philosophical Society of Washington*, in file labeled "List of Honors and Offices," HSA, box 39; cf. to alphabetical list filed with the finding aids to the Henry Papers, HSA.

16. Proper Compensation

1. This entire chapter relies on, unless otherwise indicated, Henry to [Wheatstone], 27 Feb. 1846, *Henry Papers* 6:382–85; it remains unclear whether a letter derived from either this draft or that of 26 Feb. reached Wheatstone. See also Henry's European Diary, 25 [March 1837], *Henry Papers* 3:205–6, and Diary of Mary A. Henry, 2 Dec. 1867.
2. Gray to Henry, 12 Jan. 1846, Joseph to James Henry, 31 Jan. 1846, Henry to Peter Bullions, 5 Aug. 1846, *Henry Papers* 6:368–70, 371–74, 457–62.
3. Various letters to and from Henry, between 13 and 18 Dec. 1843, *Henry Papers* 5:469–73, and between 30 Jan. and 4 March 1844, *Henry Papers* 6:13–48. Joseph to James Henry, 28 June–8 July 1844, *Henry Papers* 6:119–21. Hall, *Reminiscences,* 10. See also "Introduction," *Henry Papers* 6:xxiii–xxv.
4. Alfred Vail, *The American Electro Magnetic Telegraph* (Philadelphia: Lea & Blanchard, 1845), 115–208, esp. p. 134; Henry's copy, with passages of interest marked with vertical pencil lines, is in the Bell-Henry Library.
5. This and the next four paragraphs rely on, unless otherwise indicated, "Record of Experiments," 29 Oct. [1838], and various letters between Henry and Gale, Morse, Bache, Silliman, and James Fisher, between 23 Nov. 1838 and 17 Aug. 1839, *Henry Papers* 4:143–44, 161, 181–82, 197–99, 211–12, 214–15, 218–21, 247–49, between 11 Jan. 1842 and 26 Aug. 1843, *Henry Papers*

5:140–41, 144–45, 146–47, 150–51, 154–55, 321–23, 371, 374, 388–90, and between 24 Jan. 1844 and 30 April 1844, *Henry Papers* 6:9–10, 21–22, 63–65, 95–96. See also Coulson, *Joseph Henry*, 214–26.

6. Morse to Vail, 16 March 1847, Alfred Vail Telegraph Collection, Smithsonian Archives, record unit 7055; in a personal letter to the author of 17 March 1997, David Hochfelder included an excerpt from this letter. Mabee, *American Leonardo*, 119, 150, 309. Prime, *Life of Samuel F. B. Morse*, 16–23, 161–69.
7. This and the next two paragraphs rely on the following sources. Henry to Bache, 12 Nov. 1845, Excerpt, Diary of John R. Buhler, 21 Feb. [1846], Vail to Henry, 17 July 1846, and Morse to Henry, 17 Oct. 1846, *Henry Papers* 6:325–27, 376–77, 450–51, 520–22. Vail, *American Electro Magnetic Telegraph*, 85, 87–88. Henry, "Communication from Prof. Henry . . . Relative to a Publication of Prof. Morse," 87, and "Deposition of Joseph Henry, in the Case of Morse *vs.* O'Reilly," 111–12, 115; both reprinted in *Annual Report of the Board of Regents of the Smithsonian Institution, 1857*. Morse to Vail, 21 May 1850, Vail to Morse, 24 May 1850, Morse Papers, Library of Congress (HPP, nos. 11038 and 11040 with alternate copies 26396 and 26397). W. James King, *The Development of Electrical Technology in the 19th Century: 2. The Telegraph*, United States National Museum Bulletin 228 (Washington, D.C., 1962), 294–302.
8. This and the next paragraph rely on the following sources. Henry, MS that begins, "Mr. Morse attempts to make it appear . . . ," written sometime after 1855 in response to Morse's article in *Shaffner's Telegraphy Companion* (see below), in folder 5, HSA, box 29. Morse to Henry, 17 Oct. 1846. Morse, "The Electro-Magnetic Telegraph. A Defence Against the Injurious Deductions Drawn from the Deposition of Prof. Joseph Henry," *Shaffner's Telegraphy Companion, Devoted to the Science and Art of the Morse American Telegraph* 2 (Jan. 1855): 81, 91. Morse to Vail, 21 May 1850. Morse to Vail, 20 July 1845, 4 Aug. 1845, 8 Oct. 1845, 3 Jan. 1846, 15 Jan. 1851, and Vail to Morse, 30 July 1845, 21 Aug. 1845, 11 Jan. 1851, Alfred Vail Telegraph Collection; in a personal letter to the author of 17 March 1997, David Hochfelder included excerpts from or copies of these letters. He also pointed out Morse's warnings to Vail against revealing technical details. In his own evaluation of Morse's and Vail's disclaimers about slighting Henry's work, Hochfelder focuses on the similarity of Morse's receiving magnet to Henry's intensity electromagnet and suggests: "Both men were being disengenuous." Hochfelder, "Joseph Henry and the Telegraph," paper presented at the joint meeting of the American Physical Society and the American Association of Physics Teachers, Washington, D.C., 1997.
9. Henry to Bache, 12 Nov. 1845, Joseph to Harriet Henry, [24 Dec.] 1845, Henry to De La Rive, 8 May 1846, *Henry Papers* 6:325–27, 344, 420–23. Vail to Morse, 24 May 1850.
10. Although another article in this supplementary volume offered even a clearer means of countering Vail's book, Henry apparently chose not to exploit it.

The opportunity arose when he helped an APS colleague prepare the article on the telegraph, a companion piece to his own general entry on electricity and magnetism. In tracing the history of telegraphy, this article (also published anonymously) passed over Henry's contribution, merely saying: "Various other instruments were, from time to time, proposed, but none of any practical value until the year 1837." Moreover, the article conceded that, whereas viable systems appeared in 1837 in England, Germany, and the United States, Morse's was "the best and simplest of all the forms of the electro-magnetic telegraph." Although willing to highlight his personal achievements in his own, unsigned article on electricity and magnetism, Henry perhaps balked at appearing vain to the colleague writing the article on the telegraph; in public forums or professional exchanges with all but his closest intimates, Henry always maintained a stance of modesty and propriety. [John F. Frazer], "Telegraph," in *Encyclopædia Americana: Supplementary Volume* (Philadelphia: Lea and Blanchard, 1847), 14:575–76. See also *Henry Papers* 6:299–300 n. 14. Henry to John Frazer, 12 Sept. 1845, Henry to James W. Alexander, 14 March 1846, *Henry Papers* 6:315–16, 393–95. Henry to [Jonathan H. Lane], 22 April 1847, *Henry Papers* 7:85–86.

17. On to the Smithsonian

1. This first section of the chapter relies on, unless otherwise indicated, the following sources. Henry to Bache, [5] Sept. and 7 Sept. 1846, *Henry Papers* 6:493–500, 500–503. Henry, MS that begins, "By the advice of my friend Bache, I have concluded to note down the principal facts of my connection with the Smithsonian Institution," written sometime after mid-December 1846, in folder 1, HSA, box 30. For factual information on the development of the Smithsonian Institution, this entire chapter and the following chapter rely on, unless otherwise indicated, a March 1995 printing of the computer-based "chronology of the history of the Smithsonian through 1890," maintained by Pamela M. Henson, historian, Smithsonian Archives. For two Princeton students' recollections of serving as amanuenses for Henry as he moved through various versions and drafts of his plan for the Smithsonian—in these two cases, most likely Henry's later, more considered "Programme of Organization"—see: Hall, *Reminiscences*, 9–10, and Cameron, "Reminiscences," 170. For the later "Programme of Organization," 13 Dec. 1847, see *Henry Papers* 7:244–48.
2. *Henry Papers* 6:138 n. 6, 252 n. 7, 385 n. 9. Slotten, *Patronage, Practice, and the Culture of American Science*, 13–15.
3. Francis Markoe Jr. to Richard Rush, 19 Aug. 1846, Henry to Peter Bullions, 14 Oct. 1846, *Henry Papers* 6:482–85, 515–18.
4. Rothenberg et al., eds., "An Act to Establish the Smithsonian Institution,"

Henry Papers 6:463–71. Secondary sources on the Smithsonian's history abound, ranging from the nineteenth-century studies by William J. Rhees and G. Brown Goode to the recent sesquicentennial retrospective, James Conaway, *The Smithsonian: 150 Years of Adventure, Discovery, and Wonder* (Washington: Smithsonian Institution Press, 1995). Cynthia R. Field, Richard E. Stamm, and Heather P. Ewing, *The Castle: An Illustrated History of the Smithsonian Building* (Washington: Smithsonian Institution Press, 1993), ix–xi.

5. *Henry Papers* 6:299–300 n. 14.
6. Henry, "Biographical Memoir of Alexander Dallas Bache" (1869), *National Academy of Sciences: Biographical Memoirs* 1 (1877): 181–205. Slotten, *Patronage, Practice, and the Culture of American Science*, 6–26. Various letters to and from Henry, between 28 June and 16 Dec. 1843, *Henry Papers* 5:365, 450–57, 462–72. *Henry Papers* 6:453 n. 7. Four letters between Henry and Bache, between 12 [Feb.] and 3[–4] March 1844, *Henry Papers* 6:30–32, 32–35, 44–45, 46–48.
7. Various letters between Henry, Bache, and Lewis R. Gibbes, between 15 March and 16 Nov. 1845, *Henry Papers* 6:250–55, 259–64, 276–78, 280–84, 330–332. Henry, "The Coast Survey," *Princeton Review* (April 1845; reprint, Princeton, N.J.: John T. Robinson, 1845): 1, 16, 23–24.
8. Rothenberg et al., eds, *Henry Papers* 6:77–78 n. 10. Henry to Bache, 24 [Feb] and 16 April 1844, *Henry Papers* 6:44–45, 75–80. Exchange of letters between Bache and Henry, 30 June and 3 July 1846, *Henry Papers* 6:437–39. Series of letters from Joseph to Harriet Henry, between [11] and 21 July 1846, *Henry Papers* 6:440–49, 451–57. Henry W. Elliott, "Sketch of the Life and Work of Joseph Henry," typescript, p. 14, in file titled "Recollections of Joseph Henry," HSA, box 27. See also Slotten, *Patronage, Practice, and the Culture of American Science*, 25–41, and "Introduction," *Henry Papers* 6:xxv–xxvi. For a discerning overview of the Lazzaroni, see chap. 16 of Bruce, *Launching of Modern American Science*.
9. Henry, MS that begins, "By the advice of my friend Bache, I have concluded to note down the principal facts of my connection with the Smithsonian Institution." Henry to Peter Bullions, 14 Oct. 1846, Henry to Bache, 2 Nov. and 16 Nov. 1846, *Henry Papers* 6:515–18, 533–34, 538–39. This and the next two paragraphs also rely on Rothenberg et al., eds., "The Election of the Secretary of the Smithsonian Institution," *Henry Papers* 6:544–55.
10. Bache to Forbes, 15 Sept., Faraday to Bache, 2 Oct., Brewster to [Smithsonian Regents], 24 Oct. 1846, *Henry Papers* 6:503–4, 510–11, 527–28.
11. *Henry Papers* 6:557–58 n. 3. Markoe to Richard Rush, 19 Aug. 1846, Charles Page to Henry, 4 Dec. 1846, Benjamin B. French to Henry, 3 Dec. 1846, *Henry Papers* 6:482–85, 555–56, 568–69. William J. Rhees, ed., *The Smithsonian Institution: Journals of the Board of Regents, Reports of Committees, Statistics, Etc.*, Smithsonian Institution, Smithsonian Miscellaneous Collections, no. 329 (Washington, D.C., 1879), 1–20.

12. This and the next paragraph rely on, unless otherwise indicated, Henry to Bache, 4 Dec. 1846, *Henry Papers* 6:556–59. See also Gray to Henry, 28 Nov. [1846], *Henry Papers* 6:541.
13. Henry to Peter Bullions, 14 Oct. 1846, Henry to Amos Dean, 19 Oct. 1846, Henry to Bache, 21 Nov. 1846, Joseph to James Henry, 2[-4] Dec. 1846, Henry to [Hawley], 28 Dec. 1846, *Henry Papers* 6:515–18, 523–24, 539–40, 543–44, 610–15. See also "Introduction," *Henry Papers* 6:xxviii–xxxi.
14. Henry to Peter Bullions, 14 Oct. 1846, Joseph to James Henry, 2[-4] Dec. 1846.
15. "Election of the Secretary of the Smithsonian Institution," 550–55. Benjamin French to Henry, 3 Dec. and 4 Dec. 1846, Bache to Henry, 4 Dec. 1846, *Henry Papers* 6:568–69, 564, 564–67. Henry, "Biographical Memoir of Alexander Dallas Bache," 197–98. See also Coulson, *Joseph Henry*, 178–82.
16. Bache to Henry, 4 Dec. 1846, pp. 564–67. See also Enoch Hale to Henry, 15 Dec. 1846, *Henry Papers* 6:587–88, and Joseph to Harriet Henry, 27 Jan. 1847, *Henry Papers* 7:27–29.
17. Henry to Bache and Bache to Henry, 5 Dec. 1846, *Henry Papers* 6:569–71.
18. *National Intelligencer*, 5 Dec. 1846, p. 3, cols. 3–4; copy in binder labeled "Media on SI, Aug.-Dec. 1846," HPP. Reprinted. in *Albany Argus,* 11 Dec. 1846, and in *Memorial*, 407–9. *Henry Papers* 7:41 n. 9, 369–70 n. 2. For later, clearer perceptions of Henry as co-discoverer of mutual induction, see above chap. 6, n. 21.
19. M[orse], "Secretaryship of the Smithsonian Institute: Correspondence," *New York Observer*, 19 Dec. 1846; copy in binder labeled "Media on SI, Aug.-Dec. 1846," HPP; also reprinted in Prime, *Life of Samuel F. B. Morse*, 551–52.
20. Morse to Henry, 17 Oct. 1846. Henry, MS that begins, "Mr. Morse attempts to make it appear. . . ." Morse, "The Electro-Magnetic Telegraph. A Defence Against the Injurious Deductions Drawn from the Deposition of Prof. Joseph Henry," 81–82, 89, 95. Henry, "Communication from Prof. Henry . . . Relative to a Publication of Prof. Morse," 87, and "Deposition of Joseph Henry, in the Case of Morse *vs*. O'Reilly," 115. Vail to Morse, 24 May 1850. Mabee, *American Leonardo*, 90, 166, 310. *Henry Papers* 7:275–76, n. 4.
21. Sears Walker to Henry, 4 Dec. 1846, James Espy to Henry, 7 Dec. 1846, John Miller to Henry, 27 Dec. 1846, *Henry Papers* 6:567–68, 574–76, 609–10. Joseph to James Henry, 2[-4] Dec. 1846. Henry to Gray, 12 Dec. 1846, Henry to Eliphalet Nott, 26 Dec. 1846, *Henry Papers* 6:586, 607–9.
22. For a discussion of the "social ethics" of Henry's American milieu, see "Introduction," *Henry Papers* 6:xiv–xv, xx–xxii. For a discussion of the early nineteenth-century "culture of Whiggery" and its scientific ties, see Slotten, *Patronage, Practice, and the Culture of American Science,* 13–24. Joseph to Harriet Henry, 3 May 1847, *Henry Papers* 7:97–100.
23. Henry to Robert Hare, 29 Dec. 1846, *Henry Papers* 6:615–17. See also Jo-

seph Henry to William Henry, 7 June [1850], Family Correspondence, HSA (HPP, no. 7864). This and the next paragraph rely on, unless otherwise indicated, Henry to anonymous, 29 Dec. 1846, *Henry Papers* 6:618–19.

24. Henry, "Address to the American Association for the Advancement of Science" (1850), in *A Scientist in American Life: Essays and Lectures of Joseph Henry,* ed. Arthur Molella et al., p. 48. James Alexander to Henry, 9 Dec. 1846, *Henry Papers* 6:579–81. Slotten, *Patronage, Practice, and the Culture of American Science,* 19, 26–27.
25. Joseph to Harriet Henry, [23] Jan. 1847, and various letters to, from, and concerning Henry, between 24 May and 8 Sept. 1847, *Henry Papers* 7:22–23, 106–77.
26. Henry to Bache, 15 March 1845, Henry to [Wheatstone], 27 Feb. 1846, Joseph to Harriet Henry, 15 Dec. 1846, [19 Dec. 1846], and 22 Dec. [1846], *Henry Papers* 6:250–55, 382–85, 588–90, 600–601, 604–5. *Henry Papers* 7:54 n. 4. James Gilliss to George Marsh, 8 May 1854, Marsh Papers, Bailey/Howe Library, University of Vermont (HPP, no. 29607).
27. Joseph to Harriet Henry, 15, 17, and 22 Dec. 1846, *Henry Papers* 6:588–93, 604–5. Field, Stamm, and Ewing, *The Castle,* 12.
28. Henry to [Hawley], 28 Dec. 1846, Henry to John Ludlow, 29 Dec. 1846, Henry to James Coffin, 31 Dec. 1846, *Henry Papers* 6:610–15, 617–18, 623–24. Marc Rothenberg, "'The Man of Science Has No Country': Nationalism v. Internationalism at the Early Smithsonian," paper presented at annual meeting of the History of Science Society, Atlanta, 1996.
29. Hodge to Henry, 5 Dec. 1846, Hope to Henry, 7 Dec. 1846, Henry to Nott, 26 Dec. 1846, *Henry Papers* 6:571–73, 576–77, 607–9. "The Smithsonian Institute," *Scientific American* 2 (26 Dec. 1846): 109; copy in binder labeled "Media on SI, Aug.-Dec. 1846," HPP.
30. Henry to Trustees, 18 Dec. 1846, *Henry Papers* 6:597–98. Henry to James Carnahan, 27 June 1848, *Henry Papers* 7:344–45. Hope to Henry, 7 Dec. 1846.

Epilogue

1. [Charles Jewett] to [George Marsh], 19 Sept. 1848, *Henry Papers* 7:398–99. Henry to Mrs. Bache, 19 Feb. 1866, HSA (HPP, no. 6739).
2. "Henry, Joseph," in *The Biblical Repertory and Princeton Review: Index Volume from 1825 to 1868,* 197. Henry to Joseph Varnum Jr., 22 June 1847, *Henry Papers* 7:121–22; quoted also in Daniel J. Kevles, *The Physicists: The History of a Scientific Community in Modern America* (Cambridge, Mass.: Harvard University Press, 1987), 5.
3. William J. Rhees, *Catalogue of Publications of the Smithsonian Institution (1846–1882),* Smithsonian Miscellaneous Collections, no. 478. (Washington:

Smithsonian Institution, 1882), v–xi. See also: Curtis M. Hinsley, *The Smithsonian and the American Indian* (1981; rpt., Washington: Smithsonian Institution Press, 1994); Henry W. Elliott, "The Smithsonian Institution," *International Review* 7 (1879): 694–95; and Coulson, *Joseph Henry,* 191–92, 200. For an up-to-date list of the many books and articles on the development of the Smithsonian Institution (including Henry's role), see the computer-based bibliography maintained by Pamela M. Henson, historian, Smithsonian Archives. For a recent overview of the Institution's early evolution, see the sesquicentennial volume: Conaway, *Smithsonian,* 14–139. For a perceptive overview of the Institution's early decades, see chap. 14 of Bruce, *Launching of Modern American Science;* see also pp. 323–25.

4. Paul H. Oehser, *Sons of Science* (New York: Henry Schuman, 1949), 51–53. For similar coverage, see also chap. 8 of Oehser, *The Smithsonian Institution* (New York: Praeger Publishers, 1970), and Coulson, *Joseph Henry*, 192–94, 203–5. *Henry Papers* 7:469 n. 2.
5. Fleming, *Meteorology in America,* 75–78, 141–42, 160–62. Henry, "Meteorology in Its Connection with Agriculture" (1855–59), a five-part series, reprinted in *Scientific Writings* 2:6–402.
6. Daniel Goldstein, "'Yours for Science': The Smithsonian Institution's Correspondents and the Shape of Scientific Community in Nineteenth-Century America," *Isis* 85 (Dec. 1994): 573–99. E. F. Rivinus and E. M. Youssef, *Spencer Baird of the Smithsonian* (Washington: Smithsonian Institution Press, 1992), 41–44, 56–97. Oehser, *Sons of Science*, pp. 44–47. Henry, "Presidential Address to the National Academy of Sciences" (1876), in Molella et al., *A Scientist in American Life: Essays and Lectures of Joseph Henry,* pp. 134–36. Henry to Joseph Leidy, 23 March 1854, Academy of Natural Sciences, Philadelphia (HPP, no. 2133). Henry, 2 Dec. 1854, "Locked Book," HSA, box 39 (HPP, no. 23327). See also Joel J. Orosz, *Curators and Culture: The Museum Movement in America, 1740–1870* (Tuscaloosa: University of Alabama Press, 1990), 201–12. An earlier statement of Orosz's interpretation of the Henry-Jewett-Baird triangle met with a strong rebuttal from: S. Dillion Ripley and Wilcomb E. Washburn, "The Development of the National Museum at the Smithsonian Institution, 1846–1855: A Response to Joel J. Orosz's Article," *Museum Studies Journal* 2 (1987): 6–11. For Orosz's counter-response, see his "In Defense of the Deal: A Rebuttal to S. Dillon Ripley's and Wilcomb Washburn's 'Response'," *Museum Studies Journal* 3 (1987): 7–12. See also Diary of Mary A. Henry, 2 Dec. 1867.
7. Morse to Editors of the *Daily Times,* withdrawn (unpublished) letter to the editors, 19 July 1854, New-York Historical Society (HPP, no. 4391). Morse, "The Electro-Magnetic Telegraph. A Defence Against the Injurious Deductions Drawn from the Deposition of Prof. Joseph Henry," 6–96. Henry to Alexander, 11 April 1855, extract of letter prepared by Mary Henry, in folder

labeled "Important Inserted Papers," HSA, box 54. "Report of the Special Committee of the Board of Regents on the Communication of Professor Henry," 88–99; reprinted in *Annual Report of the Board of Regents of the Smithsonian Institution, 1857*. See also Coulson, *Joseph Henry*, 225–34. For newspaper and journal coverage of Henry's death, funeral, and the events' reverberations, this chapter relies primarily on the multitude of contemporary obituaries and articles collected in the "Henry Memorial. Scrap Book. 1878," HSA, box 47, and the smaller groups included in HSA, boxes 43, 46, 59, and 96. For Henry and the telegraph, see, e.g., the following items, copies in "Scrap Book": "Death of Joseph Henry," *New York World,* 14 May 1878; *Denver Daily News,* ca. mid-May 1878; "Death of Professor Henry," *National Republican*, 14 May 1878; and "Professor Joseph Henry," *Scientific American*, ca. May 1878.

8. Henry to Bache, 31 March 1847, *Henry Papers* 7:116. Diary of Mary A. Henry, 20 Jan. 1866. Field, Stamm, and Ewing, *The Castle*, 1–17, 26–36. Henry, "Presidential Address to the National Academy of Sciences" (1876), pp. 133–36. Henry, "Thoughts on Architecture," in Molella et al., *A Scientist in American Life,* 34.
9. Rivinus and Youssef, *Spencer Baird of the Smithsonian*, 122–40, 189–91. *New York Daily Tribune,* 18 May 1878, copy in "Scrap Book." I. Michael Heyman, "Smithsonian Perspectives," *Smithsonian* 27 (Aug. 1996): 10.
10. L. E. Chittenden, *Recollections of President Lincoln and His Administration* (New York: Harper & Brothers, 1901), 235–38. For personalized coverage of William's death, the fire, Lincoln's assassination, Henry's views on slavery, and wartime events, see Diary of Mary A. Henry, esp. 3 July 1862, 19 Oct. 1862, 6 Dec. 1862, 24 Nov. 1863, 25 Jan.–4 Feb. 1865, 15 April–19 April 1865. See also Henry, Desk Diary, 14–19 April 1865, HSA, box 14. See also Coulson, *Joseph Henry*, 235–46. For Henry's later involvement with the American Colonization Society, see "Offices and Honors," copy extracted from *Bulletin of the Philosophical Society of Washington*, in file labeled "List of Honors and Offices," HSA, box 39. Joseph to Harriet Henry, 21 April and 17 July 1865, HSA (HPP, nos. 7674 and 7676); see also Henry to James Hall, 2 Jan. 1863, Hall Collection, Manuscripts and Special Collections, New York State Library (HPP, no. 4013), Henry to Asa Gray, 7 Dec. 1860, HSA (HPP, no. 7297), and Henry to [Benjamin] Peirce, 28 March 1866, HSA (HPP, no. 12300).
11. Henry, "Address to the American Association for the Advancement of Science" (1850), 39–43, and "The Philosophy of Education" (1854), 80–87.
12. Joseph to Harriet Henry, 30 May 1862, HSA (HPP, no. 7874). Taylor, "The Scientific Work of Joseph Henry," 308–19, 345–56. Henry to Asa Gray, 6 Nov. 1852, Archives of the Gray Herbarium, Harvard University (HPP, no. 7240). Arnold B. Johnson, "The United States Lighthouse Establishment,"

Appletons' Annual Cyclopædia, 1880 (New York: D. Appleton, 1881), 435–36, 444–46. See also Coulson, *Joseph Henry,* 250–59.

13. This and the next two paragraphs rely on the following sources. Henry to L. H. Morgan, 16 Aug. 1876, HSA (HPP, no. 5099). "A Loss to Science," *New York Daily Tribune,* 14 May 1878, copy in "Scrap Book." Henry, "Opening Address to the National Academy of Sciences" and "Closing Address to the National Academy of Sciences" (1878), reprinted in *Scientific Writings* 2:536–39. Henry to Robert Patterson, 19 April 1878, HSA (HPP, no. 17611). Franklin L. Pope, "Life and Work of Joseph Henry," *Journal of the American Electrical Society* 2 (1879): 143–45; quoted in Coulson, *Joseph Henry,* 316–17. Phebe Mitchell Kendall, comp., *Maria Mitchell: Life, Letters, and Journals* (Boston: Lee and Shepard, 1896), 244–45; for the journal entry, see also Maria Mitchell, Diary, April 1878, Nantucket Maria Mitchell Association (HPP, no. 52054).
14. Gray to Baird, 7 Oct. 1882, Baird Papers, Smithsonian Archives, record unit 7002 (HPP, no. 23570).
15. For appraisals, pro and con, of Henry's Smithsonian performance, see, e.g., the following items, copies in "Scrap Book": *Providence Press,* 14 May 1878; "Prof. Baird to Succeed Prof. Henry," *Journal,* Boston, 17 May 1878; A. G. [Asa Gray?], "The Late Prof. Henry," *Boston Journal,* letter to the editor dated 20 May 1878; *New York Daily Tribune,* 18 May 1878; "The Smithsonian," *Washington National Standard,* 31 May 1878; "Professor Joseph Henry," *New York Evening Express,* 14 May 1878. See also Blodgett to Baird, 17 May 1878, Baird Papers (HPP, no. 23266). For the Nestor appellation, see, e.g., "A Loss to Science," *New York Daily Tribune,* 14 May 1878 and "Joseph Henry," *Polytechnic Review* (May 1878). In reporting a conversation overheard at the time of Henry's funeral, the correspondent to the *Cincinnati Commercial* revealed that either he or his copy editor was rusty on allusions to Greek heroes: "I heard a knot of men, of high scientific position, say, as they grouped sorrowfully in a hotel corridor: 'Henry's place can never be filled. There is no man in the world like him. He was truly the Chester of American science.'" See E.H., "The Burial of Professor Henry," correspondence dated 21 May 1878, *Cincinnati Commercial,* 26 May 1878; copy not only in "Scrap Book" but also in folder labeled "Henry's Death, Burial, Clippings, Memorials," HSA, box 46; the author of this literate, detailed account might be journalist and naturalist writer Edgar W. Howe (1853–1937).
16. "In Memory of Joseph Henry," *Philadelphia Public Ledger and Daily Transcript,* 14 May 1878, copy in "Scrap Book." "Death of Professor Henry," *Washington Evening Star,* 13 May 1878, copy in "Scrap Book." *Richmond State,* 14 May 1878 and 15 May 1878, copies in "Scrap Book." (Writing in the city that had stood as the capital of the Confederacy through most of the Civil War, the Richmond editors perhaps betrayed a lingering, postwar resentment of the Washington scientific establishment.) Tomlinson, "Professor Henry."

17. Albert E. Moyer, *American Physics in Transition: A History of Conceptual Change in the Late Nineteenth Century*, The History of Modern Physics, 1800–1950, vol. 3, eds. Gerald Holton and Katherine R. Sopka (Los Angeles: Tomash Publishers, 1983), xviii–xix. Bruce, *Launching of Modern American Science*, 104–5.
18. "Henry, Joseph," in *The Biblical Repertory and Princeton Review: Index Volume from 1825 to 1868*, 194–200. See above, chap. 1, n. 2, for evidence of Henry's authorship and for periodicals that reprinted the 1871 biographical sketch and list. To summarize Henry's scientific labors, the eulogists also frequently extracted information from a biographical essay that Frederick Barnard wrote for *Johnson's Cyclopædia.* Most of these borrowers, however, probably did not realize that Barnard, president of Columbia College, wrote his essay in consultation with Henry. For Henry's involvement in preparing the Barnard essay (and the *Princeton Review* list), see Benjamin Silliman, "Joseph Henry, LL.D.," *American Journal of Science and Arts*, 464–66. For a revised version of Barnard's essay, see Barnard, "Henry (Joseph)," *Johnson's New Universal Cyclopædia* (New York: A. J. Johnson & Son, 1876–78), 2:878–79.
19. Sherman's address at "Princeton College, N.J., June 19th 1878." See also Sherman, "Address," in *Memorial,* 117. For comments on the government honoring itself, see "Joseph Henry Memorial," *Philadelphia Public Ledger and Daily Transcript,* 15 April 1881, "The Late Joseph Henry," *Newark Daily Advertiser,* 24 April 1883, and Dall, "Prof. Henry's Statue Unveiled"; copies in William A. Henry's "Rhetoric" notebook, HSA, box 59.
20. For the relocations of the statue, see the files titled "Statue by William W. Story," HSA, box 46.
21. Dall, "Prof. Henry's Statue Unveiled." Field, Stamm, and Ewing, *The Castle,* 158. Reingold, "The New York State Roots of Joseph Henry's National Career," 142–43.

Index